Frontiers in Biotransformation

Vol. 8

Frontiers in Biotransformation Volume 8

Medicinal Implications in Cytochrome P-450 Catalyzed Biotransformations

Edited by

Klaus Ruckpaul and Horst Rein

Akademie Verlag

Editor: Prof. Dr. K. Ruckpaul
Prof. Dr. H. Rein
Max-Delbrück-Centrum für molekulare Medizin
Robert-Rössle-Str. 10
O-1115 Berlin

This book was carefully produced. Nevertheless, authors, editors and publisher do not warrant the information contained therein to be free of errors. Readers are advised to keep in mind that statements, data, illustrations, procedural details or other items may inadvertently be inaccurate.

1st edition
Editorial Director: Christiane Grunow
Copy Editor: Sabine Ulbricht
Composited, Printed and Bound in Great Britain at The Alden Press, Oxford, England
Cover design: Martina Förster

Deutsche Bibliothek Cataloguing-in-Publication data:

Frontiers in biotransformation.—Berlin: Akad.Verl.
ISBN 3-05-500456-6

Vol. 8. Medicinal implications in cytochrome P-450 catalyzed biotransformations.-1. ed.-1993

Medicinal implications in cytochrome P-450 catalyzed biotransformations/ed. by Klaus Ruckpaul and Horst Rein.-
1. ed.-Berlin: Akad.Verl., 1993
(Frontiers in biotransformation; Vol. 8)
ISBN 3-05-501366-2
NE: Ruckpaul, Klaus [Hrsg.]

Gesamt-ISBN 3-05-500456-6
Vol. 1-ISBN 3-05-500457-4
Vol. 2-ISBN 3-05-500458-2
Vol. 3-ISBN 3-05-500459-0
Vol. 4-ISBN 3-05-500460-4
Vol. 5-ISBN 3-05-500461-2
Vol. 6-ISBN 3-05-501284-4
Vol. 7-ISBN 3-05-501329-8
Vol. 8-ISBN 3-05-501366-2

Akademie Verlag is a member of the VCH Publishing Group

Printed on acid-free paper

Contents

Acknowledgment

The editors of this series wish to express their serious gratitude to several generous sponsors who enabled the publication of volume 8 which is intended to focus on basic research as source for knowledge and innovative force for developments in all fields of human life.

Bayer AG, Wuppertal/Leverkusen, Germany
Sandoz Forschungsinstitut, Wien, Austria
Dr. Karl Thomae GmbH, Biberach an der Riss, Germany

Introduction

H. Rein and K. Ruckpaul

Research on cytochrome P-450 provides an appropriate example for the long way from basic research to medicinal application either for diagnostic or therapeutic purposes. After the discovery of the endoplasmic hepatic P-450 system as being involved in induction and chemical carcinogenesis in the first half of the 50s by Miller and Miller about twenty years passed till Ames applied the unique enzymatic capability of the microsomal system to examine the carcinogenic and mutagenic potential of newly developed drugs or environmental pollutants. Kellermann in the beginning of the 70s first tried to correlate the inducibility of P-450 aryl hydrocarbon hydroxylase activity in human lymphocytes with the occurrence of cancer. Extending this line pharmacogenetic variants of P-450 isozymes (debrisoquine hydroxylase deficiency, 1977; mephenytoine hydroxylase deficiency, 1984) were taken to recognize extensive and poor metabolizers in rationalizing an individual dosage and furthermore to find a linkage between these anomalies and the biotransformation activity towards other compounds (e.g. precarcinogens).

The other line of medicinal implication was and further is to develop new therapeutic principles: either by using P-450 directed drugs as inhibitors of steroidogenic P-450 isozymes within the human organism or by inhibiting P-450 of pathogenic fungi. Present efforts in research are aimed at realising the next step in this development namely to control and influence biosynthesis of P-450 by substances interfering with transcriptional or translational processes. The aim of the present volume is to elucidate some steps in this long laborious way from basic research to medicinal application.

In the first chapter B. Kemper gives a competent overview on our present knowledge about the genomic structure of P-450 genes from 10 mammalian P-450 families, their relation to evolution, regulation of expression and to pathological states. The regulation of the Ah-inducible and steroidogenic genes has begun to be understood due to the existence of heterologously expressing cell lines. Despite different progress with regard to isozymic specificities the rapid advances in the last decade will provide an appropriate basis for a deeper understanding of molecular mechanism of the expression of several P-450 gene families. The characterization of mammalian P-450 genes may help to study structural and regulatory polymorphism at the genetic level and this way to elucidate the molecular basis of congenital diseases and individual differences in response to therapeutic drugs and environmental toxins.

Because mammalian cytochromes P-450 are membrane anchored proteins the membrane topology of the different P-450 forms is of interest to understand regulation of biosynthesis and electron transfer processes as functional basis of the biotransformation of lipophilic substrates. The expression of microsomal P-450 proceeds cotranslationally. M. SAKAGUCHI and T. OMURA give an excellent overview of the current-state-of-art. Amino-terminal segments function as signal sequence for the membrane insertion and the characteristics of this topogenic sequence is determined by the balance between the length of the hydrophobic segment and the amino-terminal charges. From the experimental data to which the authors have essentially contributed general structural features of microsomal P-450s are derived which lead the authors to detailed conclusions about the different topological orientations of the topogenic sequences. Moreover, in eukaryotic cells, co-translational insertion of P-450 molecules into the microsomal membrane by the amino-terminal signal-anchor sequence triggers their correct folding.

The clinical use of P-450 inhibitors concerns the inhibition of the synthesis or catabolism of endogenous substrates as demonstrated by the inhibition of aromatase at the treatment of estrogen-dependent breast cancer. The importance of inhibition of P-450 enzymes for therapeutic and toxicological reasons is presented in chapter 3 by M. A. CORREIA and P. R. ORTIZ DE MONTELLANO. Not only the usefulness of inhibitors for a selective chemotherapy of breast and prostate cancer is discussed but also the aspect of inhibition of xenobiotic metabolizing enzymes, altering the pharmacokinetic properties or toxicity of drugs or harmful environmental agents.

A further important aspect concerning the P-450 inhibition is evident from the chemotherapy of fungal infections. Fungal cells can be attacked by compounds exerting different mechanisms as discussed in chapter 4 by I. SCHUSTER. Inhibition of ergosterol biosynthesis has become the predominant rationale in the prevention and treatment of all kinds of fungal infections ranging from plant diseases to human and animal mycosis. Two enzymes – squalene epoxidase and 14α-lanosterol demethylase – are involved in the biosynthesis of ergosterol, the dominant sterol in most fungi and an indispensable component of their membrane structure. The inhibition of these enzymes results in changes in membrane composition with toxic consequences for the fungi. Both enzymes are oxidases but only the demethylase is a member of the P-450 superfamily. In this instructive overview not only a spectrum of antimycotic drugs based on inhibition of ergosterol synthesis is presented but also clinical aspects at application of these compounds. Thus, interference of the antimycotics with drug metabolizing enzymes of the liver and inhibition of steroid hydroxylating P-450 forms of the patients leading to side effects is reviewed, too.

Although the principal task of P-450 catalyzed biotransformations of xenobiotics is to be seen in the detoxication the initial activation step often results in a toxic intermediate. The review of Y. TERELIUS, K. O. LINDROS, E. ALBANO and M. INGELMAN-SUNDBERG is focussed on such metabolic activations which are catalyzed by specific forms of P-450. Described are especially compounds with documented hepatotoxicity at relevant levels as opposed to chemicals that are mainly considered as carcinogens. Up to now only few studies of hepatotoxicity were carried out with purified P-450 isozymes or with specific inhibitory antibodies. Therefore the authors considered also results from experiments where the link between hepatotoxicity and isozyme specificity has been obtained by specific inducers. In many cases the toxicity of a xenobiotic is strongly influenced by induction. This is exemplified by P-450 2E1 which activates numerous compounds exerting hepatotoxicity in the centrilobular region of the liver acinus where this isozyme is specifically induced by substrates, hormones and dietary factors. Because P-450 2E1 is involved in the activation of a high number of cytotoxic chemicals – more than 60 different substrates have been identified – inducers therefore strongly influence the hepatotoxic action of several kinds of compounds. Interestingly, the genetic polymorphism of several P-450 genes involving that of P-450 2E1 permits to relate a pronounced interindividual sensitivity for developing liver cirrhosis and necrosis to the expressed P-450 genotype.

Nitrate tolerance is a well known pharmacological phenomenon at a prolonged treatment of myocardial ischemia and its principal symptom angina pectoris. In the last chapter H. SCHRÖDER presents the interesting hypothesis that P-450 could be involved in nitrate tolerance. In own experimental results the author could show that the potency of organic nitrates for relaxation and second messenger stimulation (cyclic GMP) correlates with their respective degree of biotransformation by P-450. The physiological relevance of nitric oxide as endothelium-derived relaxing factor (EDRF) and endogenous activator of soluble guanylate cyclase has been emerged recently. Obviously nitric oxide from organic nitrates as the more than 100 years old therapeutically used glycerol trinitrate is formed by the reductive potential of P-450.

Chapter 1
Mammalian Cytochrome P450 Genes

B. KEMPER

1. Introduction

The first nucleotide sequence of the cDNA for a cytochrome P450 (P450) was reported only about 10 years ago (FUJII-KURIYAMA et al., 1982). Since then there has been an explosion of information on protein, cDNA, and gene sequences which has defined this class of enzymes as a superfamily. In the periodic update on P450 nomenclature, a total of 154 genes and seven pseudogenes were compiled into 27 gene families (NEBERT et al., 1991). Genes have been analyzed from mammals, birds, fish, insects, mollusks, fungi, plants, and bacteria emphasizing the ubiquitous nature of the P450s. There are 10 mammalian families most of which have been mapped to specific chromosomes in the mouse and human genomes. The first sequence of a genomic gene for P450 was reported in 1983 (MIZUKAMI et al., 1983b) by early 1991, 47 (66 if orthologous genes in different species are counted separately) mammalian genes had been at least partially sequenced (Table 1). The ultimate number of P450 genes remains a matter of conjecture. In the rat, 40 distinct genes or cDNAs have been characterized which is probably only a fraction of the total number. For example only three of an estimated 8–10 genes (ATCHISON and ADESNIK, 1983; MIZUKAMI et al., 1983a) in the rat CYP2B subfamily have been characterized. Most studies of P450 genes have concentrated on liver tissue which contains the highest concentrations of P450 but it is likely that unique P450s are expressed in other tissues. An estimate of between 100 and 200 P450 genes in a single species seems reasonable. If the average size of a P450 gene is about 10,000 nucleotides, then about 2×10^6 nucleotides or nearly 0.1% of the entire mammalian genome is devoted to encoding P450s.

A comprehensive discussion of P450 proteins, cDNAs and genes is beyond the scope of this review. The theme of this volume of "Frontiers in Biotransformation" is the clinical implications of P450s. I have focused on advances in the study of the structure of mammalian genomic P450 genes as they relate to evolution, to the regulation of gene expression, and to pathological states. The survey of the literature extends to about January, 1991 and emphasis is placed on papers written since about mid 1989. As an aside, I would strongly encourage authors of P450 papers to include CYP or P450 in the title or key words which will increase the visibility of the paper in this era of electronic searches, as well as simplify the work of reviewers. Second, I would encourage the use of the standard nomenclature for P450 (NEBERT et al, 1991), at least in a footnote if a more descriptive functional term is preferred, since it greatly facilitates comparative analysis of different P450 genes. There are a

Abbreviations used: Ah: aromatic- or aryl-hydrocarbon; BTE: basal transcription element; CAT: bacterial chloramphenicol transferase; CRE: cAMP response element; CREB: CRE binding protein; CTF: common transcription factor; CYP: cytochrome P450 gene; HNF: hepatic nuclear factor; GRE: glucocorticoid response element; LF-A1: liver transcription factor-A1; NF1: nuclear factor 1; P450: cytochrome P450; RFLP: restriction fragment length polymorphism; SDI: sex difference information; TCDD: 2,3,7,8-tetrachlorodibenzo-p-dioxin; TPA: 12-O-tetradecanoylphorbol-13-acetate; XRE, xenobiotic response element.

Table 1. Sequenced P450 genes.

	Gene	Species	5'[1]	3'[2]	Size[3]	Intron sites[4] Intron total	Database Accession #	Reference
			kilobase pairs					
1.	CYP1A1	rat	0.6	0.3	6.0	6/6*	K02246	SOGAWA et al., 1984
2.	CYP1A1	rat	0.5	–	6.1	6/6*	M26129	HINES et al., 1985
3.	CYP1A1	rat	4.4	–	–	–	M14633	SOGAWA et al., 1986
4.	Cyp1a1	mouse	0.8	1.8	6.2	6/6*	M10021	GONZALEZ et al., 1985
5.	CYP1A1	human	1.6	0.1	6.3	6/6*	X02612	JAISWAL et al., 1985
6.	CYP1A1	human	1.7	0.1	5.8	6/6*	X04300	KAWAJIRI et al., 1986
7.	CYP1A2	rat	1.2	0.5	6.9	6/6*	K03241	SOGAWA et al., 1985
8.	Cyp1a2	mouse	0.9	1.3	6.7	6/6*	M10022	GONZALEZ et al., 1985
9.	CYP1A2	human	0.1	–	7.6	6/6	L00383-9	QUATTROCHI et al., 1986
10.	CYP1A2	human	1.9	0.1	7.8	6/6	M31664-7	IKEYA et al., 1989
11.	CYP2A1	rat	4.5	1.4	12.8	8/8*	M33312	MATSUNAGA et al., 1990b
12.	CYP2A2	rat	5.5	0.5	23	8/8	M33313, M33325, M34392	MATSUNAGA et al., 1990b
13.	CYP2A3	rat	3.4	2.9	8.1	8/8*	M33190	UENO and GONZALEZ, 1990b
14.	Cyp2a4	mouse	0.5	0.3	8	8/8	M25146-7, M26202-8, J04631	LINDBERG et al., 1989
15.	Cyp2a5	mouse	0.5	0.3	8	8/8	M26204-11, J04631	LINDBERG et al., 1989
16.	CYP2B1	rat	0.1	0.1	23	8/8	M11251, L00313-20	SUWA et al., 1985
17.	CYP2B2	rat	0.1	0.8	14	8/8	J00720-8	MIZUKAMI et al., 1983b
18.	CYP2B2	rat	–	–	–	7/8		ATCHINSON and ADESNIK, 1983
19.	CYP2B2	rat	1.4	–	–	–	Y00410	JAISWAL et al., 1987
20.	CYP2B2	rat	0.8	–	–	1/8	M15947	RANGARAJAN et al., 1987
21.	CYP2B4P	rabbit	0.1	–	7	8/8	M11280-8	ZAPHIROPOULOS et al., 1986
22.	CYP2B8	rat	0.5	–	–	1/8	J04808	GIACHELLI et al., 1989
23.	Cyp2b9	mouse	1.9	0.1	38	8/8	M60267-73	LAKSO et al., 1991
24.	Cyp2b13	mouse	1.9	0.2	–	8/8	M60352-59	LAKSO et al., 1991
25.	Cyp2b?	mouse	0.7	–	–	1/8		LAKSO et al., 1991
26.	CYP2C1	rabbit	0.7	–	–	1/8	M74199	ZHAO et al., 1990
27.	CYP2C2	rabbit	0.7	–	–	6/8	M14955	GOVIND et al., 1986
28.	CYP2C3	rabbit	1.9	–	>25	8/8	M31245-54, J02901	CHAN and KEMPER, 1990
29.	CYP2C4	rabbit	0.8	–	–	5/8	M74200-03	ZHAO et al., 1990

30.	CYP2C5	rabbit	0.5	–	>21	8/8	M55664	PENDURTHI et al., 1990
31.	CYP2C6	rat	1.3	–	–	1/8	X06712, M36848	UMENO et al., 1988b
32.	CYP2C6	rat	–	–	–	4/8		KIMURA et al., 1989b
33.	CYP2C11	rat	0.9	0.1	35	8/8	M18356-63	MORISHIMA et al., 1987
34.	CYP2C12	rat	1.9	–	>35	8/8	M33544-50	ZAPHIROPOULOS et al., 1990
35.	CYP2D2	rat	0.5	3.0	4.0	8/8*	X52027	MATSUNAGA et al., 1990a
36.	CYP2D3	rat	1.4	4.3[6]	4.4	8/8*	X52028	MATSUNAGA et al., 1990a
37.	CYP2D4	rat	1.3	3.5	4.7	8/8*	X52029	MATSUNAGA et al., 1990a
38.	CYP2D5	rat	4.3[6]	2.2	4.5	8/8*	X52030	MATSUNAGA et al., 1990a
39.	CYP2D6	human	1.5	3.5	4.4	8/8*	M33388, M33189[5]	KIMURA et al., 1989c
40.	CYP2D7P	human	4.6[7]	1.5	4.3	8/8*	M33387	KIMURA et al., 1989c
41.	CYP2D8P	human	1.5	4.6[7]	5.3	8/8*	M33387	KIMURA et al., 1989c
42.	Cyp2d9	mouse	0.5	–	4.8	8/8*	M24262, J04471	WONG et al., 1989
43.	Cyp2d10	mouse	0.5	–	4.3	8/8*	M24263, J04471	WONG et al., 1989
44.	Cyp2d11	mouse	0.5	–	5.2	8/8*	M24264, J04471	WONG et al., 1989
45.	CYP2E1	rat	1.5	0.8	10.3	8/8*	M20131	UMENO et al., 1988c
46.	CYP2E1	rabbit	0.5	0.1	10	8/8	M21351-63, J03726	KHANI et al., 1988
47.	CYP2E1	human	2.8	0.6	11.4	8/8*	J02843	UMENO et al., 1988a
48.	CYP2E2	rabbit	0.5	0.1	9	8/8	M21364-72, J03726	KHANI et al., 1988
49.	CYP2G1	rat	–	–	–	8/8	M31923-31, J04715	NEF et al., 1990
50.	CYP4A1	rat	1.0	1.2	14.1	12/12*	M57718	KIMURA et al., 1989a
51.	CYP4A2	rat	2.5	0.1	10.6	11/11*	M57719	KIMURA et al., 1989a
52.	CYP4A6	rabbit		Abstract				MUERHOFT et al., 1991
53.	CYP4B1	human	0.3	–	19	10/10	X16699	YOKOTANI et al., 1990
54.	CYP7	human		Abstract				MOLOWA and CIMIS, 1991
55.	CYP11A1	human	0.6	0.1	>20	8/8	D00161-9, D00186	MOROHASHI et al., 1987
56.	CYP11A1	bovine	0.9	–	–	1/8	J05245	AHLGREN et al., 1990
57.	CYP11A1	human	2.3	–	–	–		MOORE et al., 1990
58.	CYP11A1	rat	0.9	0.1	12.5	8/8		OONK et al., 1990
59.	Cyp11a1	mouse	0.8	–	–	–	J05511	RICE et al., 1990a
60.	CYP11B1	human	0.2	–	5.0	8/8	M32863, M32878-9, J05140	MORNET et al., 1989
61.	CYP11B1	human	1.3	–	–	–		KAWAMOTO et al., 1990b
62.	CYP11B1	bovine	0.6	0.1	5.2	8/8	D00449-59	HASHIMOTO et al., 1989

Table 1. continued.

	Gene	Species	5′[1]	3′[2]	Size[3]	Intron sites[4]	Database	Reference
			kilobase pairs			Intron total	Accession #	
63.	CYP11B1	bovine	0.4	–	–	8/8		Kirita et al., 1990
64.	Cyp11b1	mouse	0.4	–	–	–		Mouw et al., 1989
65.	CYP11B2	human	0.2	–	5.0	8/8	M32864, M32880-1, J05140	Mornet et al., 1989
66.	CYP17	human	1.8	0.1	6.6	7/7*	M19489	Picado-Leonard and Miller, 1987
67.	CYP17	human	0.5	–	7.0	7/7	M31146-53	Kagimoto et al., 1988
68.	CYP17	bovine	0.5	0.1	6.2	7/7		Bhasker et al., 1991
69.	Cyp17	mouse		in press				Youngblood et al., 1991
70.	CYP17	rat		Abstract				Ivanovitch et al., 1991
71.	CYP19	human	0.9	0.2	>52	9/9	M30795-804, J05105	Means et al., 1989
72.	CYP19	human	0.7	–	>70	9/9	M32245	Harada et al., 1990
73.	CYP19	rat	0.5	–	–	–		Hickey et al., 1990
74.	CYP19	human	0.5	–	>70	9/9	X55983	Toda et al., 1990
75.	Cyp21a1	mouse	0.1	0.1	3.0	9/9*	M15009	Chaplin et al., 1986
76.	CYP21A1	bovine	3.0	0.1	3.5	9/9*	M11267	Chung et al., 1986
77.	CYP21A1	porcine	0.1	–	–	2/9		Geffrotin et al., 1990
78.	CYP21A1P	human	0.2	–	3.2	9/9*	M13935	White et al., 1986
79.	CYP21A1P	human	1.7	0.2	3.2	9/9*	M12793	Higashi et al., 1986
80.	CYP21A2	human	0.2	–	3.2	9/9*	M13936	White et al., 1986
81.	CYP21A2	human	1.7	0.2	3.2	9/9*	M12792	Higashi et al., 1986
82.	CYP21A2	human	0.7	0.1	3.2	9/9*	M26856-7	Rodriques et al., 1987
83.	Cyp21a2p	mouse	0.1	0.1	2.9	9/9*	M15008	Chaplin et al., 1986

[1] Size of the 5′ flanking region sequenced.
[2] Size of the 3′ flanking region sequenced.
[3] Size of the gene from the RNA initiation site to the polyadenylation site.
[4] The number of intron sites determined over the total number of introns. An asterisk indicates that the total sequence of the exons and introns was reported.
[5] Accession number for the CYP2D6(44M) polymorphic gene (Hanioka et al., 1990).
[6] Distance between the CYP2D3 and CYP2D5 genes.
[7] Distance between the CYP2D7P and CYP2D8P genes.

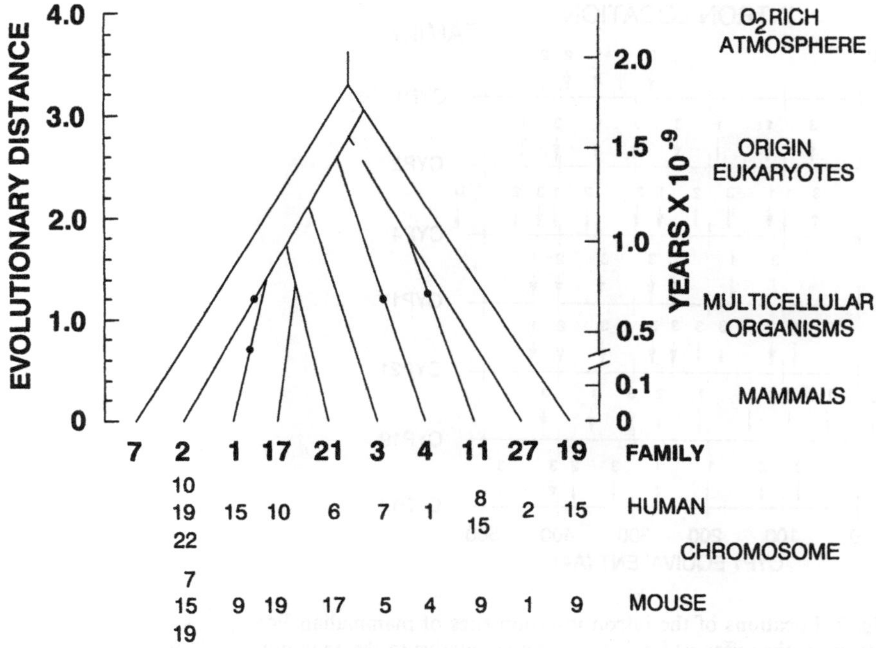

Fig. 1. Evolution of the mammalian P450 families. The evolutionary distance between the mammalian families calculated by the unweighted-pair-group method of analysis is shown. The approximate time in years that correspond to the evolutionary distance is taken from NELSON and STROBEL (1987). The approximate times of several critical evolutionary events are shown at the right. The small circles in the tree represent the time of the first gene duplication within the family leading to the formation of subfamilies and, thus, represent the ancestral gene for each family. At the bottom of the figure the chromosomal locations of the various families are shown. Listing of multiple chromosomes indicates that subfamilies are on different chromosomes. The evolutionary distances are based on the data in NEBERT et al. (1991).

large number of excellent reviews on various aspects of P450 molecular biology: comprehensive reviews (ADESNIK and ATCHISON, 1986; NEBERT and GONZALEZ, 1987, 1989; GONZALEZ, 1988, 1990; GOTOH and FUJII-KURIYAMA, 1989), categorization and nomenclature (NEBERT et al., 1987, 1989, 1991), regulation (WHITLOCK, 1987, 1989, 1990; NEBERT and JONES, 1989; WHITLOCK et al., 1989; OKEY, 1990, SIMPSON et al., 1990; KOOP and TIERNEY, 1990), evolution (NELSON and STROBEL, 1987; NEBERT et al., 1989; GONZALEZ and NEBERT, 1990) and genetic polymorphisms (GUENGERICH, 1988, MEYER et al., 1990).

2. Gene structure and evolution

2.1 Genomic sequences

P450 genes for which genomic sequences have been reported are listed in Table 1.

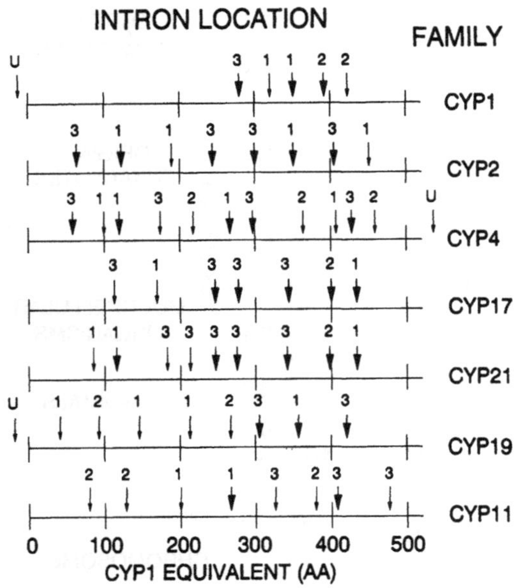

Fig. 2. Locations of the intron insertion sites of mammalian P450 genes. The positions of introns of the different families are shown relative to the analogous position in the CYP1A1 gene product. The protein sequences were aligned according to GOTOH and FUJII-KURIYAMA (1989). The large arrows indicate locations that are present in two or more gene families and the small arrows indicate unique sites. The numbers indicate the nucleotide in the codon after which the intron insertion occurs. The "U" represents intron locations in the untranslated regions of the mRNA.

There is remarkable diversity in the structures of these genes. The sizes vary from the CYP21 genes which are 3.2 kb long to the human aromatase genes (CYP19) which are greater than 70 kb. Large differences in size are even seen within a single family particularly the CYP2 family in which the length of 2C and 2D genes are 35 kb and 4.5 kb, respectively. Since the sizes of the mRNAs for each P450 gene are similar, changes in the lengths of the introns are responsible for the difference in size of the genes. For example, a large change in a single intron in the closely related rat CYP2B1 and CYP2B2 genes is responsible for the change in size of these genes (SUWA et al., 1985). The number of introns varies from 6 in the CYP1 family to 12 in the CYP4 family.

2.2. Chromosomal location

The locations on mouse and human chromosomes have been determined for all the mammalian P450 families except CYP7 (Fig. 1, see NEBERT et al., 1991 for references). During the evolution of this superfamily, the genes have spread to nine different chromosomes in humans. Genes in different subfamilies within the same family often are on different chromosomes, although there is some conservation, for example, three of the CYP2 subfamilies and the two CYP4 subfamilies are on the same chromosome. Within subfamilies, all the genes are on the same chromosome. The genes within a subfamily cluster presumably have remained near each other after gene duplication. Direct evidence for adjacent P450 genes has been obtained in the CYP2D and CYP21 families. Four genes in the rat CYP3D locus are present within 60 kb in the order 2D2, 2D3, 2D5, and 2D4 as determined by sequencing and Southern analysis (Matsunaga et al., 1990a). In the human CYP2D locus, two pseudogenes are separated by 4.6 kb (Kimura et al., 1989c). The CYP21 locus resulted from a duplication that included the C4 histocompatibility gene and the two CYP21 genes are separated by 24 kb in humans (Higashi et al., 1986; White et al., 1986). Gene subfamilies may also remain close together. The close linkage of the CYP2A and 2B subfamilies resulted in the initial erroneous identification of the coumarin-7-hydroxylase as a CYP2B gene rather than a CYP2A gene (Miles et al., 1990b).

2.3. Intron locations

There is both remarkable diversity and conservation of the locations of the intron insertion sites in the genes (Fig. 2). Between gene families there are generally only a small fraction of the intron sites conserved while within a gene family the locations of the introns are completely conserved. The only exception to this latter observation occurs in the CYP4 family. An intron in the 3′ untranslated region of CYP4A1 is not present in CYP4A2 or CYP4B1 and the fourth intron in CYP4A1 is not present in CYP4B1. The position of the first intron of CYP4A1 differs by two codons from that CYP4B1 and is probably an example of intron migration. There are some intron positions which are common between gene families based on the criteria that the intron interruptions are after the same nucleotide in a codon and are within 10 codons of each other. The first criteria is included since two independent mutations at either end of the intron are required to shift the insertion site within a codon and would be rare occurrences. The latter criteria is rather arbitrary to allow for some intron migration which might result from a point mutation that alters the donor or acceptor site for splicing. The CYP17 and CYP21 families share five common intron locations and a sixth, intron 2, is at the same location but splits different codon nucleotides. Families CYP2 and CYP4 share three intron sites and several families share one or two sites. Altogether there are 29 intron locations unique to individual gene families and 12 positions that occur in more than one of the seven families characterized.

2.4. Evolution

The evolution of the P450 genes has been analyzed by the unweighted-pair-group method of analysis (Nelson and Strobel, 1987; Nebert and Gonzalez, 1987; Gotoh and Fujii-Kuriyama, 1989; Nebert et al., 1991) and an abbreviated tree of the 10 mammalian familes is shown in Figure 1. The times shown in Figure 1 are based on estimates of Nelson and Strobel (1987) and are dependent on assuming a constant rate of acceptance of mutations. Divergence between bacterial and eukaryotic P450s is set to 1.4 billion years. A linear relationship from 0 to 1.4 billion years leads to estimates for the time of mammalian radiation that are not consistent with paleontological data so a break in the scale is made at the bird–mammalian divergence. Conservation of sequences for functional interactions and gene conversions can result in nonlinear rates of acceptance of amino acid change and thus the times shown should be considered only a rough guide. Gene conversions were first observed in CYP2B genes (Atchison and Adesnik, 1986) and appear to be a common phenomenon within several gene families (Higashi et al., 1988a; Matsunaga et al., 1990a, 1990b; Lakso et al., 1991). The neighbor-joining method, which makes no assumptions about a constant rate of change produces a similar tree and provides some assurance that qualitatively the relationships shown in Figure 1 are correct (Nebert et al., 1991). With those provisos in mind, it is interesting that the ultimate P450 ancestral gene arose about the same time that its substrate, oxygen, began to be present in the atmosphere. By about the time of the development of multicellular organisms, all the major families of P450 had evolved. The massive gene duplication in families one to four leading to subfamilies is largely a phenomenon occurring in multicellular eukaryotes. The rapid expansion of the xenobiotic metabolizing genes within these familes about 500 million years ago occurred during the period when land-dwelling vertebrates began to appear. This has led to the hypothesis that this expansion was an adaptive response to the inability of land animals to easily eliminate toxic chemicals in plants that were ingested, a form of plant-animal warfare (Gonzalez and Nebert, 1990).

2.5. Intron origins

The evolution of introns in the P450 genes and indeed, in genes in general, remains a matter of conjecture and controversy. The main argument has centered on whether introns are ancient or relatively new (see Belfort, 1991 for a short review). In the first case ancestral genes would contain many introns, perhaps obtained during formation of the gene by the fusion of multiple functional domains. Complete loss of these introns to increase efficiency would have occurred in rapidly dividing cells like bacteria, the genes of which do not contain introns, while a partial retention of introns in eukaryotic genes might provide some selective advantage. In the recent intron scenario, ancestral genes, like bacterial genes, would not contain introns and introns would have arisen because of the evolution of mobile insertion elements in eukaryotes. The case for recent development of introns is supported by the presence of conserved mobile self-splicing introns in widely separated phyla. The case for ancient introns received additional strong support by the discovery of structurally similar group 1 introns at identical locations in $tRNA^{Leu}$ genes in cyanobacteria and chloroplasts (KUHSEL et al., 1990; XU et al., 1990). It seems likely that these two mechanisms are not exclusive and that both contribute to the structure of present day mammalian genes.

The interesting differences in intron structure in P450 genes also seem most compatible with both loss and gain of introns. The similarity of structure that is apparent in comparing protein sequences of P450s from bacteria to humans strongly supports the hypothesis that this gene superfamily was generated from a single ancestral gene by gene duplication and divergent evolution. This is further supported by the presence of up to 12 sites that are probably common in genes in different families (Fig. 2). The probable presence of common sites in CYP19 and CYP2 genes which diverged about 1.7 billion years ago is consistent with the ancient theory of introns. On the other hand, there are 29 intron sites unique to individual families and the presence of 41 intron sites in the P450 genes characterized so far would require a very large number of introns in the ancestral gene. This number is presumably much larger than 41 since loss of introns has resulted in near unique structures of some P450 genes, for example CYP11 genes which contain only two common intron locations with other genes. If intron loss is random, a very large number of ancestral introns would have to be postulated to explain with a reasonable probability the existence of genes such as CYP11 and CYP19 which contain a total of 20 introns with no common locations. A more reasonable conclusion is that intron loss has been accompanied by insertion of introns at random sites within the genes.

Regardless of the mechanisms which led to present intron locations in the P450 genes, the rate at which such changes were accepted remains an enigma. Within gene families, the intron location is invariant, yet presumed ancestral genes separated by similar or lesser evolutionary distances have almost completely different intron structures. For example, within the CYP2 family, the CYP2D genes separated from the other CYP2 subfamilies about 0.7 billion years ago (1.4 billion years apart including both legs, Fig. 2) yet identical intron locations are present in these two gene subfamilies. This conservation of intron location occurred in spite of rather dramatic gene duplications generating the multiple members of these two subfamilies, and genetic deletions and/or insertions in the introns as evidenced by the 10-fold difference in sizes of the introns in CYP2D genes and CYP2C genes. In contrast, the distance between the ancestral CYP2 gene and the ancestral CYP1 gene is only about 0.55 billion years, considerably less than the 1.4 billion years for the CYP2 subfamilies, yet only one of the 13 introns in these two postulated ancestral genes is common. These considerations would imply that either the estimated evolutionary distances are greatly in error or that the rates of intron location change have greatly decreased relative to that of amino acid change. Major factors that might affect the estimate of the rate of amino acid changes betwen P450s are gene conversion and accelerated change by selective pressure. There is compelling evidence for gene conversion which has been reviewed in some detail (ADESNIK and ATCHISON, 1986; GONZALEZ and NEBERT, 1990). Gene conversion would likely reduce the rate of change between related P450 genes and thus would result in underestimates of evolutionary distance. Since massive gene duplication has occurred recently, gene conversion would be expected to be an important factor. However, this would bias evolutionary time estimates in the wrong direction to explain the recent conservation of intron locations within families. For many of the subfamilies, selective pressure for amino acid changes might exist in order to provide a broader protection against toxic chemicals that are ingested. This might lead to an overestimate of the evolutionary distance for recent times. While this would be in the expected direction to explain intron location conservation, structure-activity studies have shown that single point mutations can dramatically alter substrate specificity of P450s (LINDBERG and NEGISHI, 1989; KRONBACH et al., 1989). It seems unlikely that such a selective drive could explain the large differences in estimates of evolutionary distances required to explain the differences in the apparent rates of intron location change.

The rate of intron location change then appears to have changed dramatically relative to the rate of amino acid change. From the estimated evolutionary distances, the evolution of the P450 families occurred before the appearance of multicellular eukaryotic organisms, while the evolution of the subfamilies occurred largely after this time. In the early single cell organisms, introns may have been primarily self-splicing mobile elements which would have permitted a high rate of change of intron location as a result of deletion and insertion. In eukaryotes, in which the P450 subfamilies developed, the nature of introns presumably evolved toward the modern higher mammalian introns which have lost the self-splicing activity and are no longer mobile. These introns might be expected to be more stable and, thus, the rate of intron location change would be greatly reduced in more recent times in agreement with the observed conservation of intron location within families.

The occurrence of common intron sites in some cases roughly correlates with the evolutionary relationships determined by the unweighted-pair-group method of analysis of protein sequences which has led to the suggestion that this property be considered in evaluating evolutionary relationships (NEBERT et al., 1991). The relatively closely related CYP17 and CYP21 genes share five common intron sites while CYP11 genes, which are distantly related to the other gene families, share only two sites with all other genes. On the other hand, CYP2 and CYP4 genes which separated about 1.4 billion years ago contain three common sites while CYP2 and CYP1 which separated about 0.8 billions years ago, as did CYP17 and CYP21, share only one common site. The CYP19 genes are most distantly related to the other P450 genes and contain three common sites, two with CYP2. Even though the designation of common sites in distantly related genes invariably requires postulation of some intron migration, disconcordance seem too frequent and the number of common sites too small to contribute greatly to the determination of evolutionary relationships.

3. Regulation of gene expression

3.1. Chromatin structure

Changes in chromatin structure are often associated with activation of the expression of a gene (ESSIENBERG et al., 1988; GROSS and GARRAD, 1988). Usually the entire region near an active gene becomes selectively sensitive to nuclease, but discrete regions of about 200 base pairs, termed hypersensitive sites or regions, may exhibit extreme sensitivity to nuclease digestion. Hypersensitive regions are thought to represent nucleosome-free regions accessible for binding of transcriptional regulatory proteins. Several studies have examined the DNase 1 or restriction endonuclease sensitivity of P450 genes. In general, changes in nuclease sensitivity have not correlated well with the induction of P450 gene expression.

3.1.1. DNase 1 sensitivity

Genes which are induced by polycyclic aromatic hydrocarbons have been analyzed in the most detail. The P450m gene is expressed constitutively but is induced about 4-fold by treatment of rats with 3-methylcholanthrene (EINCK et al., 1985). Digestion of rat liver nuclei with DNase 1 followed by slot blot analysis showed that bulk DNA, insulin gene DNA, and somewhat surprisingly the albumin gene, were equally sensitive to nuclease digestion. Treatment with 3-methylcholanthrene had no effect. In contrast, the 5′ flanking sequence of the P450m gene was more sensitive than bulk DNA and was also unaffected by drug treatment. The central exon region in untreated samples was similar to bulk DNA but sensitivity was increased by 3-methylcholanthrene treatment of the rats. By indirect Southern analysis, with an intron three probe, three DNase 1 hypersensitivity sites in the intron regions and 1 in the 5′ flanking sequence were identified, regardless of treatment with 3-methylcholanthrene. Similar experiments with thymus nuclei revealed no hypersensitive sites. Treatment of the nuclei with micrococcal nuclease produced a nucleosomal ladder for bulk DNA but not for the central exon region of P450m, indicating that at least this region of the gene was not organized into nucleosomes. These data indicate that the expression of the P450m gene is associated with increased DNase 1 sensitivity and the presence of discrete hypersensitive regions, which are not changed by treatment with an inducer.

The rat CYP1 genes, which in contrast to P450m are not expressed constitutively, were also studied by the same group (EINCK et al., 1986). Micrococcal nuclease digestion indicated that these genes also were not in a nucleosomal structure. By indirect Southern analysis with a probe detecting both CYP1A1 and CYP1B1 genes, two hypersensitive sites the 3′ region of the gene and one site in intron 1 were detected in uninduced samples. In 3-methylcholanthrene treated samples, the intron site was not detected and a new site in the 5′ flanking region near the promoter was observed. Judging from the intensities of the bands these sites are probably in CYP1A1. Using a 200 bp probe near the 5′ promoter region of the rat CYP1A1 gene, FOLDES and BRESNICK (1989) detected a distal hypersensitive site at about −3 kb in both treated and untreated samples. The position of the probe was the same as that of the induction dependent hypersensitive site mapped by EINCK et al. (1986) which, therefore, could not be detected.

In the mouse CYP1A1 gene, 2,3,7,8-tetrachlorodibenzo-p-dioxin (TCDD) treatment did not alter any DNase 1 hypersensitive sites (DURRIN and WHITLOCK, 1989). General increases in sensitivity to restriction nucleases after TCDD treatment were detected from −1200 bp to −167 bp in the 5′ flanking region as well as the +800 region. These changes were dependent on functional interactions of the Ah receptor with the gene since no changes were seen in a variant cell line with defective receptor binding to the gene. The changes were detected within five minutes of TCDD treatment and were independent of protein and RNA synthesis as assessed by inhibitor studies. These studies in the rat and mouse suggest that activation of the CYP1 genes results in general increased nuclease sensitivity and an increase in a DNase 1 hypersensitive site in the proximal promoter region.

A comparative analysis of DNase 1 hypersensitive sites in phenobarbital-responsive and unresponsive rabbit P450IIC genes revealed two major hypersensitive regions, a proximal site near the promoter and a distal site at about −2 kb (KIM and KEMPER, 1991). Like the results with rat CYP1A1 gene, the proximal site correlated well with gene expression. It was present in CYP2C2 and CYP2C3 genes in both control and phenobarbital-treated liver samples and only in treated samples for the CYP2C1 gene which is consistent with the pattern of expression of these genes (LEIGHTON and KEMPER, 1984). Only CYP2C2 is expressed in kidney and the proximal site was present for this gene in kidney samples. In contrast, the −2.0 kb site was found only in the phenobarbital-responsive genes, CYP2C1, CYP2C2, and CYP2C4 but not in the constitutive CYP2C3 gene. This site was present, however, in both treated and untreated samples in liver and kidney. The presence of the distal hypersensitive site in the phenobarbital-responsive genes and the observation that DNase 1 sensitivity was increased in this distal region by phenobarbital treatment suggest that it may play a role in regulation of the gene. Such regulation by interaction between upstream sequences and sequences proximal to the promoter is a common phenomenon in other genes including the CYP1A1 gene (see below).

In another study, two DNase 1 hypersensitive sites were detected in the 5′

flanking region of the rat P4502C12 gene which is predominantly expressed in the female liver (ZAPHIROPOULOS et al., 1990). There were, however, no sex-dependent differences in the hypersensitive sites. The study of hypersensitive sites in several P450 genes, therefore, has not identified discrete regions of the genes which are responsible for the induction of P450 gene expression. However, as expected from observation in other genes, hypersensitivity near the promoter of P450 genes and general sensitivity to nuclease correlate with gene expression.

3.1.2. Methylation of genes

The methylation of genes has been shown to correlate in some cases with the expression of the genes and actively expressed genes are usually hypomethylated (CEDAR, 1988). In mammalian cells, the dinucleotide CpG can be methylated. The methylation status of a gene may be studied by exploiting restriction enzymes such as Hpa II, Hae II, and Hha I that contain CpG in their recognition site but do not cleave at sites that are methylated. Msp I has the same recognition site as Hpa II but does cleave at methylated sites so that differences between Hpa II and Msp I digestions reveal methylation patterns. The relationship between methylation and gene expression has been studied in detail in CYP2D and CYP2E genes.

Rat CYP2E1 expression is undetectable at birth and increases to high levels within a week of birth (SONG et al., 1986). Methylation of the gene as a function of age was studied with the enzymes Hpa II and Hha I (UMENO et al., 1988c). Since the sequence of the gene was known, the location of each site could be determined and sites in the 5′ flanking region through intron 6 were studied. Prior to birth all sites were methylated and within one day after birth the site in the 5′ flanking region was partially demethylated. By one week demethylation of a second site near exon 2 was also observed. After 10 weeks the first two sites were completely demethylated as was the next site toward the 3′ end in exon 3. The early demethylation of sites near the 5′ end of the gene correlates well with the increased transcription during the first week, but the later progressive demethylation is unrelated to transcription changes.

The CYP2D3 gene is not expressed until about two weeks of age in the rat after which expression increases until puberty, while the CYP2D5 gene is expressed at peak levels within the first week. Methylation of the CYP2D3 and CYP2D5 genes was studied by using Hha I, Hae II, Hpa II, and Msp I (MATSUNAGA and GONZALEZ, 1990). Eight sites in CYP2D3 from the 5′ flanking region to intron 6 and one site in the 3′ flanking region were studied as a function of age. All sites remained >95% methylated except for two sites in exon 1 which were methylated at one week but were <60% methylated at 10 weeks. Similar studies in CYP2D5 revealed that all sites were >95% methylated at 20 days of gestation but that sites in exon 1 and intron 1 were partially demethylated by seven days and <40% methylated by 10 weeks. One site in exon 1 was partially demethylated as early as

```
             -200                -180                -160                -140                -120                -100
R1A1.1   GTGTGTGTGTGTGTGTGTGTGTGTGCCTGCCTGTACCTACATCAGGTGTCCCCATTTGGCTTCTTGCCTATCTCATCCATCCCCCTCCTGCCTAGAGCACT
M1A1.4   TCTTTTTCTTTTCTTATGTTTTTTTGTGCCTGTACCTACATCAGGTATCCGGTATGGCTTCTTGCCTATCTCCCCCCTCCCCCCTGCCTAGAGCACTCCCT
                                                                     SP1                 CBP
H1A1.5   TGTACAGGCACCGAGATGTGTCCCATAGTGGGTTCTTGCCCACCCGACCCCCCACCCCCGCCGCCCTCCGCCACCTTTCTCTCCAATCCCAGAGAGACCAG
            HNF4A            HNF4A
R1A2.7   CCATGGACCCAGAACCACTTCTGACCCAAAGAGCTATTCCTAAGCCCCAAGTGAAATGTGGAACTGAGGGATCATGGCTTGAAGCCAAGAATTGATCCTTT
                              HNF4a     CBPa    HNF4A
M1A2.8   TCTCCATGAATATTTATCTCCATGGACCCCAAGCCACTTCTGACCCCAAGTGAAAGGTGGAACTGAGGAGCCATGGCTTGAAGCCAAGAATTGATCCTTTC
                                           CRE                                    HNF4a
H1A2.10  CCCAGGGACCCCACCCCTTCTGGCCTATCCCCAAAGAGTCACCCTGGGTCTTAGGTAGTAGGTGGAGCTGAGGGATAATGGCCCAAGGCCAAGAGTTGACT
                                                                                              HNF4a
R2A1.11  AAGCACACCGAGGCTTTAAGAATTTTGAAGTAAGCCTGCCACCCAGAGGTCTTCCTATTTGCCCATGTTGTGGGTGTTGCAACAAAGACAGGGTCAGTGTT
                                                                                              HNF4a
R2A2.12  AAAAAAAAAAAAAAAAAAAAAAAAGAATTTTGAAGTAAGCCTGCCACCTTTCTTCCTATTTGCCCATGTTGTGGGTGTTGCAACAAAGACTGGGTCAATGTT
                                                                                                    CRE
R2A3.13  TTCCTCCTTAGACACCACAAACTCAGTCTACTCCCTGTTAATCTGTCATGGAGCCCTCCTCCTCCTTGAGTGTTGGCTATGTCCCAAACTAGGTGCATGAC
M2A4.14  ACTTTCTCCTTAGCTACCACAGATTCAGTCTACTCCCTGTTAATCTGTCATAGGGCCCTCCTCCTCCTCCAGTGTTGGTAATGTCCCAAACTTGGTGCACA
M2A5.15  ACTTTCTCCTTAGCTACCACAGATTCAGTCTACTCCCTGTTAATCTGTCATAGGGCCCTCCTCCTCCTCCAGTGTTGGCAATGTCCCAAACTTGGTGCACA
                                                  GRE                 GRE
R2B1.16  ACCCATCCCCATAAAACAAGAGGCCTAAGTCCCAGTGCCCTTTTGTCCTGTGTATCTGTTTCGTGGTGTCCTTGCCAACATGTATGGTGTGGGTAAGGGAA
                                                  GRE                 GRE
R2B2.19  ACCCATTTACATAAAACAAGAGGCCTAAGTCCCAGTGCCCTTTTGTCCTGTGTATCTGTTTCGTGGTGTCCTTGCCAACATCTATGGTGTGGGTAAGGGAA
                                                              GRE
R2B8.22  CTATCTCATTTCATGATGTCTTTGCCAACATCTAAGGTTGGGATAAAGGAATGGGGAGTGAACAGGTCACCAGGAGCCCAAACAGGAAGTATGGGTGCTCA
                         GRE
M2B12.23 GCATGTATCTGCCCATGTCCTTGCCAAAGTCTAGGGTGGTTGTAATGGAATAGCGAATGAGCAGGTCACCAGCAGCCCAGGTAGGAACTATGGTGCTCACA
                         GRE
M2B13.24 GCATGTATCTGCCCATGTCCTTGCCAAAGTCTAGGGTGGTTGTAATGGAATAGTGAATGAGCAGGTCACCAGCAGCCCAGGTAGGAACTATGGTGCTCACA
               HNF4a     GRE
M2B14.25 GTATCTGGTTCATGGTGTCCTTGCCAAAGTCTAGGGTGGTTGTAATGGAATAGTGAATGAGCAGGTCACCAGCAGCCCAGGTAGGAACTATGGTGCTCACA
                         CBP              CBP                                           HNF4a HNF4b
r2C1.26  AGAATTTCTGGGTTGGATTGGGGATGAGGAAAGTATTGGAAAGTAGAGCAAGGGGAAAGACAGTTTATTTTCATTGTTTCAGTGGACCAAAGTCCATCCAG
         GRE                                                    GRE                         HNF4a HNF4b
r2C2.27  AGAACAGGACCCCTGCTCTTGGGTGGAGGAGGGGAGCACTGGAAAACAGAGCAAGAGGAAAGAACATCTTATTTTCACCTTTTTCAGTGGTCCAAAGTCCA
r2C3.28  GACACCTTTTAGAAAAAAAAAAAAAACCTTTTTTTGGAGAAGTGAAAGGTAGTATTTAGTGGGAATTTCAACATTTAAAATGATTGTTTTTTTTTCCTACC
r2C4.29  CATAGGAGCAACCCAGTACAACGAAATTTCCATTTCCAAGTGGGCACAGGAAAGAATGCAATGCGACTAAAGAATATCTTGTTTTCATCTATGTCATGGAG
r2C5.30  GGAGCAACCCAGTACAATGAAATTACCATTTCCCGAGTTGGGCACAGGAAAGAATGCAATGCGACTAAAGAATATCTTATTTTCATCTCTGTCGTGGAGGC
                                                                                                  HNF4b
R2C6.31  CCTTGGGAAGGTCAAGCAGATAAGTACACAGTTTGAGTTATATCATGAGATGTTTAAGAACGAAGAAGAGCTTATTTTTGTTATTGTCCCTAGGATAAAGT
                                                     GRE                                HNF4a
R2C11.33 GGTCCCACCCCTGGTTATTACACAGATTTGTGGTAGTAGAAAGAGGAAGAACAGTTTTCACTTGTGTATTTTAACAGGTCAGGGTCCACAAAGAAGAAATA
                                                                                             HNF4b
R2C12.34 TAACTCAATAAAATTAAGCCAAAATCTCTTACAGACTTCCAACAAAAATGACAAAGTTAAAGGAGCTATTTTTGTCTTATCATTGAGACAAAGTTCATTTG
R2D2.35  GATTTCAATGCCACTCCCATGGACAAGCACCACCGGAGAGAAGAAAGGGCAAGAACCTCTGATGTTTATTTTCCCTGGACTTTTCTGACACCCCTTCCCCC
                        HNF4a                                                           HNF4a
R2D3.36  CCGTCTTGAAGCAGGGGTCCACTGGACATCTGCCTACCATGCTCAGCCCAACAGAATCACAATGCTCAGCTCTGGAAAGAAGTGAACCTCACTGGCCCCTC
                 XRE                                                               HNF4a
R2D4.37  ATGCACACACGCATGCACAAACAGAACTTAGAGAAGGAGGGATCTCTAAGACCAAATGCCCCTGGCCTGCCTTGAACCTTCCTTGCACTCTGCATGGTCAG
                        HNF4a                                                           HNF4a     HNF4a
R2D5.38  CCGTCTTGAAGCAGGGGTCCACTGGACATCTGCCTACCGTGCTCAGCCCAACAGAATCACAATGCTCAGCTCTAGTAGGAACTGAACCTCACTGGCCCCTC
```

Fig. 3. Sequences of the proximal promoter regions of the mammalian CYP genes. Sequence is shown from 200 bp upstream of the RNA initiation site (−200) to the first nucleotide in the mRNA (+1). Sequence numbers in the figure may not agree precisely with the numbers for sequences used in the text which are taken from authors' papers. In some cases the assignment of the RNA initiation site differs between laboratories. The gene is indicated on either side of the sequence with the species indicated for rat (R), rabbit (r), mouse (M), human (H), bovine (B), and porcine (P), followed by the name of the gene. The number after the period refers to the number of the sequence as listed in Table 1. Lines over the sequences indicate regions that were protected from DNase 1 digestion by nuclear proteins. Underlined are motifs that were

```
                  -80                 -60                 -40                 -20                 +1
                   •                   •       BTE         •          TATA     •                   •
CCCTAAGGCTGTCCCTCCCTCAGTCCCACCACTGGGCTCAGAGAAGGAGGCGTGGCCAACAGACAGTCATATAAAGGTGGTGGTGCCTTCACCCTAACCA R1A1.1
                   •                   • BTE               •          TATA     •                   •
AAGGCTGTCCCTCCCTCGGTCCCACCACTGGGCTCAGATAAGGAGGCGTGGCCAACAGACACAGAGTCCTATAAAGGTGGTGGTGCCTTCACCCTAACCC M1A1.4
     HNF4a         •                   •        BTE        •              TATA•                    •
CCCGGTTCAGGCTGCTTCTCCCTCCATCTCAGCTCGCTCCAGGGAAGGAGGCGTGGCCACACGTACAAGCCCGCCTATAAAGGTGGCAGTGCCTTCACCC H1A1.5
                   •                   •                   • BTE        TATA   •                   •
TAGCTTTGTGCAGTATCCAGCTTTGATATCAAGTGACCTGGGCAGCTGAGCCAGTCCAGAGGAGGTGGTATTATAAAAGGCCCCTTGTCAGAACCCAGAA R1A2.7
                   •             HNF4a •     HNF4A         •          TATA     •                   •
AGCTTTGTGCAGCATCCAGCTATGATATCAAGTGACCCGGGCACCTGGGCCAGTCCAGAGTGGGTGGTGTTATAAAAGGCCCCTCGTCGGAACCCAGAAG M1A2.8
                   •                   •           CBP •     SP1          TATA•                    •
TCCAACTTTTTCAGTGATCCAGCTTTCATATCAGGTGATCAGGACAACCAGGCCAATCTGATGGGGGCGGTGTTTATAAAAAGGCCTCACCTAGAGCCAG H1A2.10
      CBP          •                   •     HNF4b         •           TATA •                   •
AGGAGATAGGATTGGAGGGTAAAAGACTCAACTAGACAAACAGGAGCAAAGGCCATCCTGTGTCCCTGGGAGTATAAAGGTACTATCTCAGCCTTGGCTA R2A1.11
                   •                   •     HNF4b         •           TATA •                   •
AGAAAATAGGGTTGGGAGGCAAAAGACTCAACTAGACAAACAGGAGCAAAGGCCATCCTGTGTCCCTGGGAGTATAAAGGTACTATCTCAGCCTTGGCTA R2A2.12
.                  •                   •                   •           TATA •                   •
GGGGGTAGTCTGGAGGTAAAATTAGACAGTTGCATAATCAAGACCAAAGGCCAGCCTTCTATCTCAGAATGTATAAAAACAAGCTAATTCAGCCATCACT R2A3.13
                   •                   •                   •        TATA       •                   •
ATAGTGGGTAGTCTGGAGGTGAAATAGTTGCATAATCAAGACCAAAGTCCGGCCTTCTGTCTCTGGATGTATAAAAGCAAGCTAATTCAGCCATCACTGT M2A4.14
                   •                   •                   •        TATA       •                   •
ATAGTGGATAGTCTGGAGGTGAAATAGTTGCATAATCAAGACCAAAGTCCGTCCTTCTGTCTCTGGATGTATAAAAGCAAGCTAATTCAGCCATCACTGT M2A5.15
             B.m.  •   BTE             •                   • SP1       TATA •                   •
TGAGGAGTGAATAGCTAAAGCAGGAGGCGTGAACATCTGAAGTTGCATAACTGAGTGGAGGGGCGGATTCAGCATAAAAGATCCTGCTGGAGAGCATGCA R2B1.16
             B.m.  •   BTE             •                   •           TATA •                   •
TGAGGAGTGAATAGCCAAAGCAGGAGGCGTGAACATCTGAAGTTGCATAACTGAGTGTAGGGGCAGATTCAGCATAAAAGATCCTGCTGGAGAGCATGCA R2B2.19
                   •                   •                   •    SP1          TATA•                 •
CATACCTGGACAGTGATGAGACCGGGAGGTGGGAACAGCTGGAGTTGCTTAAGTGAGTAGGGAGGCGGGCTCATGATAAAAGATCCTGCTGGAGTGTTCC R2B8.22
                   •                   •   TATA            •                •GRE               •
TTCCTGGACAGTGTGAGACAGGAGGTGGGAACAGTTGGAGTTGTATAACCTAGTGGGGAGGTGGCTTAAGCATCAAAAGTCCTGCTGGACTGTGCACTGA M2B12.23
                   •                   •   TATA            •                •GRE               •
TTCCTGGACAGTGTGAGACAGGAGGTGGGAACAGTTGGAGTTGTATAACCTAGTGGGGAGGTGGCTTAAGCATCAAAAGTCCTGCTGGACTGTGCACTGA M2B13.24
                   •                   •   TATA            •                •GRE               •
TTCCTGGACAGTGTGAGACAGGAGGTGGGAACAGTTGGAGTTGTATAACCTAGTGGGGAGGTGGCTTAAGCATCAAAAGTCCTGCTGGACTGTGTACTGA M2B14.25
                   •                   •HNF4a              •           TATA •                   •
AAGAAACACCAAAGTACATCAATCCACTGGGTACTTTGTGGGTCAAAGACACACATCAATAATAGCGAGTATTATAAAAACCTGCGATTGCCAGCTCTCA r2C1.26
                   •         CBP       •                   •              TATA•                    •
CTCAGAGAGAATGAAGTACACCTAGTTCATTGGGTACTTTTTAGGTCAAAGGACACAAAATTAGCAGGAGTGTTATAAAAGCCTGGCTTGCAGCTTCCA r2C2.27
        HNF4b      •         HNF4a     •                   •           TATA •                   •
GTAAGACCAAAGTGCAATGTTTATTGAAGGTTCAAAGATTAGTACATATTTGGTGATAAGATTGTTGTGCATTTATAAAGTCTTTGAATGCTCAACTTTA r2C3.28
   HNF4b           •                   • HNF4a CBP         •              TATA•                    •
GCAAAGTCCACTCAGGAGAGAGGATAAAGTACATCGTTGCTTGACCCGTCAGAGAAGCATATTAGCAAGAGTGGTATAAAAGCCTGTGCTTCCCAGCTCA r2C4.29
HNF4b              •                   •                   •           TATA •                   •
AAAGTCCACTCAGGAGAGAGGATAAAGTGCACTGTTGCTTGACCCGTCAGAGAAGCATATTAGCAAGAGTGGTATAAAGCCTGTGCTTCCCAGCTCACAG r2C5.30
                   •                   •                   •        TATA       •                   •
TCATGTAAAAGAAAAAGATCAGCAGGTAGCTGATTAGTCAATTATGCAATATTGATTCAGCAGCCACATGATAAAAGCTCTATGCTTCCTGGCAGAGCTG R2C6.31
                   •CBP                •           CBP     •        TATA       •                   •
TTAAAGCATATCTAGTTGATTGGTCACTTAGGTATCAGAAGCTCATGTTGAATTGGCAGCTAGCTAGCATTATAAAAGTCCTGGACAGCAAGCTCACAGG R2C11.33
                   •                   •                   •           TATA •                   •
AAAGAGAGATAAACAGTGGCCAGATGGCTGTGCAAAGTTTGCCAAAAGCAAAATATTGATTTTTATGGTGTTATATAAAATCTCTGGAGTGCCTGATAAC R2C12.34
                   •                   •       GRE  HNF4b •          TATA      •                   •
ATACAGGCTCTGTTTGAACATTAACACAGCTACTTGGCTCAATTTAGAGAACAAAGTGCAGCTTTACCCCCTTTTAAGGGACCAGGTTCCCAGTCTGTTG R2D2.35
                   •         CBP HNF4a •    GRE  GRE HNF4b•          TATA      •           CBP     •
CCTACACTTGACCAAACAAAACACAGGATTGGCCCTGGAGGCAGAGGACAAAGGACAACCCTGGCTCTCCTTTATAAGGGGACAGGCTCCAAGATTGGGA R2D3.36
     CBPa          •                   •                   •      TATA         •                   •
TGCAGCCACTTGGCTTAAGGCAGAGGAGAAAGGGCAGCTCGGTCCCTTATTGAGTGCCCAGGGCCTGATAATGCTATCGTTCCTTGGTCCTAGGTGGCTG R2D4.37
                   •         CBP HNF4a •       HNF4b       •          TATA     •                   •
CCTACACTTGACCAAACAAAACACAGGATTGGCCCTGGAGGCAGAGGACAAAGGACAACCCTGGCTCTCCTTTATAAGGGGACAGGCTCCAAGCTTGGGA R2D5.38
```

selected for inclusion in the figure because of interest for P450 expression, that appeared frequently or that were described in P450 papers. The University of Wisconsin Genetic Computer Group program "Findpatterns" was used to search for sequences in the P450 sequences that match motifs in the Eukaryotic Transcription Factor Recognition Sites file(release 2.1.) compiled by Dr. David Ghosh. Sites from the transcription factor file include: TATA, TATA box sequence; SP1, Spl general transcription factor; CBP, usual CCAAT motif; CBPa, variant of the CCAAT motif; HNF4a, consensus sequence for hepatic nuclear factor-4 (Hardon et al., 1988); CRE, cAMP responsive element; GRE, half site for the glucocorticoid hormone receptor response element; OCT, octamer binding

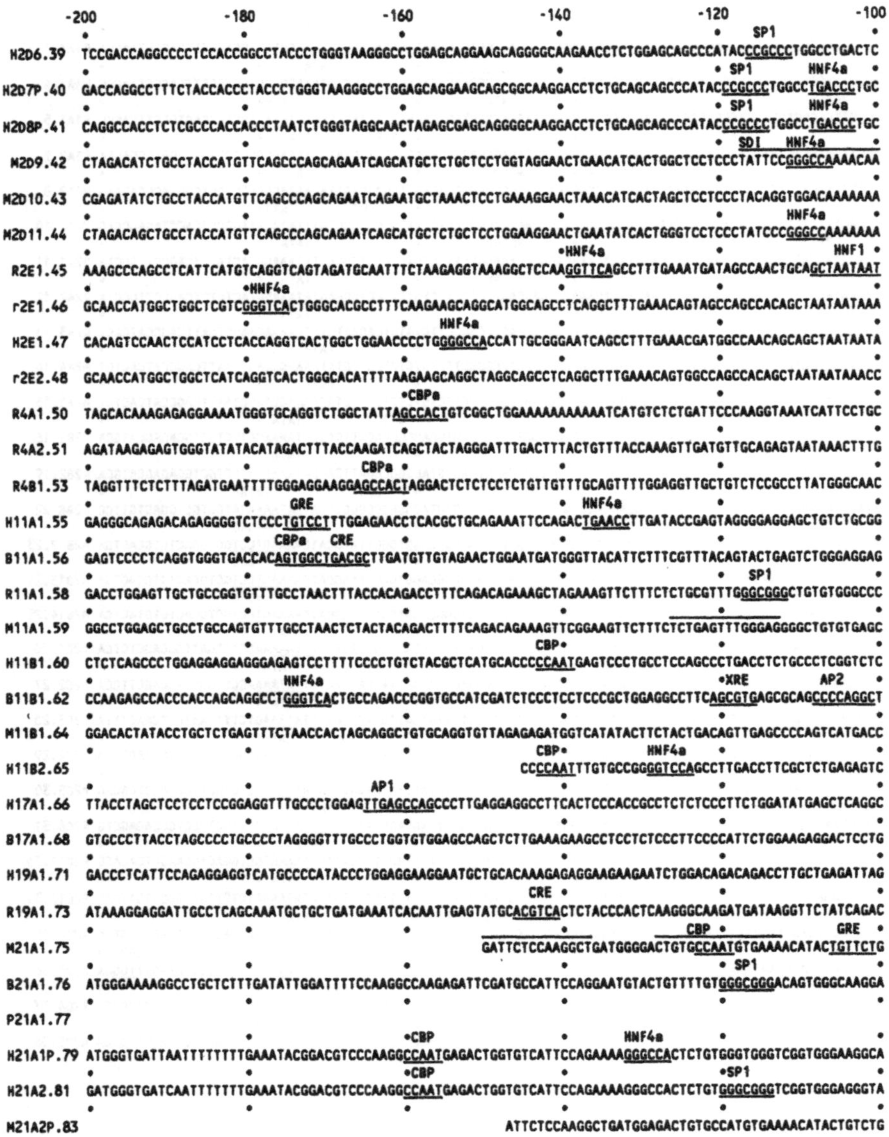

protein motif; XRE, xenobiotic response element; AP1 and AP2, activating proteins 1 and 2. Other symbols are: BTE; basal transcription element (Yanagida et al., 1990); HNF4b, HNF-4-like sequences conserved in rabbit CYP2C genes (Zhao et al., 1990); B.m., conserved sequence in *Bacillus megaterium* and CYP2B1/2 genes (He and Fulco, 1991); CTE, common transcription element (Yoshioka et al., 1990), Ad1 and Ad2, regulatory elements in CYP11B1 (Morohashi and Omura, 1990), C21R, sequences in Cyp11a1 similar to regulatory elements in Cyp21a1 (Rice et al., 1990).

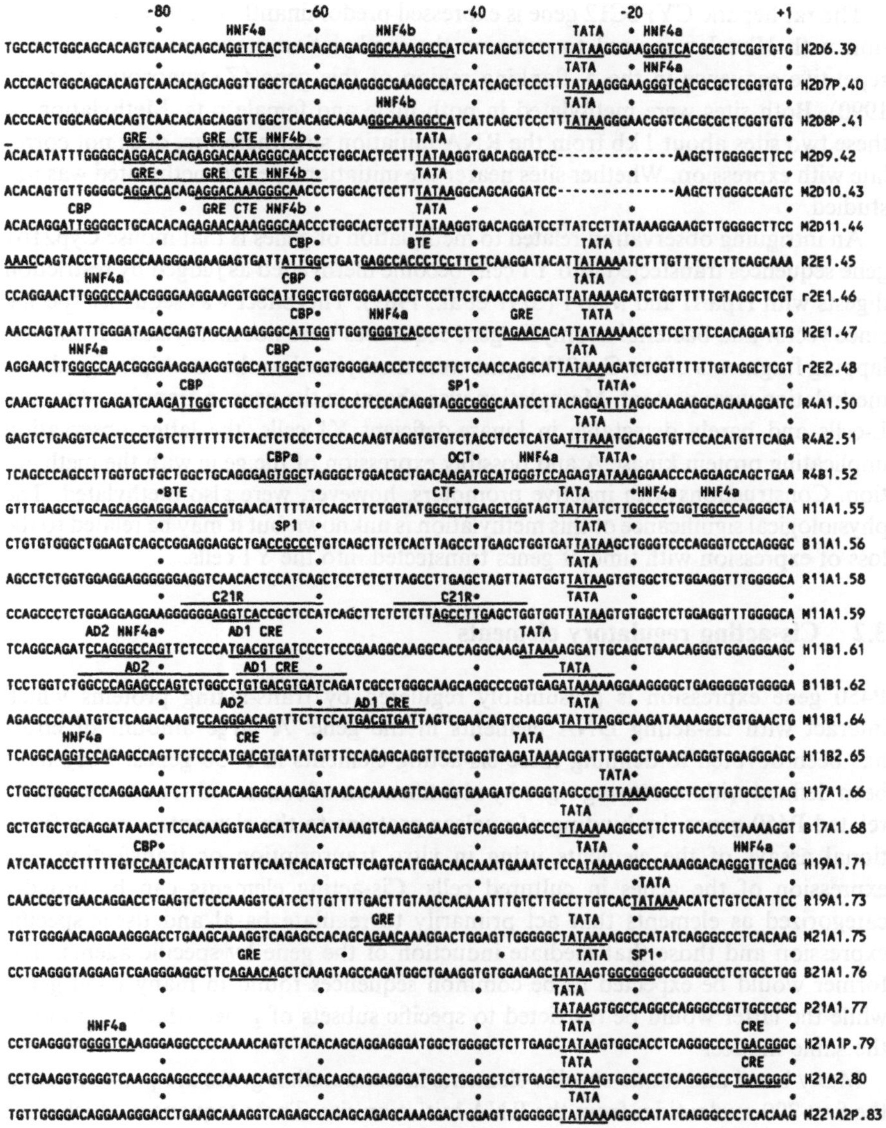

Fig. 3. Continued.

one day after birth. These results suggest that in the CYP2D genes like the CYP2E gene, demethylation of sites near the RNA initiation site but not in other regions of the genes, is associated with gene expression. Whether the demethylation is responsible for the activation of expression or a result of active expression is not yet clear.

The rat hepatic CYP2C12 gene is expressed predominantly in the female. Digestion with Hha I was used to examine the methylation status of two sites in a repetitive sequence in the 5′ flanking region of this gene (ZAPHIROPOULOS et al., 1990). Both sites were methylated in both male and female rats. Methylation of these two sites about 1 kb from the RNA initiation site, therefore, does not correlate with expression. Whether sites nearer the initiation site are methylated was not studied.

An intriguing observation related to methylation of genes is that mouse Cyp21b1 gene sequences transfected into Y1 cells become methylated as judged by restriction digests with Hpa II and Msp I (SZYF et al., 1989). This effect was sequence specific since vector and bacterial neomycin gene sequences were not methylated. Nonoverlapping fragments of the Cyp21b1 gene were methylated so that multiple signals for methylation were present. Methylation was absent in the gene in transfected mouse L-cells and barely detectable in kinase-deficient Y1 cells, the latter observation implicating protein kinase A and possibly expression of the gene with the methylation. Constructions with inactive promoters, however, were also methylated. The physiological significance of this methylation is unknown but it may be related to the loss of expression with time of genes transfected into the Y1 cells.

3.2 Cis-acting regulatory elements

P450 gene expression is presumably regulated by trans-acting proteins which interact with cis-acting DNA elements in the gene. A large amount of effort has been devoted to defining these cis-acting elements in P450 genes. They have been defined with increasing rigor by identification of conserved sequences among related P450 genes, by binding of nuclear proteins to the elements, and by functional assays of the elements using in vitro transcription or transfection and expression of the genes in cultured cells. Cis-acting elements can be broadly categorized as elements that act primarily to regulate basal and tissue-specific expression and those that mediate induction of the gene by specific agents. The former would be expected to be common sequences found in many P450 genes while the latter would be restricted to specific subsets of genes which respond to the same inducer.

Many basal and tissue-specific elements in mammalian genes are present within the first 200 nucleotides from the RNA initiation site. These sequences which have been determined for the P450 genes are shown in Figure 3. Elements that mediate induction on the other hand often have properties of enhancers and may act at a large distance from the initiation site in an orientation independent manner.

3.2.1. Common basal elements

The only common regulatory region shared by all P450 genes is an AT rich region about 20 to 25 nucleotides from the RNA initiation site. The consensus for this sequence is TATAAA and it is referred to as the TATA sequence. As shown in Figure 3, all P450 genes have an AT rich region in the appropriate position and most have at least the sequence ATAA. Exceptions are the rat CYP4A genes and the CYP17 genes which contain nonconsensus AT rich regions.

The CAAT sequence is also a common cis-acting element. A protein binding to these sequences has also been found to be important for liver-specific expression of genes (LANDSCHULZ et al., 1988). As can be seen in Figure 2, only about 40% of the sequences have a form of CAAT sequence and with the exception of the CYP2E subfamily the positions of the motif are not conserved. This sequence may influence transcription in some P450 genes but is not generally required.

A consensus sequence for LF-A1 (HNF-4), G–G–A/T–C/T–C–A, (HARDON et al., 1988) occurs frequently in most of the gene families including the adrenal P450s. This is a partial consensus binding sequence for a factor that is enriched in liver and kidney. As discussed below, sequences homologous to other HNF-4-like sequences are found in many of the CYP2 genes. These HNF-4-like sequences bind to nuclear proteins and in one case have been shown to be important regulatory sequences (see below). The position of G–G–A/T–C/T–C–A consensus motif is not well conserved among the P450 genes and functional assays will be required to establish their importance.

3.2.2. Functional proximal promoter elements

The functional activities of sequences in the proximal promoter region and binding of proteins to these regions have been studied in the CYP1A, CYP2A, CYP2C, CYP2D, CYP2E and the adrenal P450 genes. The proximal promoter in the adrenal genes is intimately associated with induction by cAMP and will be discussed in the section on cAMP induction.

3.2.2.1. CYP1A1 gene

Three groups have independently shown that the proximal region other than the TATA sequence is required for both basal and induced expression of the CYP1A1 gene. Induced expression is also dependent on xenobiotic response elements (XRE) present about 1 kb upstream of the RNA initiation site. Deletion analysis of the rat CYP1A1 promoter demonstrated that a region from –53 to –44 (basal transcription element-BTE, Fig. 3) was required for basal expression in transfected cells in the absence of multiple XRE's required for inducible expression (YANAGIDA et al., 1990). If this region was deleted in constructions containing the normal distal XRE's, activity was reduced by 100-fold but relative inducibility was only slightly reduced. A similar reduction in activity was seen if the SV40 enhancer was substituted for the CYP1A1 upstream sequences. Nuclear proteins from both hepatic and non-hepatic cell lines that bind to this region were detected by both DNase 1 footprinting and gel retardation analyses. Although this region contains sequences similar to the Sp1 and NF1 binding sites, oligonucleotides containing these motifs did not compete for the binding of factors to the BTE. Similar sequences are present in several other P450 genes (see Fig. 3, CYP2B1/2, rat CYP2E1, human CYP11A) regulated by different inducers as well as in other liver genes. The CYP2B1 BTE was shown to compete for the binding of nuclear proteins to the CYP1A1 sequence. This element, therefore, may be required for basal as well as induced expression of many genes.

The region –245 to –50 has been shown to be essential for the expression of the mouse Cyp1a1 gene in transfected cells as determined by internal deletions which retain the XRE's (NEUHOLD et al., 1989). Multiple complexes with nuclear proteins were formed by this region independent of TCDD treatment. In other studies on the mouse gene, the 5′ flanking region up to –419 was shown to be inactive by itself, but if combined with a functional domain containing XRE's, inducible gene expression was observed (JONES and WHITLOCK, 1990). Deletion of sequences from –166 to –55 reduced expression about 10-fold but the residual activity was still inducible. Only complete deletion of the promoter past the TATA sequence resulted in the loss of inducible expression. The sequences –45 to –59 and –110 to –130 (Fig. 3), which contained a possible CTF/NF1 or a G-box adjacent to a second possible CTF/NF1 site, respectively were protected from DNase 1 digestion by nuclear proteins. No effects of TCDD treatment on the binding patterns were observed. Mutation of the proximal region, which is analogous to the rat BTE sequence, reduced expression 80% and eliminated binding of the protein. Mutation of either the G-box or the CTF/NF1-like sequence in the distal region individually had little effect on expression but mutation of both reduced expression 50%. These results indicate that even though the BTE sequence has little or no promoter activity in the absence of the XRE's it is required for expression in the presence of the XRE's. Other factors in the proximal region may play a more modest role.

3.2.2.2. Mouse Cyp2d genes

The mouse Cyp2d9 gene is expressed specifically in males while the closely related Cyp2d10 gene is expressed in both males and females. Potential differences in the proximal promoter region of these two genes were studied by DNase 1 footprint analysis and in vitro transcription (YOSHIOKA et al., 1990). DNase 1 footprinting of the 2d9 and 2d10 5′ flanking regions revealed that regions from –86 to –102 (SDI in Fig. 3) and –44 to –67 (CTE in Fig. 3) were protected in *2d9* but only the proximal region was protected in *2d10*. Therefore the proximal region was designated the common transcription element (CTE) and the upstream region was identified as the sex difference information (SDI) region. There was, however, no difference in the footprints with male and female extracts. The promoters of Cyp2d9/10/11 were each capable of being templates for transcription in cell free mouse liver extracts when fused to the G-free cassette. Deletion analysis from the 5′ end of the Cyp2d9 gene showed that removal of half of the SDI sequence reduced activity several fold. Deletion of both the SDI and the CTE reduced activity to near background. Dramatic losses of activity in transcription were observed when the altered base pairs in the Cyp2d10 sequence were substituted for the Cyp2d9 SDI establishing unequivocally that the SDI is a regulatory region for in vitro transcription. The SDI region is not well conserved among the CYP2D genes nor the other family 2 genes but sequences similar to the CTE sequence are well conserved among many of the family 2 genes (Fig. 3). It is likely that the CTE is equivalent to an HNF-4-like sequence and is important for the expression of these genes in liver and kidney. Since the SDI binding protein is present in both sexes it is not yet clear whether this regulatory region is important in the sex specific expression of the 2D9 gene or is a more general regulatory factor.

3.2.2.3. Rabbit CYP2C genes

In the rabbit CYP2C genes there is a conserved nonamer sequence at about −100 bp (Zhao et al., 1990; Chan and Kemper, 1990) to which a rabbit liver nuclear protein binds. The sequence is very similar to the mouse Cyp2d9 CTE sequence described above and to the HNF-4 motif which has been shown to be important for transcription of other liver genes (Sladek et al., 1990). The HNF-4 factor is also present in the kidney (Xanthopoulos et al., 1991) and thus may be important for the expression of CYP2C2 in that tissue although the other rabbit 2C members are not expressed in the kidney. Sequences similar to HNF-4 are found in other P450's particularly the CYP2 family (Fig. 3, HNF4b). The conservation of this sequence and the position in the 5′ flanking region suggests that this region is important for the liver-specific expression of these P450s. This hypothesis is further strengthened by the demonstration in the mouse Cyp2d9 gene that the equivalent CTE motif is a strong regulatory sequence (Yoshioka et al., 1990).

3.2.2.4. Rat CYP2E1 gene

The rat CYP2E1 gene, which is activated within one day after birth, has been analyzed by in vitro transcription. The expression of this gene was studied in adult hepatic nuclear extracts and transcripts were detected by primer extension

(UENO and GONZALEZ, 1990a). Beginning with a fragment extending from −4,600 to +1,410, deletion from the 3′ end to +190 had little effect on transcription. Deletion of the 5′ end to −2,600 resulted in a 30% increase in activity. Progressive deletion of sequences from the 5′ end identified the regions −274 to −170 and −113 to −87 as containing positive regulatory elements. No activity was seen if the TATA sequence was deleted. Footprint analysis revealed no binding in the −274 to −170 region but three regions within −127 to −56 were protected (see Fig. 3). Binding of liver nuclear proteins to the region −127 to −89 was observed and DNA containing known HNF-1 motifs sites competed for the binding. HNF-1 is present in liver and is important for the expression of several liver genes (BAUMHUETER et al., 1990). These oligonucleotides containing HNF-1 sequences also decreased the cell-free transcription of the CYP2E1 gene in liver extracts. A sequence from −108 to −96 in the rat 2E1 gene resembled an HNF-1 site (Fig. 3) and its deletion is probably responsible for the loss of transcription observed when the −113 to −87 region was deleted. The role of the HNF-1 factor in the developmental expression of the CYP2E1 gene, however, is not clear. Nuclear extracts from fetal and new born rats, which do not express CYP2E1, produced DNase 1 footprints very similar to the adult extracts. By gel retardation analysis, the same protein-DNA complex was also observed in fetal and newborn rats except that a second higher mobility complex with similar binding characteristics was present. The role of this complex in transcription remains to be established. In mice homozygous for deletion of a region of chromosome 7 both HNF-1 and CYP2B1 expression are absent (GONZALEZ et al., 1990). Neither of the structural genes for these two proteins is deleted so that another gene that regulates HNF-1 expression is presumably absent. These results are consistent with an HNF-1 requirement for CYP2E1 expression but a direct effect of a deleted gene on CYP2E1 expression cannot be excluded. These results suggest that an HNF-1 factor is important for expression of the CYP2E1 gene. It is likely that other factors are important in establishing the developmental pattern of its expression.

3.2.2.5. Rat CYP2A genes

The rat CYP2A1 gene is expressed predominantly in adult female rats while the CYP2A2 gene is expressed exclusively in adult males. The CYP2A1 and CYP2A2 genes were analyzed by in vitro transcription and expression of each gene was the same in extracts from male and female rat livers (MATSUNAGA et al., 1990b). Either factors missing from the extract or sequences not present in the gene templates (−6 kb to +1.5 kb) are responsible for the sex-dependent expression. The CYP2A1 promoter was about 5-fold as active as the CYP2A2 promoter even for short templates of −74 to +169 in which the CAAT sequence specific to 2A1 is removed.

3.2.3. Inducible response elements

3.2.3.1. Polycyclic aromatic hydrocarbon induction

3.2.3.1.1. Xenobiotic response element (XRE)

The regulation of the CYP1A1 gene by polycyclic aromatic hydrocarbons has been studied more intensively than any of the other P450 genes at the molecular, cellular, and genetic levels. This gene is expressed in the liver only after treatment with polycyclic aromatic hydrocarbons and is also expressed in many other tissues. Initial evidence for an aromatic hydrocarbon (Ah) receptor was the demonstration of a cytosolic binding protein for the high affinity ligand TCDD (POLAND et al., 1976). Strong evidence for the functional significance of the Ah locus was the correlation between Ah inducibility and the presence of the Ah receptor in inbred strains of mice (POLAND and KNUTSON, 1982). Continuous cell lines in which CYP1A1 expression could be induced permitted the selection of several classes of regulation-deficient variants, some of which were defective in the Ah receptor (HANKINSON et al., 1985). Studies of the regions of the genes important for regulation were greatly facilitated by continuously cultured cell lines in which transfected CYP1A1 genes could be expressed.

The mechanism of Ah induction which emerged from the early studies is analogous to that of the steroid hormones. The inducer first binds to the Ah receptor, and this complex undergoes a temperature-dependent activation for binding to specific sequences in the 5′ flanking region of CYP1A1 genes variously termed drug, dioxin, arylhydrocarbon or xenobiotic response elements (designated XRE's in this review). Once activated, these binding regions have the properties of enhancers and activate gene expression by unknown mechanisms. Three major regulatory regions were defined in the 5′ flanking region of the rat and mouse CYP1A1 gene. The Ah responsive region contains multiple copies of the XRE centered at about −1 kb. An inhibitory region is present at about −500 bp, and a second positive regulatory element is in the proximal promoter as described above. The regulation of this gene, therefore, involves complex interactions between proximal and distal positive and negative elements. For details of this earlier work, readers are referred to the several recent reviews of this area (NEBERT and GONZALEZ, 1987; WHITLOCK, 1987, 1989, 1991; WHITLOCK et al., 1989; GOTOH and FUJII-KURIYAMA, 1989; NEBERT and JONES, 1989, GONZALEZ, 1988, 1989; OKEY, 1990).

A consensus sequence for binding of the Ah receptor to the XRE has been determined by comparison of the multiple XRE's in the rat and mouse genes. The importance of the individual nucleotides has been analyzed by studying the binding of the receptor to natural and mutated XRE's by gel retardation and footprinting analyses. In the rat system, an extended consensus sequence, G–N–N–C/C–C/A–N–G–N–N–C–N–T–N–T–C–A–C–G–C–A/T–A, was initially proposed (FUJISAWA-SEHARA et al., 1988). Footprinting analysis using the small molecule, o-phenanthroline/copper, for cleavage, indicated that the receptor protected about 15 bp centered on the core sequence (SAATCIOGLU et al., 1990a). Use of a small molecule for cleavage allows a better determination of the length of the protected region. A less extended consensus sequence for the binding, C–X–A/T–X–A/T–C–A–C–G–C–A/T–A/T (SAATCIOGLU et al., 1990) was proposed based on the regions protected by the receptor. Six strong Ah receptor binding sites and two weakly binding sequences, which still contained a C–A–C–G–C core, were detected in the mouse Cyp1a1 5′ flanking region (DENISON et al., 1989). A consensus binding sequence of A/T–N–N–C–A–C–G–C–N–A–G/C–N–Py–N–G/C–N–Pu was proposed.

The xenobiotic responsive region in the mouse gene was determined by transfection experiments to be in the region –1302 to –820. This region contained four XRE's and a core consensus sequence of these four sequences was C–A–C–G–C–N–A (FISHER et al., 1990). This core sequence is common in both the rat and mouse consensus sequences. Although there are multiple XRE core consensus sequences in both the rat and mouse genes, three XRE's are conserved at –1066, –991, –903 in the mouse gene, at –1047, –976, and –889 in human and at –1090, –1014, –899 in the rat (NEBERT and JONES, 1989). The conservation of these three XRE's suggests that they are functionally important. A second possibly important sequence is an Sp1 binding site, G–G–G–C–G–G, which is conserved between the second and third XRE's in each species. Inspection of the consensus sequences reveals that there is no symmetry in the XRE binding site, in contrast to the responsive elements for other members of the steroid receptor superfamily. The receptor appears to bind as a monomer to the XRE as determined by gel retardation analysis of labeled receptor and DNA (DENISON et al., 1989). This contrasts with the steroid hormone receptors that dimerize before binding to the DNA but is consistent with the lack of symmetry in the binding sequence.

Methylation interference, methylation protection, and mutational analyses have been used to study the requirements of the individual base pairs in the XRE for binding to the Ah receptor. In the mouse, gel retardation analysis demonstrated that TCDD treatment resulted in a new protein-DNA complex with the region –901 to –1071 in wild type but not receptor deficient mutants (NEUHOLD et al., 1989). Methylation interference experiments indicated that methylation of each of the three G's in the GC base pairs of the core sequence, C–A–C–G, interfered with binding. Mutational analysis with dinucleotide substitutions in oligonucleotides also demonstrated that the CACG core of the conserved sequence of the XRE was required for binding (NEUHOLD et al., 1989). WHITLOCK'S group also studied the binding requirements of the XRE by methylation protection and interference studies (SHEN and WHITLOCK, 1989). Methylation interference results were the same as NEUHOLD et al., (1989) except that the G in the 3′ GC base pair in the C–A–C–G–C core was also protected.

Methylation of two core G's was inhibited by binding of the receptor to the XRE while methylation of two flanking G's on one strand and the core G on the other strand were weakly inhibited. These results suggest that the receptor binds to the major groove of the DNA in this region. Transfection of plasmids containing methylated G's in the XRE resulted in about a 4-fold reduction in expression (Shen and Whitlock, 1989). CpG is a potential methylation site in mammalian genes so the inhibition of induction by methylation of this dinucleotide could have regulatory significance. However, the unliganded receptor binds to the XRE which should protect it from methylation and prevent inhibition of induction. This is consistent with the universal inducibility of the CYP1A1 gene in various tissues.

Binding of the rat receptor to the XRE has also been analyzed. Mutant XRE oligonucleotides were used to compete for the binding of the Ah receptor to the XRE (Saaticioglu et al., 1990b). Mutations in the C–A–C–G–C core had the greatest effect on competitive binding while changes in the nucleotides flanking the core had smaller effects in agreement with the mouse studies. Methylation of each of the GC base pairs in the core sequence was shown to interfere with the binding of the Ah receptor to the XRE. In these studies, a role for Zn^{++} in binding of the receptor was suggested by the inhibition of the binding by chelators of divalent ions, such as 1,10-phenanthroline. Co-incubation with excess Zn^{++} or removal of 1,10-phenanthroline by dialysis in the presence of Zn^{++} could prevent or reverse the inhibition. Since steroid hormone receptors are Zn^{++} fingers proteins, it is reasonable to expect that the Ah receptor also requires Zn^{++} for binding as shown in these studies. However, isolation and characterization of the Ah receptor will be required to establish this point.

The studies on the binding of the Ah receptor to the XRE indicate that it may be a zinc finger factor similar to steroid hormone receptors. In contrast to the steroid hormone receptors, the Ah receptor appears to bind as a monomer to a nonsymmetrical DNA motif. The importance of the core C–A–C–G–C is well established for binding but the role for the conserved flanking sequences is not. Finally, although much work has been done on binding characteristics, the functional significance of each base pair for the induction of XRE containing genes has not yet been established by studying the activity of the modified XRE's in transfected cells.

3.2.3.1.2. Functional XRE analysis

The functional significance of multiple XRE's in the responsive region was addressed by studying the four responsive elements in the mouse gene (Fisher et al., 1990). Transient expression in transfected cells showed that inducible expression was proportional to the number of responsive elements present. Elimination of each of the elements individually by block substitution produced the same reduction in expression for each element. A single XRE was largely inactive but addition of a second XRE resulted in a dramatic increase in expression. Smaller additional increases were observed by adding three or four copies. The Sp1 binding site at –943 had no activity by itself but inclusion of it with an XRE increased activity about 10-fold. Block substitution of the Sp1 site confirmed its regulatory role. These results indicate that binding the ligand–Ah receptor complex to the multiple XRE's results in a synergistic increase in CYP1A1 expression and the common factor, Sp1, plays an important complementary role with the XRE's.

An important recent result has been the development of a cell-free system in which regulated expression of the Cyp1a1 gene has been obtained (WEN et al., 1990). Fractionated nuclear extracts were required since an inhibitor was present in complete extracts. Increased transcription with extracts from TCDD-treated cells was observed that was dependent on the presence of at least two XRE's in the template. A functional Ah receptor was required since transcription was not induced in extracts from receptor defective cell lines. This assay should greatly facilitate analysis of factors involved in Ah induction of the CYP1A1 gene.

The mechanism by which Ah receptor binding to the XRE leads to expression of the CYP1A1 gene is not known. One possibility is that a conformational change is induced in the DNA which leads to activation of the gene. Bending of DNA is known to alter the electrophoretic mobility of a linear DNA fragment in gels. The change in the mobility is dependent on the position of the bend in the linear fragment with the greatest change occurring when the bend is in the middle. ELFERINK and WHITLOCK (1990) studied a series of linearized plasmids in which the position of the XRE was systematically changed. The results clearly showed that bending of the DNA occurred when the Ah receptor bound to the XRE. The relationship of this bending to the function of the gene is unknown.

3.2.3.1.3. Ah receptor accessory proteins

Several studies have addressed the role of accessory proteins in the binding of the Ah receptor to the XRE. The receptor is present in two forms in the Hepa 1c1c7 cell line, a larger 9S form and a smaller 6S form. The 90 kd heat shock protein is associated with the larger form but not the 6S form (WILHELMSSON et al., 1990). The 9S form did not bind to the XRE as assessed by gel mobility shift studies indicating that the heat shock protein must be displaced before binding to the DNA can occur. Analogous to the steroid hormone receptors, activation of the Ah receptor also requires at least two steps, first, binding of ligand to the receptor and followed by dissociation of the heat shock-90 protein.

The gene for a second protein that interacts with the Ah receptor was identified by exploiting the ability of human DNA to complement the receptor translocation defect in the group C variants of the mouse Hepa-1 cell line (HOFFMAN et al.,

1991). This property was used to isolate a complementing normal gene by successive transformations of the defective cells with human DNA. The gene, termed the arnt for Ah receptor nuclear translocation gene, is a 87 kd protein which does not bind ligand but is required for translocation of the receptor ligand complex to the nucleus. The arnt protein contains sequence similar to the DNA-binding helix-loop-helix motif in transcription factors. It is interesting to speculate that this protein forms a complex with the Ah receptor which is required for the nuclear translocation and DNA binding of the receptor.

Negative regulation of the expression of the rat CYP1A1 gene has been reported by the cotransfection of vectors expressing E1a (SOGAWA et al., 1989). Repression was dependent on the presence of the XRE in the CYP1A1 gene, but occurred with or without 3-methylcholanthrene in the medium. Since E1a does not bind to DNA, but mediates its effects by protein interactions (KOVESDI et al., 1982), the inhibitory effect is probably mediated by binding of E1a to the Ah receptor. This result suggests that an E1a-like factor may be responsible for the negative regulation of Cyp1a1 genes (HANKINSON et al., 1985). A putative repressor, which may be an Ela-like protein, is thought to be generated in a Hepa 1 cell line with a dominant mutation that is deficient for Ah receptor mediated regulation of the Cyp1a1 genes.

The gene products of the Cyp1a1 and Cyp1a2 genes may act as repressors of Ah regulated genes. Expression of functional Cyp1a1 or Cyp1a2 proteins can repress the transcription of endogenous Cyp1a1 genes and another Ah inducible gene that encodes NAD(P)H: monadione oxido reductase (RAYCHAUDHARI et al., 1990). This autoregulation may proceed by effects on transcriptional factors. Sequences in the Cyp1a1 and Cyp1a2 genes that might mediate this effect have not been identified.

Two proteins other than the Ah receptor have been shown to bind to the XRE in the rat (SAATCIOGLU et al., 1990b). These proteins, designated XF1 and XF2, can be separated from the Ah receptor by changing the conditions for extraction of nuclear proteins. The binding of these proteins to the XRE is different from that of the receptor based on footprinting and methylation interference. Cycloheximide treatment results in the activation of CYP1A1 expression (ISRAEL et al., 1985) and concurrently the rapid disappearance of XF1 and XF2 but not the Ah receptor. The possibility that XF1 and XF2 are inhibitory regulatory factors has been postulated to explain the cycloheximide effect.

Effects of TPA on the TCDD receptor have been studied because TPA dramatically inhibits the induction of CYP1A1 in mice. TCDD-induced increases in transcription of CYP1A1 and CYP1A2 were dramatically decreased by co-administration of TPA (OKINO and TUKEY, 1991). A corresponding decrease in the binding of the Ah receptor to the XRE was observed. The mechanism by which TPA affects function Ah receptor levels is not clear but phosphorylation via protein kinase C or other modifications of proteins that interact with the Ah receptor may be important.

3.2.3.1.4. CYP1A2 gene

The CYP1A2 gene is also induced by polycyclic hydrocarbons, but differs from the CYP1A1 gene in that it is also constitutively expressed and its expression is largely restricted to the liver (see QUATTROCHI and TUKEY, 1989). Initially it was thought that the induction of the CYP1A2 gene might be different from the CYP1A1 gene since no XRE sequences were present in 1.9 kb of 5′ flanking region of the human CYP1A2 gene (IKEYA et al., 1989). However, transfection of 3.7 kb of 5′ flanking region into HepG2 cells resulted in expression of the gene and 2-fold induction with 3-methylcholanthrene treatment (QUATTROCHI and TUKEY, 1989). A sequence at −2903 contained a hexamer sequence conserved in the XRE's in the CYP1A1 gene. The fragment from −3202 to −1595 conferred 3-methylcholanthrene inducibility on the SV40 promoter but only in hepatic cells in contrast to the CYP1A1 XRE's which are active in many cell types. These results indicate that the mechanism of weak transcriptional activation of the CYP1A2 by 3-methylcholanthrene is similar to that of the CYP1A1 gene. Additional liver-specific elements may be present in the CYP1A2 enhancer which would account for its restricted expression in hepatic cells in contrast to the widespread pattern of CYP1A1 expression.

3.2.3.2. Phenobarbital induction

Phenobarbital is one of the classical inducers of P450. The major forms of phenobarbital inducible P450s are in the CYP2B subfamily, although P450s in the CYP2C (LEIGHTON and KEMPER, 1984; ZHAO et al., 1987) and CYP3A (GONZALEZ et al., 1986) subfamilies also contain P450s that are inducible by phenobarbital. A P450 expressed in *Bacillus megaterium* is also induced by phenobarbital (FULCO, 1991). The mechanism by which this drug induces gene expression remains largely unknown. In contrast to the CYP1A1 genes, no receptor for phenobarbital has been identified and continuous cell lines that express the phenobarbital-inducible P450's have not been available. The recent development of primary cultures of hepatocytes which continue to express these P450's in a phenobarbital-dependent manner provides hope that cis-acting and trans-acting regulatory elements for these genes will soon be identified (KOCAREK et al., 1990).

Although no functional analyses of mammalian phenobarbital-inducible P450 genes have been possible, indirect evidence for nuclear factors that regulate the CYP2B1/2 genes has been obtained (RANGARAJAN and PADMANABAN, 1989). Transcription of a fragment of the CYP2B2 gene from −179 into the first exon was increased several-fold in freeze-thawed nuclei from phenobarbital-treated rats compared to controls. Gel retardation analysis showed that a protein bound to a fragment containing this region of the 5′ flanking region and that phenobarbital-treatment altered the mobility of the protein-DNA complex (RANGARAJAN and PADMANABAN, 1989; RAO et al., 1990). A protein of 85 kd was bound to this region as determined by binding of labeled DNA to blots of nuclear proteins (RAO et al., 1990). Again phenobarbital treatment greatly increased the amount of this 85 kd protein that bound. The region from −57 to −89 was protected from DNase 1 digestion in a footprint analysis (Fig. 3). These results suggest that phenobarbital induction might be regulated through binding of this protein to the proximal promoter region.

An interesting system that may serve as a model for induction of mammalian P450 by phenobarbital is the bacterium, *Bacillus megaterium*. This bacterium contains three P450's that are induced by phenobarbital including one, $P450_{BM-3}$ or CYP102, that is fused to the P450 reductase (FULCO and RUETTLINGER, 1987). Structurally this P450 is more similar to microsomal P450s than to bacterial or mitochondrial forms. Fusions of the 5′ flanking region of CYP102 with a reporter gene directed the pentobarbital-regulated expression of the gene and required about 1.3 kb (1.1 kb to +0.2 kb) of sequence for full activity and induction (WEN et al., 1989). Fragments of 1.0 kb had very low activity with or without barbiturate treatment indicating that a regulatory element was present between −0.8 and −1.1 kb. A second property of this regulatory region was the suppression of endogenous CYP102 expression presumably by competition of a positive element for a trans-acting factor. Deletions of the 5′ flanking region from the 3′ end, demonstrated that a second region within a 300 bp fragment near the RNA initiation site was required for this repression effect. This gene, therefore, appears to contain a distal barbiturate response element and a proximal positive element.

To try to define the phenobarbital regulatory region, the 5′ flanking regions of two genes induced by phenobarbital in *Bacillus megaterium*, CYP102 and CYP106, and the rat phenobarbital-inducible CYP2B1 and CYP2B2 genes were compared (HE and FULCO, 1991). A 17 bp sequence, −76 to −89 in CYP2B1/2 (Fig. 3), was found to be similar in all four genes with some invariant positions, A–N–N–N–A–A–A–G–C–N–G–G–N–G. By gel retardation assay, a protein in extracts of both *Bacillus megaterium* and rat liver nuclei bound to this sequence in both the bacterial and mammalian genes. Treatment with barbiturates had opposite effects on the two binding activities. In the bacteria, pentobarbital treatment eliminated the binding activity, while phenobarbital treatment of rats increased the binding of the protein. In vitro incubation of bacterial or liver extracts with phenobarbital was able to induce the same effect on protein binding as in vivo treatment. Deletion of 24 bp containing the conserved sequence at −301 in the CYP106 gene transfected into *Bacillus megaterium* resulted in the elimination of the repression of the endogenous gene expression and the loss of barbiturate stimulation of the transfected gene. These results suggest a remarkable conservation of a DNA-binding protein mediating phenobarbital induction in bacteria and mammals. In spite of this conservation, the phenobarbital treatment has opposite effects on the binding of this protein. The bacterial factor represses transcription while the analogous rat protein stimulates transcription. Therefore, caution must be exercised before concluding that this protein is the primary mediator of phenobarbital induction of the mammalian P450 genes. To establish the significance of this protein in eukaryotes, functional analyses in mammalian or vertebrate systems will be required. Regardless of the outcome the remarkable parallels so far promise that the results from continued comparison of the bacterial mechanism with that of the mammalian systems, will be fascinating.

The first successful study of transfected phenobarbital-inducible genes has been reported in the chicken system (HAHN et al., 1991). The 5′ flanking region of a chicken phenobarbital-inducible gene fused to a reporter gene was transfected into primary cultures of chicken hepatocytes. Induction of expression by phenobarbital was observed, and the induction was dependent on sequences at about −1 kb. A fragment containing the putative phenobarbital-responsive region when placed in front of the enhancer-less SV40 promoter conferred phenobarbital inducibility of 14-fold, which is similar to induction in vivo. These results provide strong evidence that phenobarbital acts by activating initiation of RNA synthesis through a phenobarbital-responsive enhancer-like sequence.

It has also been proposed that phenobarbital regulates the rat CYP2B1/2 genes by effects on the elongation of RNA rather than initiation (PADMANABAN et al., 1990). This hypothesis is based on run-on transcription analyses in which exon 1 probes show high levels of transcription in untreated samples and low inducibility, and downstream exon probes show low control levels and high inducibility. Although this is an interesting hypothesis, the present evidence is only suggestive since this type of experiment is replete with potential artifacts. As exemplified by the high level of antisense transcription in these experiments, there is probably considerable transcription of segments of the gene that are not due to normal transcription driven by the RNA polymerase II promoter of the CYP2B2 gene. Earlier studies in which a fragment of only 5′ flanking sequence and part of exon 1 was transcribed in freeze-thawed nuclei in a phenobarbital responsive manner are not consistent with this hypothesis (RANGARAJAN and PADMANABAN, 1989). Furthermore, the ability of the 5′ flanking region of the chicken CYP2H1 gene to confer phenobarbital inducibility in transfected cells indicates that regulation occurs at the RNA initiation step (HAHN et al., 1991). While there is certainly precedent for regulation of RNA elongation, and both initiation and elongation may be involved in phenobarbital regulation, the present data are most consistent with primary effects of phenobarbital on initiation of RNA synthesis.

Little information is available on the expression of genes other than the CYP2B members which are induced by phenobarbital. The genes in the rabbit CYP2C subfamily contain both constitutive and phenobarbital-responsive genes and the phenobarbital effects are at least in part at the level of gene expression. The rabbit CYP2C2 gene in transfected human HepG2 cells is efficiently expressed but no effect of phenobarbital is observed (P. VENEPALLY, D. CHEN, and B. KEMPER, unpublished). Interestingly, deletions of distal 5′ flanking sequences result in a progressive increase in expression suggesting that inhibitory sequences are present in the distal region. Comparisons of the members of this family which vary in their response to phenobarbital may provide insight into the mechanism.

Finally, it is possible that phenobarbital effects on different P450 subfamilies may be mediated by different mechanisms. In primary cultures of rat hepatocytes, a five to 20 fold difference in the Kd for phenobarbital induction of CYP2B and CYP3A genes was observed (KOCAREK et al., 1990). These large differences suggest that either different receptors or different aspects of RNA synthesis or stability may be regulated in the different genes.

3.2.3.3. Heme regulation of P450 expression

The role of heme in the regulation of phenobarbital-inducible P450s has been an area of controversy. Padmanaban's group has reported that inhibitors of heme biosynthesis block the induction of both CYP1A and CYP2B genes by their respective inducers. Treatment of animals with $CoCl_2$, a heme biosynthesis inhibitor blocks in vitro phenobarbital-induced transcription of the CYP2B2 gene in freeze-thawed nuclei and eliminates the binding of the phenobarbital-dependent protein to the 5′ flanking region of the gene (RANGARAJAN and PADMANABAN, 1989). These effects can be reversed by the addition of heme in vitro. These results suggest that heme is required for the phenobarbital induction and that both heme and phenobarbital effects are mediated by the same cis-acting element and nuclear protein. In contrast, the heme inhibitor, succinylacetone, at concentrations that elevated mRNA for the heme regulatable enzyme, 5-aminolevulinate synthase, had no effect on the induction of CYP2B1/2 by 2-allyl-2-isopropylacetamide, a phenobarbital-like inducer (MAU and HANSEN, 1989; SRIVASTAVA et al., 1989). The administration of heme to rats was shown to reduce the drug-induced increases in the mRNA of a number of enzymes, including CYP2B1/2, and this effect was mediated at the level of transcription (SRIVASTAVA et al., 1990). In primary hepatocyte cultures, phenobarbital increased the levels of CYP2B1/2 mRNA and the addition of succinylacetone failed to block induction (SINCLAIR et al., 1990). As in the intact animal, 5-aminolevulinate synthase mRNA concentrations were increased by the inhibitor and addition of exogenous heme reversed the effect of the inhibitor. The conclusion from these experiments is that heme has no effect on the induction of the CYP2B1/2 genes by phenobarbital. It is not clear how to resolve the conflicts in these data. Different inhibitors were used by the two groups so that differential levels of heme inhibition could cause differing results. In the studies with succinylacetone, the effects on 5-aminolevulinate provide a control that heme levels were actually being affected by the inhibitor so at least changes sufficient to affect that enzyme have no effect on CYP2B1/2 gene expression. All of these studies suffer from the same problem of using inhibitors which may have side effects not related to heme effects. A more detailed understanding of the mechanisms regulating CYP2B gene expression will be required to resolve the role of heme.

3.2.3.4. Glucocorticoid induction

The induction of P450 genes by glucocorticoids may be relatively common. Glucocorticoids have been shown to directly regulate the expression of CYP2B and CYP3A genes (SCHUETZ and GUZELIAN, 1984; SIMMONS et al., 1987) and to potentiate the induction of CYP1A1 genes by polycyclic hydrocarbons (MATHIS et al., 1986; HINES et al., 1988; SHERRATT et al., 1989). In the case of CYP2B and CYP1A1 genes functional glucocorticoid response elements (GRE) have been demonstrated.

The role of the glucocorticoid potentiation of the rat CYP1A1 gene was studied by transfection of the 5′ flanking and intron 1 sequences fused to a chloramphenicol acetylase transferase (CAT) reporter in liver hepatoma HepG2 cells (HINES et al., 1988). Multiple GRE-like sequences are present in intron 1 of the human and rat CYP1A1 genes. Deletion of this region did not affect induction by 3-methylcholanthrene alone, but eliminated the potentiation of the induction of CYP1A1 expression by glucocorticoids. These results indicate that at least some of the GRE-like sequences in intron 1 are functional and mediate the potentiation effect.

The sequence, G–G–T–C–A–A–A–G–T–G–T–T–C–A, which is similar to the GRE is present about 1.3 kb upstream of the RNA initiation site in the rat CYP2B2 gene. Transfection of H4II cells with plasmids in which CYP2B2 DNA was inserted upstream of a thymidine kinase promoter, resulted in expression that was increased about 20-fold if dexamethasone was present in the media (JAISWAL et al., 1990). Deletion analysis indicated that the GRE was between −1406 and −1255. Insertion of multiple copies of a 25 bp DNA segment from this region that contained the putative CYP2B2 GRE also conferred glucocorticoid responsiveness. A protein in rat liver bound specifically to this GRE element. These results demonstrate that this GRE is functional but its significance in vivo remains uncertain since both induction and slight repression of CYP2B2 by glucocorticoids have been reported (YAMAZOE et al., 1987; SIMMONS et al., 1987).

Evidence has been presented that dexamethasone potentiates phenobarbital induction of CYP2B1/2 mRNA in spite of a decrease in the transcription of the genes (RAO et al., 1990). Run-on transcription assays of endogenous CYP2B2 genes in isolated nuclei or transcription of exogenous CYP2B2 genes in freeze-thawed nuclei was greatly reduced by dexamethasone. Surprisingly, this effect was not observed for the endogenous gene in the freeze-thawed nuclei as it was for transcription in isolated nuclei. The induction of the 85 kd DNA-binding protein that is increased by phenobarbital treatment, as described above, was blocked by dexamethasone. The stability of cytoplasmic CYP2B2 mRNA was not affected by dexamethasone. These data suggest that dexamethasone acts by stabilizing nuclear precursors of the CYP2B2 mRNA. This conclusion does not require a functional GRE as described above which would imply a positive transcriptional effect. Resolving these conflicting hypotheses of dexamethasone regulation will require additional experiments.

3.2.3.5. Cyclic AMP induction

The synthesis of steroids in the adrenal glands and gonads is regulated by polypeptide hormones which act by activation of adenylate cyclase. Cyclic AMP rapidly induces secretion of steroid hormones but also has long term inductive effects on the biosynthesis of the steroidogenic enzymes (SIMPSON et al., 1990). There are five steroidogenic P450's: cholesterol side chain cleavage, $P450_{SSC}$ (CYP11A1), 11β-hydroxylase, $P450_{11\beta}$ (CYP11B1), 17α-hydroxylase $P450_{17\alpha}$ (CYP17A1), 21-hydroxylase, $P450_{C21}$ (CYP21B1) and aromatase, $P450_{arom'}$ (CYP19A1). The first four are present in the adrenal and are involved in the synthesis of cortisol and other adrenal steroids and the last three are also present in the gonads and synthesize the gonadal steroids. All these P450s are encoded by nuclear genes although CYP11A1 and CYP11B1 are mitochondrial enzymes rather than microsomal P450's. In the adrenal, cAMP coordinately induces the expression of each of the four P450 genes.

The study of the mechanism by which cAMP induces adrenal P450 gene expression has been facilitated by a continuous cell line, Y1, derived from a mouse adrenal tumor. This cell line retains the capability to respond to cAMP. Each of the four adrenal P450 genes has been analyzed in this cell line by transfection of 5′ flanking regions fused to reporter genes. The most remarkable finding has been that DNA-regulatory elements that are responsive to cAMP are different in each gene. Only the CYP11B1 gene appears to have an element similar to the classic cAMP responsive element, CRE which binds the cAMP responsive element binding protein, CREB (MONTMINY et al., 1986). Other cAMP-responsive genes also lack a typical CRE, some of which contain an AP1 or AP2 site which conveys cAMP responsiveness (IMAGAWA et al., 1987).

Two other themes run through the studies of steroidogenic genes. The first is the role of protein kinase C stimulators which are generally inhibitory to the adrenal genes. The second is the effect of the protein synthesis inhibitor, cycloheximide, which inhibits cAMP induction of endogenous CYP21B1, CYP17A1 and CYP11A1 genes in cultured cells. Cycloheximide did not inhibit the expression of any of these genes transfected into cultured cells (KAGAWA and WATERMAN, 1990; LUND et al., 1990; AHLGREN et al, 1990). These results indicate that either cycloheximide effects are not directly on transcription or that the binding element for the cycloheximide sensitive factor is not present in the constructions used.

3.2.3.5.1. CYP11A1 genes

In the human CYP11A1 gene, the distal 5′ flanking region contains positive cAMP-responsive and cell specific elements. Transfection of human CYP11A1 5′ flanking regions fused to the CAT reporter genes demonstrated that in Y1 cells, a 2500 bp fragment stimulated expression 10-fold and that deletion to −573 reduced expression to background (Chung et al., 1989). These data suggest that positive-regulatory elements are present in the distal region. The role of the distal element was cell-specific since the 10-fold stimulation with the full fragment was not seen in Hela, CHO, COS1 or JEG-23 cells. The basal activity after deletion of the distal region was not cell-specific. More recent results indicate that in JEG-23 cells a strong inhibitory sequence is present between −257 and −152 that masks a strong positive element between −152 and −87 (Moore et al., 1991).

In similar studies with the human CYP11A1 gene fused to the CAT reporter, unstimulated expression decreased when the regions −2327 to −605 were deleted (Moore et al., 1990). Deletion of the −2327 to −605 region resulted in a large decrease in the induction by forskolin from 10-fold to about the same as the increase in thymidine kinase promoter control. In a third study of human CYP11A1 5′ flanking regions fused to the CAT reporter and transfected into Y1 cells, deletion of the fragment from −1733 to −1621 resulted in the loss of cAMP induction with little effect on basal expression (Inoue et al., 1991). The fragment from −1697 to −1523 could confer cAMP responsiveness to either a basal CYP11A1 promoter or an SV40 promoter. Deletions from either end reduced the response to cAMP but duplication or triplication of the remaining sequence could restore responsiveness. These results indicate that multiple elements in this region are important for cAMP responsiveness. The region from −1662 to −1647 is identical to the cAMP-responsive region of the rat somatostatin gene except for one nucleotide. Two other regions were present in the −1697 to −1523 fragment that were similar to this sequence and each contained sequences with four of five matches to the CRE consensus sequence C–G–T–C–A. Point mutations within each of the three regions independently reduced cAMP induction. An adrenal nuclear protein interacted with this region but could not be competed by the CRE from the somatostatin gene. These results indicate that the human CYP11A1 gene contains a positive cAMP responsive region in the distal upstream region centered at about −1600. This region contains sequences similar to the classical CRE but proteins that bind to this region are different from CREB.

The regions between the distal promoter region and the proximal promoter seem to have slight negative regulatory elements present which have not been defined. The proximal promoter contains an additional positive regulatory region. Deletions from –236 to –76 (Chung et al., 1989) or from –152 to –89 (Moore et al., 1990) reduced basal activity of the human CYP11A1 promoter. Deletion of the region –56 to –44 also reduced basal expression (Inoue et al., 1991). A sequence near –44 was also required for inducible expression as shown by a large reduction in the induction ratio when –44 to –40 were deleted in a construction that retained the distal cAMP-responsive (–1697 to –1523) fragment. The sequence in the –44 region is similar to the CTF/NF-1 binding site (Fig. 3). These results indicate that multiple upstream cAMP responsive elements and a proximal promoter region are required for cAMP induction of human CYP11A1 expression.

Proximal promoter elements also seem to be responsible for regulation by protein kinase C activators. Basal and induced expression of the CYP11A1 gene could be inhibited by >85% by addition of TPA and the calcium ionophore, A23187. Progressive deletions from –343 to –89 eliminated the inhibitory effect in steps indicating that multiple elements in this region mediate the effects of these protein kinase C activators.

Transfection of 5′ flanking regions to –896 of bovine CYP11A1 fused to either CAT, or β-globin reporter genes in Y1 cells resulted in forskolin-inducible expression of the genes of 3-fold and 9-fold, respectively (Ahlgren et al., 1990). The β-globin constructions contain the globin TATA region and the 3′ boundary of the CYP11A1 5′ flanking region was at –32. Deletions to –186 had little effect on expression but further deletion to –101 greatly reduced expression without changing the ratio of induction. Deletion to –50 eliminated both basal and forskolin-inducible expression. The region from –101 to –50, therefore, contains a cAMP inducible region. Fusion of the region –183 to –83 in either orientation to the minimal globin promoter resulted in inducible expression comparable to the –186 deletion. This indicates that this region contains an element with the properties of an inducible enhancer. Neither the consensus CRE nor AP2 sequences are present in this region so that a novel cAMP responsive element is responsible for the induction. The bovine gene in contrast to the human gene has a cAMP responsive element in the proximal promoter region. The possibility of another distal cAMP responsive element in the bovine gene analogous to that in the human gene has not been examined.

Expression in pools of stably transfected Y1 cells of 1500 bp of the 5′ flanking region of the mouse Cyp11a1 gene fused to a growth hormone reporter gene demonstrated that 8-Br-cAMP induced expression about 5-fold (Rice et al., 1990a). Transient expression and inducibility was much less in MA-10 cells compared to Y1 cells in spite of equivalent expression of the endogenous Cyp11a1 gene in these two cell lines. Deletion analysis indicated that regions from –424 to –327 and –219 to –77 had the largest effects on basal expression. Induced expression progressively decreased as more 5′ flanking region was deleted indicating that multiple elements mediated the induced response. Three regions of the gene from –126 to –34 (Fig. 3) and one region from –325 to –309 were protected from DNase 1 digestion by nuclear extract proteins. The sequence T–G–A–G–T–C–A at –319 matches the AP-1 consensus site. Sequences at –70 and –40 are similar to sequences important for the expression of the mouse Cyp21 gene (Rice et al., 1990b). The fourth sequence at –120, G–A–G–G–T–T–T–G–A–G–T–C–T, did not match known regulatory sites. By gel retardation assays, a protein present in steroidogenic cell lines produced a complex with the –70 region that had a different mobility than that observed with Hela or PC12 cell extracts. The binding could be competed by the similar sequence at –210 in Cyp21. Mutation of each of the individual binding regions showed that the –319 element did not affect basal or induced expression. For all other regions, basal expression was significantly reduced, but each remained responsive to cAMP.

The expression of the rat CYP11A1 gene has been characterized less extensively. Fragments extending to –940 can direct cAMP induced expression in the Y1 cells or in rat granulosa cells (Oonk et al., 1990). However, the promoter region does not contain any classical CRE elements.

3.2.3.5.2. CYP11B1 genes

The mouse and rat CYP11B1 genes have been studied in most detail. In each case multiple regulatory regions are present in the proximal promoter region to about –400 bp. In contrast to the other steroidogenic P450 genes, CYP11B1 genes contain a classical CRE. A 1.5 kb 5′ flanking region of the mouse Cyp11b1 gene fused to a growth hormone reporter gene was expressed efficiently in both transiently and stably transfected Y1 cells (Mouw et al., 1989). Expression was about 25-fold more efficient in Y1 cells than in MA-10 cells derived from a mouse Leydig tumor. Expression was also 20-fold less in kin-8 cells which are Y1 mutant cells with a defective cAMP-dependent protein kinase. These results indicate that the expression is both cell-specific and dependent on a functional cAMP regulatory pathway. In the Y1 cells deletion to –400 had little effect and additional deletions to –338, –123, and –40 resulted in progressive loss of basal expression indicating that multiple regulatory elements were present in this region.

Transfection of the mouse Cyp11b1 gene-growth hormone reporter gene into Y1 cells also resulted in about a 4-fold induction by cAMP with 1.5 kb of flanking region and 2.5-fold induction remained with a 123 bp fragment (Rice et al., 1989). Further deletion to –40 resulted in the loss of induction. The deleted region contains a sequence similar to the CRE and specific deletion of this sequence from the 1.5 kb fragment resulted in the loss of induction (Mouw et al., 1989). Insertion of 1 to 3 copies of the region from –65 to –42 in front of thymidine kinase or chorionic gonadotropin α-subunit promoters conferred inducibility (Rice et al., 1989). This region from –65 to –42 was protected by Y1 nuclear proteins from DNase 1 cleavage (Fig. 3) and CRE-containing fragments competed for the binding. Regulatory regions from the mouse Cyp21 gene (RICE et al., 1990b) did not compete even though they contain some similarity to the CRE. Southwestern analysis demonstrated that a protein with the same molecular weight as CREB interacted with the Cyp11b1 sequence. These results indicate that the mouse gene contains a functional CRE at about -65 which is responsive to cAMP induction. A decrease in the induction ratio when the fragment from −1500 to −123 is deleted suggests that other elements may also contribute to inducibility.

Upstream regions in the mouse gene were analyzed by footprint analysis. Regions protected from DNase 1 digestion were −362 to −380 and −300 to −320 with extracts from Y1 cells and only the latter region with extracts from MA-10 Leydig cells (BOGERD et al., 1990). Block mutations of fragments centered on −310 to −370 resulted in > 50% reduction in both basal and cAMP stimulated activity. Fragments from −385 to −358 or from −323 to −296 fused in either orientation to the −40 minimal Cyp11b1 gene stimulated basal expression several-fold but did not confer cAMP inducibility. Interestingly, at their normal distance from the TATA region when the CRE was deleted, these regions had little activity (MOUW et al., 1989). The −385 to −358 region was about 3-fold more active if inserted in its normal orientation. In contrast to activation of expression by the CRE element, these upstream elements increased expression in cell lines deficient for protein kinase A. The −310 element contains the sequence C–C–A–G–G–T–C–T which is similar to sequences present in the Cyp21 and Cyp11a1 genes. This sequence is also conserved in the bovine CYP11B1 gene. These results indicate that the positive regulatory elements in the −300 to −400 region appear to affect only general transcription and do not respond to cAMP. One of these elements is conserved in several adrenal P450 genes and is likely to be a common adrenal regulatory element. There are probably additional positive regulatory elements in the −300 to −120 region which have not yet been defined.

Transfection of bovine CYP11B1 and three related pseudogene 5′ flanking regions fused to CAT demonstrated that each was capable of directing cAMP-dependent expression in Y1 cells (KIRITA et al., 1990). Fragments as small as 1 kb resulted in 10-fold induction by cAMP in the normal genes. A 100 bp fragment induced expression only about 2-fold. In contrast to CYP11A1 genes, the CYP11B1 genes were not expressed in I–10 cells derived from Leydig tumor cells. These results indicate that tissue-specific expression is retained in these cell lines since the CYP11A1 gene is expressed in all steroidogenic tissues while CYP11B1 is expressed only in the adrenal cortex.

More detailed analyses revealed that four regions, designated Ad1 to Ad4, in the bovine CYP11B1 gene are important for transcription. Each of these regions is conserved among the bovine, mouse and human genes. Two other nonconserved regulatory regions, Ad5 and Ad6, were also identified. Ad1 and Ad2 were defined by cell-free in vitro transcription in extracts from bovine adrenal cortex with constructions containing the G-free cassette (MOROHASHI and OMURA, 1990). Deletions from the 5′ end from −300 to −93, which deletes Ad3, Ad4, Ad5, and Ad6, had little effect on transcription, but deletion from −93 to −81 (Ad2), −68 to −61 (Ad1), and −39 to −18 (TATA) each progressively reduced expression. The Ad1 region contains the sequence, T–G–A–C–G–T–G–A–T, which is similar to the CRE and identical to the functional cAMP-responsive element in the mouse gene (RICE et al., 1989) while the Ad2 site contains the sequence C–C–A–G–N–G–N–C–A–G–T, which does not correspond to a known regulatory binding site. Oligonucleotides containing either the Ad1 or Ad2 site inhibited in vitro transcription by about 50% and both inhibited the activity about 90% in accord with the deletion studies. Binding of proteins to both the Ad1 and Ad2 regions was demonstrated by both gel retardation and DNase 1 footprint analysis (Fig. 3) and competition with the individual oligonucleotides showed that binding to each site was independent of each other. Transcription of the CYP11B1 gene was higher in adrenal extracts than in liver extracts and the activity in the liver extracts could be increased by addition of small amounts of adrenal extract. This indicates that the Ad1 or Ad2 region provides some tissue specificity, but the anomalous expression in the liver suggests that other tissue-specific elements are present which are not active in the in vitro transcription studies. These cell-free transcription studies are useful for examining basal but not cAMP inducible transcription. They clearly identify Ad1 and Ad2 as positive regulatory regions. On the other hand, no decrease in expression is observed when Ad3 through Ad6 are deleted. As discussed in the next paragraph these regions are elements that mediate inducible (Ad3, Ad4) and basal (Ad5, Ad6) expression.

The analysis of these regions was continued by transfection of Y1 cells with the 5′ flanking regions of bovine CYP11B1 fused to a CAT reporter gene (HONDA et al., 1990). Basal activity was low with 5.6 kb of 5′ flanking region while deletion to −1.5 kb stimulated basal activity 9-fold. Additional deletions progressively decreased the basal activity. Induction by forskolin or dibuturyl cAMP of about 10 to 13-fold was observed with fragments extending to −5600 bp and was retained in deletions to −400. Deletion of sequence from −333 to −287 resulted in a decrease in induction from 9-fold to about 3-fold and deletion from −125 to −60, which includes the CRE sequence analogous to that in the mouse gene abolished the remaining cAMP response. DNase 1 footprint analysis revealed four regions protected in a fragment from −407 to −145. Two of these, Ad3 and Ad4, were present in the −333 to −287 region and represent conserved regions in the human, bovine and mouse genes. Small block mutations within the Ad3 and Ad4 regions decreased the cAMP induction ratio by about 50% in agreement with the deletions studies. Ad4 contains the sequence C–C–A–A–G–G–T–C and is similar to the −310 regulatory sequence defined in the mouse gene (BOGERD et al., 1990). Sequences that are similar to Ad4 are present in the other steroidogenic P450 genes, CYP11A1, CYP21 and CYP17, indicating that this sequence is likely to be important in the regulation of these genes. Ad3 on the hand was not well conserved in steroidogenic P450 genes other than the other CYP11B1 genes. In addition to the Ad3 and Ad4 sequences, two regions, Ad5 and Ad6, from −253 to −126 were protected from DNase 1 digestion by the adrenal proteins. Deletion of the segment from −253 to −126 did not alter the induction ratio by cAMP but did reduce activity by 60%. These regions are not well conserved in the CYP11B1 genes of different species; Ad6 contains a consensus sequence for AP-2. The Ad4 sequence competed for the binding to Ad5 and Ad6 which may reflect either binding of the Ad4 protein to these regions or an interaction between the sites. Competition with the Ad3 sequence located between the Ad4 and Ad5/6 sequences had no effect on binding to the Ad5 and Ad6 sites.

These results indicate that the regions containing the Ad1 through Ad4 sites are important for both basal and cAMP induced expression of the bovine CYP11B1 gene and Ad5 and Ad6 may be general transcription activators. The studies of the mouse gene emphasize the role of the CRE (Ad1) sequence in cAMP responsiveness while those of the bovine gene emphasize that of the upstream region. However, in both the bovine and mouse genes, deletion of upstream sequences reduced the fold-induction of cAMP by about 50% and deletion of the CRE eliminated the remaining 50%. In the mouse, the upstream region was between −1500 and −123 and was not defined further. In the bovine, the Ad3 and Ad4 sequences are the upstream inducible elements. Ad3 is conserved in the mouse gene centered at −310, a region that affected basal but not cAMP inducible expression. This discrepancy in the activity of the bovine and mouse Ad3/−310 region may reflect species differences or more trivial differences in the constructions used in the assay. In the mouse gene the effect of deletion of the −310 region on induction without deleting the CRE was not reported as it was with the bovine gene. In the bovine gene, the activities of Ad3 and Ad4 regions were not tested in the absence of the CRE (Ad1) so that it is not clear if the induction contributed by these sites is dependent on the CRE sequence as was observed in the mouse gene. Additional experiments may narrow the differences in the estimations of the relative importance of the upstream and CRE sequences in cAMP induction in the CYP11B1 genes.

Similar less extensive studies with the human CYP11B1 gene showed that 1 kb of 5′ flanking sequence resulted in 10-fold induction of expression by cAMP and that successive deletions progressively decreased expression (Kawamoto et al., 1990b). No expression was observed in Hela cells. A CRE analogous to that in the mouse gene, is present at −71 but fragments containing this sequence exhibit no promoter activity. Additional studies will be required to define the cAMP responsive element in the human gene.

3.2.3.5.3. CYP17 genes

Expression of the bovine and human CYP17 genes have been analyzed. Transfection into Y1 cells of 5′ flanking fragments of the bovine CYP17 gene fused to a

CAT gene or to a minimal globin promoter were analyzed (LUND et al., 1990). The 3′ boundary of the CYP17 5′ flanking region in the globin constructions was −40. A fragment extending only to −243 was able to drive cAMP inducible expression of the gene (LUND et al., 1990). Both basal and inducible expression were reduced slightly by longer fragments up to −437. Deletion of the region −243 to −107 or −80 reduced the induction ratio by 50% with little effect on basal expression. These results indicate that two regions from −80 to −40 and −243 to −107 are involved in the cAMP-stimulated expression of the gene. Gel retardation assays indicated that three regions −323 to −297, −243 to −225, and −89 to −25 bound to nuclear proteins from either Y1 cells and Hela cells. Oligonucleotides containing the CRE sequence effectively competed for the protein binding to the −243 to −225 region but not for that binding to the −89 to −25 region. As determined by south-western blotting, the classical CRE binds to CREB, a 43 kd protein, but the −243 to −225 region binds to a 47 kd protein. This result is consistent with the fact that this region contains no sequences similar to the CRE. Since the CRE competes for the binding, it is likely that a regulatory protein other than CREB is present in the adrenal cell extracts which can bind to the −243 to −225 region as well as CRE sequences. This region does have a sequence T–G–A–T–G, which is similar to a core CRE sequence and to the cAMP responsive sequences postulated for the CYP11A1 gene (INOUE et al., 1991). The ability of this region to act independently was established by insertion of just the −243 to −225 sequence in front of the minimal globin promoter which conferred cAMP-responsiveness. The effect was increased if two copies were inserted; three or four copies did not increase the expression further.

Transfection of 5′ fragments of the human CYP17 gene into Y1 cells resulted in maximal expression and about a 3-fold induction with cAMP for a fragment extending to −310 (BRENTANO et al., 1990). Basal and induced activity were reduced about 3-fold when the sequence from −235 to −184 was deleted. Further deletions to −104 and −82 led to progressive loss of cAMP inducibility. Thus, cAMP responsive elements are located between −82 and −184 and an element for general transcriptional activity is present between −235 and −184. In contrast, forskolin-stimulated expression was inhibited by TPA and the inhibitory element mapped between −35 and −184. This finding is consistent with inhibition of induction of the endogenous CYP17A1 gene in primary cultures of adrenal cells. The promoter was inactive in JEG-103 cytotrophoblast cells and had low activity in MA-10 cells. Four GRE-like elements in the 5′ flanking region are apparently not functional since dexamethasone did not affect expression. In the human genes, therefore, regulatory regions for cell-specific expression, cAMP induction, and TPA inhibition are present in the proximal promoter region. Different regions of the promoter are responsible for these basal, inducible, and inhibitory functions. As in the bovine CYP17 gene, distal regions of the 5′ flanking region are apparently not required for expression.

3.2.3.5.4. CYP19 genes

The expression of the human CYP19 gene has not been characterized in detail, but expression of 5′ flanking regions fused to the CAT reporter gene was studied in BeWo choriocarcinoma cells (TODA et al., 1990). Maximal expression was obtained with a fragment extending to −242. Deletion of the region −500 to −361 resulted in an 8-fold increase in activity. Deletion of the region from −242 to −182 resulted in a reduction in expression of 15-fold and further deletion to −138 reduced expression 10-fold more to background levels. Treatment with TPA or cAMP greatly increased endogenous expression of the CYP19A1 gene but cAMP did not affect expression of the transfected gene and TPA increased expression only 2-fold. Elements mediating these effects may not be present in the 500 bp of 5′ flanking region that was analyzed. These results demonstrate that inhibitory elements are present between −500 and −361 and that positive regulatory elements for basal transcription are present between −242 and −138.

3.2.3.5.5. CYP21 genes

The regulation of the mouse Cyp21 gene has been studied in detail by PARKER'S group. In early studies it was shown that only 330 bp of 5′ flanking region were required for cell-specific and hormonally inducible expression in Y1 cells (HANDLER et al., 1988, 1989). Normal promoter activity was observed with the sequence −330 to −150 in either orientation. Deletion analysis revealed that sequences between −230 and −180 were required for basal expression (PARKER et al., 1986) and the sequence −230 to −150 was able to stimulate the expression of the thymidine kinase promoter.

More recent detailed studies have revealed that this region is probably a complex regulatory region with multiple cis-acting elements (RICE et al., 1990b). DNase 1 footprinting with nuclear extracts from Y1 cells revealed 8 discrete regions that bound to proteins and were protected from DNase 1 (Fig. 2). One sequence at −120 contains the known regulatory sequence, C–C–A–A–T. Sites at −210 and −65 share the heptamer, A–G–G–T–C–A–G, and cross-competition experiments and the pattern of complexes formed in gel retardation assays indicate that the same protein binds to both elements. These two elements also contain five or six nucleotides in common with the CRE, but fragments containing known CRE sequences did not compete for binding. Proteins extracted from nonsteroidogenic cells, Hela and PC12, also protected all regions except for that centered at −140 and weaker protection was observed with the sites at −210 and −65. Gel retardation assays demonstrated that a unique complex formed with extracts from steroidogenic cells with the −210 and −65 regions that was not present in extracts from the other cell lines.

The functional role of each of these binding regions was tested by transfecting mutants of the intact Cyp21a1 gene containing 1.7 kb of 5′ flanking region into Y1 cells. Multiple substitutions were introduced into each site and expression was detected by primer extension so that only accurately initiated mRNA was assessed. Mutations at each site reduced the level of expression by 90% or more except for the site centered at −255 which resulted in a 60% decrease. Interestingly, mutation of either the −210 or −65 site, which probably interact with the same protein, resulted in >90% inhibition. These results demonstrate that regulation of the mouse Cyp21a1 gene probably involves a complex interaction of multiple factors which is consistent with the regulation of the gene by several stimuli. The protein binding studies suggest that cell-specific expression may depend on elements at −140, and possibly −65 and −210. The role of these elements in cAMP induction of this gene remains to be determined.

The 5′ flanking region of the human CYP21A2 gene was analyzed in fusions with a minimal globin promoter or with its natural promoter and a CAT reporter (Kagawa and Waterman, 1990). In the globin promoter constructions, the 3′ terminus of the CYP21A2 sequence was −96. Incremental deletions of fragments from −1305 to −840 and −612 resulted in a progressive loss of both basal and forskolin-stimulated expression whereas further deletion to −389 and then to −250 resulted in increases in basal and stimulated expression. Deletion of the segment −250 to −232 resulted in a large decrease of both basal and stimulated activity. An increase in the stimulated but not basal activity was observed when the fragment −210 to −129 were deleted. The remaining CYP21A2 sequence from −129 to −96 retained cAMP responsiveness of 10-fold, greater than that of any of the other constructions. The fragment −100 to −30 was inactive for both basal and induced expression. The positive element between −250 and −225 was not active when fused directly to the globin promoter and apparently requires sequences between −225 and −96 for its action, possibly the cAMP-responsive region from −129 to −96. Similar results were seen with the natural CYP21A2 promoter and CAT reporter except that basal expression was undetectable. These results indicate that regions upstream of −612 in the human CYP21A2 gene may contain negative regulatory elements and multiple positive regulatory regions are present from −389 to −232. The region between −210 and −129 appears to contain a negative element that selectively inhibits inducible expression, while the sequence from −120 to −96 contains the cAMP-responsive element. Nuclear proteins that bind to the −250 to −225 and the −129 to −96 regions were present in extracts from the Y1 cells as well as bovine adrenocortical cells in primary culture and human fetal adrenal tissue. Neither a fragment containing a known CRE nor the cAMP-responsive region from bovine CYP17 competed for binding of the protein to the cAMP-responsive region −129 to −96. These results indicate that the protein binding to the cAMP element is different from both CREB and the adrenal protein that binds to the CYP17 gene.

4. Adverse human P450 gene polymorphisms

Alterations of human P450 genes which affect the level or activity of a P450 enzyme lead to predictable adverse effects. Polymorphisms in the steroidogenic P450 genes can lead to dramatic alterations in the endocrine status of an individual while polymorphisms in xenobiotic metabolizing P450s may lead to abnormal responses to either therapeutic drugs or environmental toxins. The correspondence of polymorphic change in the CYP1A1 and CYP2D6 genes with risk of cancer has also been proposed. Variations in the catalytic activity of at least 18 different human P450s have been reported (GUENGERICH, 1988). The basis for some of these variations have not been established but in six cases changes in the structure of the genes have been documented. Polymorphisms involving the CYP21A2 gene and the CYP2D6 gene have been studied most intensively. In each of these cases, tightly linked pseudogenes are present and for the CYP21A2 gene the high incidence of polymorphisms may result from frequent gene conversion events in which detrimental mutations in the pseudogene are transferred to the normal gene.

The most common disorder associated with polymorphisms of P450 genes is congenital adrenal hyperplasia. This disorder is associated with a defect in the synthesis of cortisol which results in a hypertrophy of the adrenal glands because of a compensatory release of corticotropin from the pituitary. There are six known enzyme deficiences that lead to cortisol insufficiency, four of which are P450s. By far the most common is deficiency of CYP21A2 which is required for biosynthesis of adrenocorticoids and mineralocorticoids. Less frequent deficiencies are associated with CYP17A1, which is required for cortisol, androgen and estrogen synthesis, and with CYP11B1, which is required for the last step of cortisol synthesis and the conversion of deoxycorticosterone to corticosterone in the aldosterone pathway. The fourth P450 is the side chain cleavage enzyme, CYP11A1, which is required for synthesis of all adrenal and gonadal hormones. Abnormalities associated with this enzyme are so devastating that survival is rare and little information about biochemical and genetic changes is available.

There are a variety of possible mechanisms for adverse polymorphic changes which include regulatory as well as structural changes. Regulatory changes have not yet been clearly documented in human adverse effects. Structural changes may result from point mutations that result in decreased activity, inactive full length enzymes or truncated enzymes due to formation of a premature termination codon. Insertions or deletions cause the same effects or may cause a frameshift which leads to miscoding of the sequence of the protein. Either of these types of mutations may lead to changes near the intron/exon junctions which can result in altered splicing of the mRNA. The demonstration of an abnormally spliced mRNA in an individual does not necessarily establish splicing as the mechanism of the polymorphism. Splicing of normal genes can be imprecise so that both normal mRNA's and abnormally spliced mRNA's may coexist with no phenotypic effect (Miles et al., 1989, 1990; Lacroix et al., 1990). With a human population of several billion, it is not surprising that examples of all the possible mechanisms have been found for P450 genes.

4.1. CYP1A1 gene

The CYP1A1 gene can metabolize a number of potential mutagens and carcinogens so that changes in its activity might alter carcinogenic risk in an individual (Gonzalez, 1989). The CYP1A1 gene has a restriction fragment length polymorphism (RFLP) with the enzyme, Msp I, and the rare allele is homozygous in 11% of the population (Kawajiri et al., 1990). In cancer patients this homozygous allele occurred about three times more frequently. This indicates that a structural change near CYP1A1 correlates with cancer risk, but the nature of the change has not been determined and a causal relationship has not been established.

4.2 CYP2D6 gene

P4502D6 catalyzes the metabolism of debrisoquine and a large number of other drugs, such as dextromethorphan, propranolol and bufuralol. The metabolism of debrisoquine is polymorphic in humans because of an autosomal recessive mutation that leads to decreased metabolism in 5% to 10% of the population. The poor metabolizer is characterized by the absence of the CYP2D6 enzyme. Several RFLP's of the CYP2D6 gene, with Xba I have been observed. A 29 kb fragment is the most frequent and is usually associated with extensive metabolizers while 44 kb and 11.5 kb fragments are thought to represent mutant alleles (Skoda et al., 1988). The use of these RFLP's predicted the phenotype in 25% of poor metabolizers indicating that other mutations must also be causative that are not detectable by the Xba I RFLP analysis.

Initial studies based on cloned cDNA indicated that defective splicing resulting in the retention of intron 5 might be responsible for the enzyme deficiency (Gonzalez et al., 1988). More recent studies have used the polymerase chain reaction for amplification of genomic DNA to study changes reported in the cDNA (Gough et al., 1990) or have analyzed genes isolated from individuals with the 29 kb or 44 kb phenotype (Kagimoto et al., 1990; Hanioka et al., 1990). There are several point mutations in the genes including a conserved proline in exon 1 which inactivates the enzyme (Kagimoto et al., 1990). The absence of CYP2D6 enzyme, however, can be explained only by a G to A mutation at the junction of intron 3 and exon 4 which results in altered splicing. The new mRNA that is formed contains a single base pair deletion in the mRNA resulting in a frameshift mutation. The change also eliminates a BstN1 site. The use of the polymerase chain reaction in this region to test for the loss of the BstN1 site predicted the correct phenotype in 65% to 82% of poor metabolizers (GOUGH et al., 1990; HANIOKA et al., 1990). The G to A mutation in intron 3 is not present in the two related pseudogenes so that in this case gene conversion cannot be the basis for the high frequency of this polymorphic change.

Additional less frequent mutations have been described for the CYP2D6 genes. The Xba I 11.5 kb genotype is associated with a deletion of the CYP2D6 gene (see HEIM and MEYER, 1990) and a second 29 kb genotype has a single base pair deletion in exon 5 (KAGIMOTO et al., 1990). Two polymerase chain reaction amplifications have been designed to detect each of these polymorphic changes and the phenotype of 97% of poor metabolizers could be predicted (HEIM and MEYER, 1990). The remaining small fraction of poor metabolizers not predicted are presumably due to rare mutations, such as that resulting in an Xba I 16 + 9 kb RFLP and other undetected mutations (HEIM and MEYER, 1990).

A linkage between extensive metabolizers and cancer risk has been suggested by the low frequency of poor metabolizers in cancer patients (see SUGIMURA et al., 1990 for references). In a study of 45 lung cancer patients, a normal distribution of the various Xba I polymorphic fragments was observed, but none of these patients was a poor metabolizer (SUGIMURA et al., 1990). Poor metabolizers were present in healthy individuals and in patients with non-neoplastic pulmonary disease or tumours in tissue other than lung. Other recent studies have confirmed the low incidence of poor metabolizers in lung and bladder cancer patients (GOUGH et al., 1990). These results suggest that the CYP2D6 enzyme may be important in activating environmentally derived carcinogens such as are found in cigarette smoke. More extensive studies are required to establish the correlation and eliminate the possibility that a closely linked gene is responsible. A carcinogenic metabolite must be also identified to conclusively establish the role of CYP2D6 in carcinogenesis.

4.3. CYP11B1 gene

Specific defects in the CYP11B1 gene have not been reported but a Msp I RFLP in the gene is associated with deficiency in the enzymatic activity corticosterone methyl oxidase II (GLOBERMAN et al., 1988b). This enzyme catalyzes the final three reactions in the synthesis of aldosterone and was thought to be encoded by the CYP11B1 gene. However, classical $P450_{11\beta}$ deficiency has different clinical symptoms. This suggests that point mutations in the enzyme may have different effects on the synthesis of aldosterone and cortisol or alternatively, that the association results from linkage with the mutated form of a nearby related gene. This latter possibility is supported by the isolation of cDNA's closely related to CYP11B1 from rat (IMAI et al., 1990) and human (KAWAMOTO et al., 1990a) that preferentially catalyze aldosterone formation. The human cDNA seems to correspond to the CYP11B2 gene (Mornet et al., 1989).

4.4. CYP17A1 gene

The CYP17A1 gene product, $P450_{17\alpha}$ catalyzes both steroid 17-hydroxylation and 17,20-lyase (desmolase) reactions. These metabolic steps are essential for the formation of the glucocorticoids, androgens and estrogens, but not the mineralocorticoids. Deficiency of the CYP17A1 gene is extremely rare. In affected males, the failure of androgen production by the fetal testis prevents development of external genitalia (Liddle et al., 1981). In females, deficient estrogen production leads to a failure of the development of secondary sexual characteristics at puberty. The overproduction of the deoxycorticosterone produces a low renin hypertension. Treatment with glucocorticoids eliminates the adrenal hyperplasia and the overproduction of deoxycorticosterone which corrects the hypertension. Estrogen therapy beginning at the time of puberty is required for normal sexual development. The deficiency of the CYP17A1 gene product may be complete or partial and examples of genetic changes in the gene for each type have been observed. For complete deficiency, one defect is a four-base pair duplication in the CYP17A1 gene of two Mennonite families that leads to a frameshift and miscoding of the C-terminal 26 amino acids (Kagimoto et al., 1989). In Japanese patients, two independent changes have been found. In the first case a point mutation producing a stop codon at codon 17 resulted in a truncated inactive enzyme (Yanase et al., 1988). In the second, a seven bp duplication in exon 2 leads to a frameshift mutation leading to miscoding and a premature termination codon (Yanase et al., 1990). As might be expected for the partial deficiency, a more subtle change of the deletion of phenylalanine at position 53 or 54 was observed (Yanase et al., 1989). Expression of CYP17A1 with this mutation in COS cells produced an enzyme with 37% and 8% of the hydroxylase and lyase activities, respectively. On the basis of different defects in each patient of this small group, it is likely that additional defects will be found.

4.5. CYP21A2 gene

There are two CYP21 genes which resulted from a duplication of a 30 kb fragment in the HLA region of chromosome 6. This fragment also contained the C4 histocompatibility gene of the HLA locus resulting in an alternating C4A-CYP21A1-C4B-CYP21A2 gene organization. The human CYP21A1 gene is a pseudogene that contains an 8 bp deletion within codons 110–112 that leads to a frameshift mutation and a premature termination codon at codon 130 (White et al., 1986; Higashi et al., 1986). Interestingly, the mouse Cyp21a2 gene is the pseudogene and Cyp21a1 is the functional gene (CHAPLIN et al., 1986). These genes are about 98% homologous in the protein coding region suggesting a very recent duplication event, but two CYP21 genes are also present in other mammalian species with the same organization. The CYP21 genes should have diverged more than 2% if they formed by a duplication before the mammalian radiation so that frequent gene conversions are thought to be responsible for the conservation of the sequences of the two genes within a species.

Classical forms of congenital adrenal hyperplasia caused by CYP21A1 deficiency are salt-wasting, in which cortisol and androsterone are deficient and androgens overproduced, and simple virilizing, in which aldosterone production is presumably near normal (LIDDLE et al., 1981). Nonclassical congenital adrenal hyperplasia is milder and may have a late onset. The incidence of these diseases is about 1 in 5,000 to 15,000 and 1 in 3,000 for the classical and nonclassical forms, respectively (SPEISER et al., 1985). Severe forms result in marked masculinization of the external genitalia of female fetuses and the salt-wasting form is often fatal without appropriate intervention. Milder forms may be asymptomatic but usually are manifested as rapid early growth in children between two and 10 years and by the lack of development of secondary sex characteristics in females. Glucocorticoid and/or mineralocorticoid administration is effective therapy for the disease.

A large number of different mutations in the CYP21A2 gene have been associated with congenital adrenal hyperplasia. In about 25% of cases, deletion of the CYP21A2 gene is present (RUMSBY et al., 1986, 1988; WERKMEISTER et al., 1986; MOREL et al., 1989; OWERBACH et al., 1990; MORNET et al., 1991). The remaining 75% of cases are associated with point mutations that lead to aberrant splicing of the mRNA or decreased activity of the enzyme. In nearly all cases the mutations that are observed are present in the CYP21A1 pseudogene and probably arose by gene conversion. In some cases the sequence of the entire CYP21A2 gene is converted and in others a portion of the gene is changed. The presence of the pseudogene and the high rate of gene conversion explains the high incidence of congenital adrenal hyperplasia caused by CYP21A2 deficiency.

As expected from the multiple differences between the two CYP21 genes, multiple point mutations have been described and relative frequencies have been studied in large samples of patients (OWERBACH et al., 1990; MORNET et al., 1991). A frequent change is a GC substitution in intron 2 that leads to aberrant splicing of the mRNA (HIGASHI et al., 1988b). In exon 8, an 8 bp deletion producing a frameshift leads to an inactive protein and is associated with the severe forms of the disease (OWERBACH et al., 1990; MORNET et al., 1991). An isoleucine to asparagine change at position 172 in exon 4 (AMOR et al., 1988) is observed at low frequency in all forms of the disease (MORNET et al., 1991). In exon 6, a cluster of three mutations (HIGASHI et al., 1988b) resulting in amino acid changes is also associated with all forms of the disease. In exon 7, a valine to leucine substitution at position 281 (SPEISER et al.,

1988) is correlated with the milder nonclassical form of the disease. Substitution of a stop codon for glutamine at position 318 (GLOBERMAN et al., 1988a) and substitution of arginine at 356 with tryptophan in exon 8 have been detected in defective genes (CHIOU et al., 1990). A potential gene conversion change of proline at position 30 to leucine has not been detected in natural genes but in vitro mutagenesis has shown that this change results in a partially active enzyme (TUSIE-LUNA et al., 1991). The change is potentially a cause of the milder nonclassical form of congenital adrenal hyperplasia. All of the changes discussed above can be explained by gene conversion of CYP21A2 by the CYP21A1 gene. Two changes, serine at 269 to threonine and asparagine at 494 to serine, have been described in a defective CYP21A2 gene which are not present in the CYP21A1 gene and could not have been produced by gene conversion. The former of these is apparently a neutral mutation (OWERBACH et al., 1990; TUSIE-LUNA et al., 1991).

Nearly all the mutations leading to CYP21A2 related to congenital adrenal hyperplasia have now been described. In principle it should be possible to combine RFLP and polymerase chain reaction techniques to provide prenatal testing with high accuracy. This remains a difficult task in practice because multiple possible mutations must be assessed and the presence of the closely related CYP21A1 gene will result in misdiagnosis if specific hybridization conditions are not stringently followed (MORNET et al., 1991).

4.6. CYP27A1 gene

A rare disease, cerebrotendinous xanthomatosis, is a sterol storage disorder which leads to neurological dysfunction and tendon xanthomas. A study of two patients with this disease identified two different point mutations in the CYP27A1 gene (CALI et al., 1991). The first was an arginine to cysteine change at position 362 in the adrenodoxin binding region of the P450 enzyme. The second was a substitution of cysteine for arginine at position 446 which is two codons from the conserved cysteine at 444 that is a ligand for the heme moiety. The rate of 27-hydroxylation activity catalyzed by mutant CYP27A1 cDNA's transfected into COS cells was dramatically decreased compared to the wild type enzyme establishing these changes as causes of the disease.

5. Reflections

Great strides have been made in the characterization of P450 genes, but much remains to be done. The P450 superfamily is a rich source for the study of molecular evolution. It contains genes essential to life, like CYP21A2, and those that would seem to be dispensable on an individual basis such as members of the large xenobiotic metabolizing subfamilies. There may be selective pressure for new mutations to develop novel substrate specificities in the latter. Comparative analysis of the molecular evolution of these diverse but related genes should reveal interesting differences. An explanation for the evolution of the introns in P450 genes remains elusive. Additional gene sequences from lower vertebrate and nonvertebrate species will contribute to the understanding of intron origins of P450 and genes in general.

The regulation of the Ah-inducible and the cAMP-inducible genes have begun to be understood, but the regulatory proteins that mediate these effects have yet to be isolated and characterized in detail. Studies on the regulation of other inducible P450s are less advanced. To understand the regulation, receptors need to be identified. Cell lines, transgenic mice, or in vitro systems need to be developed in which expression can be assayed. Studies on the Ah-inducible and cAMP-inducible genes have been facilitated by the existence of cell lines in which the genes can be expressed. New methods for culture of hepatocytes are likely to lead to rapid advances in the near future, particularly for the phenobarbital-inducible forms. It is likely that P450s with novel tissue and ligand-dependent regulation have yet to be identified, particularly in nonhepatic tissues. The rapid advances in the last decade provide confidence that such challenges will be met and detailed understanding of the molecular mechanisms of the expression of several P450 gene families will be achieved in the 90's.

The study of the relationship of polymorphic changes in P450 genes to disease is in its infancy. Although the CYP21A2 and CYP2D6 genes have been analyzed in detail, most other structural and regulatory polymorphisms are not understood at the genetic level. These variations are important for understanding many congenital diseases and individual differences in the response to therapeutic drugs and environmental toxins. The tantalizing correlation between polymorphisms and risk of cancer remains to be proven. Ultimately these studies may lead to procedures that can predict the response of an individual to drugs and toxins so that rational individual drug doses can be administered and adverse risks can be minimized.

Acknowledgement

Thanks to the members of the Kemper laboratory who provided advice on this manuscript. Research on P450 in the author's laboratory is supported by N.I.H. grants GM39360 and GM35897.

6. References

Adesnik, M. and M. Atchison, (1986), CRC Crit. Rev. Biochem. **19**, 247–305.

Ahlgren, R., E. R. Simpson, M. R. Waterman, and J. Lund, (1990), J. Biol. Chem. **265**, 3313–3319.
Amor, M., K. L. Parker, H. Globerman, M. I. New, and P. C. White, (1988), Proc. Natl. Acad. Sci. USA **85**, 1600–1604.
Atchison, M. and M. Adesnik, (1983), J. Biol. Chem. **258**, 11285–11295.
Atchison, M. and M. Adesnik, (1986), Proc. Natl. Acad. Sci. USA **83**, 2300–2304.
Baumhueter, S., D. B. Mendel, P. B. Conley, C. J. Kuo, C. Turk, M. K. Graves, C. A. Edwards, G. Coustois, and G. R. Crabtree, (1990), Genes Dev. **4**, 372–379.
Belfort, M., (1991), Cell **64**, 9–11.
Bhasker, C. R., B. S. Adler, A. Dee, M. E. John, M. Kagimoto, M. X. Zuber, R. Ahlgren, X. Wang, E. R. Simpson, and M. R. Waterman, (1989), Arch. Biochem, Biophys. **271**, 479–487.
Bogerd, A. M., A. Franklin, D. A. Rice, B. P. Schimmer, and K. L. Parker, (1990), Mol. Endocrinol. **4**, 845–850.
Brentano, S. T., J. Picado-Leonard, S. H. Mellon, C. C. D. Moore, and W. L. Miller, (1990), Mol. Endocrinol, **4**, 1972–1979.
Cali, J. J., C. -L. Hsieh, U. Francke, and D. W. Russell, (1991), J. Biol. Chem. **266**, 7779–7783.
Cedar, H., (1988), Cell **53**, 3–4.
Chan, G., and B. Kemper, (1990), Biochemistry **29**, 3743–3750.
Chaplin, D. D., L. J. Galbraith, J. G. Seidman, P. C. White, and K. L. Parker, (1986), Proc. Natl. Acad. Sci. USA **83**, 9601–9605.
Chiou, S. H., M. C. Hu, and B. C. Chung, (1990), J. Biol. Chem. **265**, 3549–3552.
Chung, B. C., K. J. Matteson, and W. L. Miller, (1986), Proc. Natl. Acad. Sci. USA **83**, 4243–4247.
Chung, B. C., M. C. Hu, C. C. Lai, and C. H. Lin, (1989), Biochem. Biophys. Res. Commun. **160**, 276–281.
Denison, M. S., J. M. Fisher, and J. P. Whitlock, Jr., (1989), J. Biol. Chem. **264**, 16478–16482.
Durrin, L. K. and J. P. Whitlock, Jr., (1989), Mol. Cell. Biol. **9**, 5733–5737.
Einck, L., J. Fagan, and M. Bustin, (1985), Biochemistry **24**, 5269–5275.
Einck, L., J. Fagan, and M. Bustin, (1986), Biochemistry **25**, 7062–7068.
Elferink, C. J. and J. P. Whitlock, Jr., (1990), J. Biol. Chem. **265**, 5718–5721.
Essienberg, J. C., I. L. Cartwright, G. H. Thomas, and S. C. R. Elgin, (1988), Ann, Rev. Genet. **19**, 485–536.
Fisher, J. M. L. Wu, M. S. Denison, and J. P. Whitlock, Jr., (1990), J. Biol. Chem. **265**, 9676–9681.
Foldes, R. L. and E. Bresnick, (1989), Biochem. Pharmacol. **38**, 4425–4428.
Fujii-Kuriyama, Y., Y. Mizukami, K. Kawajiri, K. Sogawa, and M. Murmatsu, (1982), Proc. Natl. Acad. Sci. USA **79**, 2793–2797.
Fujisawa-Sehara, A., M. Yamane, and Y. Fujii-Kuriyama, (1988), Proc. Natl. Acad. Sci. USA **85**, 5859–5863.
Fulco, A. J. and R. T. Ruettlinger, (1987), Life Sci. **40**, 1769–1775.
Fulco, A. J., (1991), Annu. Rev. Pharmacol. Toxicol. **31**, 177–203.
Geffrotin, C., P. Chardon, D. F. de Andres-Cara, R. Feil, C. Renard, and M. Vaiman, (1990), Anim. Genet. **21**, 1–13.
Giachelli, C. M., J. Lin-Jones, and C. J. Omiecinski, (1989), J. Biol. Chem. **264**, 7046–7053.
Globerman, H., M. Amor, K. L. Parker, M. I. New, and P. C. White, (1988a), J. Clin. Invest. **82**, 139–144.
Globerman, H., A. Rosler, R. Theodor, M. I. New, and P. C. White, (1988b), N. Engl. J. Med. **319**, 1193–1197.

Gonzalez, F. J., S. Kimura, and D. W. Nebert, (1985), J. Biol. Chem. **260**, 5040–5049.
Gonzalez, F. J., B. J. Song, and J. P. Hardwick, (1986), Mol. Cell Biol. **6**, 2969–2976.
Gonzalez, F. J., (1988), Pharmac. Rev. **40**, 243–288.
Gonzalez, F. J., R. C. Skoda, S. Kimura, M. Umeno, U. M. Zanger, D. W. Nebert, H. V. Gelboin, J. P. Hardwick, and U. A. Meyer, (1988), Nature **331**, 442–446.
Gonzalez, F. J., (1990), Pharmac. Ther. **45**, 1–38.
Gonzalez, F. J., and D. W. Nebert, (1990), Trends. Genet. **6**, 182–186.
Gonzalez, F. J., S. -Y. Liu, C. A. Kozak, and D. W. Nebert, (1990), DNA Cell Biol. **9**, 771–776.
Gotoh, O. and Y. Fujii-Kuriyama, (1989), in: Frontiers in Biotransformation, (K. Ruckpaul and H. Rein, eds.), Vol **1**, Taylor and Francis, New York 195–243.
Gough, A. C., J. S. Miles, N. K. Spurr, J. E. Moss, A. Gaedigk, M. Eichelbaum, and C. R. Wolf, (1990), Nature **347**, 773–776.
Govind, S., P. A. Bell, and B. Kemper, (1986), DNA **5**, 371–382.
Gross, D. S. and W. T. Garrad, (1988), Ann. Rev. Biochem. **57**, 159–197.
Guengerich, F. P., (1988), Trends in Pharmacol. Sci. **101**, 107–109.
Hahn, C. N., A. J. Hansen, and B. K. May, (1991), J. Biol. Chem., 17031–17039.
Handler, J. D., B. P. Schimmer, T. R. Flynn, M. Szyf, J. G. Seidman, and K. L. Parker, (1988), J. Biol. Chem. **263**, 13068–13073.
Handler, J. D., B. P. Schimmer, T. R. Flynn, M. Szyf, D. A. Rice, and K. L. Parker, (1989), Endocr. Res. **15**, 31–47.
Hanioka, N., S. Kimura, U. A. Meyer, and F. J. Gonzalez, (1990), Am. J. Hum. Genet. **47**, 994–1001.
Hankinson, O., R. D. Anderson, B. W. Birren, F. Sander, M. Negishi, and D. W. Nebert, (1985), J. Biol. Chem. **260**, 1790–1795.
Harada, N., K. Yamada, K. Saito, N. Kibe, S. Dohmae, and Y. Takagi, (1990), Biochem. Biophys. Res. Commun. **166**, 365–372.
Hardon, E. M., M. Fain, G. Paonessa, and R. Cortese, (1988), EMBO J. **7**, 1711–1718.
Hashimoto, T., K. Morohashi, and T. Omura, (1989), J. Biochem. **105**, 676–679.
He, J. -S. and A. J. Fulco, (1991), J. Biol. Chem. **266**, 7864–7869.
Heim, M. and U. A. Meyer, (1990), Lancet **336**, 529–532.
Hickey, G. J., J. S. Krasnow, W. G. Beattie, and J. S. Richards, (1990), Mol. Endocrinol, **4**, 3–12.
Higashi, Y., H. Yoshioka, M. Yamane, O. Gotoh, and Y. Fujii-Kuriyama, (1986), Proc. Natl. Acad. Sci. USA **83**, 2841–2845.
Higashi, Y., A. Tanae, H. Inoue, and Y. Fujii-Kuriyama, (1988a), Am. J. Hum. Genet. **42**, 17–25.
Higashi, Y., A. Tanae, H. Inoue, T. Hiromasa, and Y. Fujii-Kuriyama, (1988b), Proc. Natl. Acad. Sci. USA **85**, 7486–7490.
Hines, R. N., J. B. Levy, R. D. Conrad, P. L. Iversen, M. L. Shen, A. M. Renli, and E. Bresnick, (1985), Arch. Biochem. Biophys. **237**, 465–476.
Hines, R. N., J. M. Mathis, and C. S. Jacob, (1988), Carcinogenesis **9**, 1599–1605.
Hoffman, E. C., H. Reyes, F. -F. Chu, F. Sander, L. H. Conley, B. A. Brooks, and O. Hankinson, (1991), Science **252**, 954–958.
Honda, S., K. Morohashi, and T. Omura, (1990), J. Biochem. **108**, 1042–1049.
Ikeya, K., A. K. Jaiswal, R. A. Owens, J. E. Jones, D. W. Nebert, and S. Kimura, (1989). Mol. Endocrinol. **3**, 1399–1408.
Imagawa, M., R. Chiu, and M. Karin, (1987), Cell **51**, 251–260.
Imai, M., H. Shimada, Y. Okada, Y. Matsushima-Hibiya, T. Ogishima, and Y. Ishimura, (1990), FEBS Let. **263**, 299–302.
Inoue, H., N. Watanabe, Y. Higashi, and Y. Fujii-Kuriyama, (1991), Eur. J. Biochem. **195**, 563–569.

Israel, D. I., M. G. Estolane, D. R. Galeazzi, and J. P. Whitlock, Jr., (1985), J. Biol. Chem. **260**, 5648–5653.
Ivanovitch, J. D., U. M. Zanger, M. R. Waterman, and H. R. Fevold, (1991), Fed. Proc. **5**, A1161.
Jaiswal, A. K., F. J. Gonzalez, and D. W. Nebert, (1985), Nucleic Acids Res. **13**, 4503–4520.
Jaiswal, A. K., E. Rivkin, and M. Adesnik, (1987), Nucl. Acids Res. **15**, 6755.
Jaiswal, A. K., T. Haaparanta, P. V. Luc, J. Schembri, and M. Adesnik, (1990) Nucleic Acids Res. **18**, 4237–4242.
Jones, K. W. and J. P. Whitlock, Jr., (1990), Mol. Cell. Biol. **10**, 5098–5115.
Kagawa, N. and M. R. Waterman, (1990), J. Biol. Chem. **265**, 11299–11305.
Kagimoto, M., M. Heim, K. Kagimoto, T. Zeugin, and U. A. Meyer, (1990), J. Biol. Chem. **265**, 17209–17214.
Kagimoto, K., M. R. Waterman, M. Kagimoto, P. Ferreira, E. R. Simpson, and J. S. Winter, (1989), Hum. Genet. **82**, 285–286.
Kawajiri, K., J. Watanabe, O. Gotoh, Y. Tagashira, K. Sogawa, and Y. Fujii-Kuriyama, (1986), Eur. J. Biochem. **159**, 219–225.
Kawajiri, K., K. Nakachi, K. Imai, A. Yoshii, N. Shinoda, and J. Watanabe, (1990), FEBS Lett. **263**, 131–133.
Kawamoto, T., Y. Mitsuuchi, T. Ohnishi, Y. Ichikawa, Y. Yokoyama, H. Sumimoto, K. Toda, K. Miyahara, I. Kuribayashi, K. Nakao, K. Hosoda, Y. Yamamoto, H. Imura, and Y. Shizuta, (1990a), Biochem. Biophys. Res. Commun. **173**, 309–316.
Kawamoto, T., Y. Mitsuuchi, K. Toda, K. Miyahara, Y. Yokoyama, K. Nakao, K. Hosoda, Y. Yamamoto, H. Imura, and Y. Shizuta, (1990b), FEBS Lett. **269**, 345–349.
Khani, S. C., T. D. Porter, V. S. Fujita, and M. J. Coon, (1988), J. Biol. Chem. **263**, 7170–7175.
Kim, J. and B. Kemper, (1991), Biochemistry **30**, 10287–10294.
Kimura, S., H. Hanioka, E. Matsunaga, and F. J. Gonzalez, (1989a), DNA. **8**, 503–516.
Kimura, H., K. Sogawa, Y. Sakai, and Y. Fujii-Kuriyama, (1989b), J. Biol. Chem., **264**, 2338–2342.
Kimura, S., M. Umeno, R. C. Skoda, U. A. Meyer, and F. J. Gonzalez, (1989c), Am. J. Hum. Genet. **45**, 889–904.
Kirita, S., T. Hashimoto, M. Kitajima, S. Honda, K. Morohashi, and T. Omura, (1990), J. Biochem. **108**, 1030–1041.
Kocarek, T. A., E. G. Schuetz, and P. S. Guzelian, (1990). Mol. Pharmacol. **38**, 440–444.
Koop, D. R. and D. J. Tierney, (1990), Bioessays **12**, 429–435.
Kovesdi, I., R. Reichel, and J. R. Nevins, (1982), Cell **45**, 219–228.
Kronbach, T., T. M. Larabee, and E. F. Johnson, (1989), Proc. Natl. Acad. Sci. USA **86**, 8262–8266.
Kuhsel, M. G., R. Strickland, and J. D. Palmer, (1990), Science **250**, 1570–1573.
Lacroix, D., M. Desrochers, M. Lambert, and A. Anderson, (1990), Gene **86**, 201–207.
Lakso, M., R., Masaki, M. Noshiro, and M. Negishi, (1991), Eur. J. Biochem. **195**, 477–486.
Landschulz, W. H., P. E. Johnson, E. Y. Adashi, B. J. Graves, and S. L. McKnight, (1988), Genes Dev. **2**, 786–800.
Leighton, J. K. and B. Kemper, (1984), J. Biol. Chem. **259**, 11165–11168.
Liddle, G. W., (1981), in: Textbook of Endocrinology, (R. H. Williams, ed.). W. B. Saunders Co., Philadelphia, 249–292.
Lindberg, R., B. Burkhart, T. Ichikawa, and M. Negishi, (1989), J. Biol. Chem. **264**, 6465–6471.
Lindberg, R. L. P. and M. Negishi, (1989), Nature, **339**, 632–634.

LUND, J., R. AHLGREN, D. H. WU, M. KAGIMOTO, E. R. SIMPSON, and M. R. WATERMAN, (1990), J. Biol. Chem. **265**, 3304–3312.
MATHIS, J. M., R. A. PROUGH, R. N. HINES, E. BRESNIK, and E. R. SIMPSON, (1986), Arch. Biochem. Biophys. **246**, 439–448.
MATSUNAGA, E. and F. J. GONZALEZ, (1990), DNA Cell. Biol **9**, 443–452.
MATSUNAGA, E., M. UMENO, and F. J. GONZALEZ, (1990a), J. Mol. Evol. **30**, 155–169.
MATSUNAGA, T., M. NOMOTO, C. A. KOZAK, and F. J. GONZALEZ, (1990b), Biochemistry **29**, 1329–1341.
MAY, B. K. and A. J. HANSEN, (1989), Clin. Exp. Pharmacol. Physiol. **16**, 497–500.
MEANS, G. D., M. S. MAHENDROO, C. J. CORBIN, J. M. MATHIS, F. E. POWELL, C. R. MENDELSON, and E. R. SIMPSON, (1989), J. Biol. Chem. **264**, 19385–19391.
MEYER, U. A., R. C. SKODA, and U. M. ZANGER, (1990), Pharmacol. Ther. **46**, 297–308.
MILES, J. S., A. W. MCLAREN, and C. R. WOLF, (1989), Nucleic Acids Res. **17**, 8241–8255.
MILES, J. S., A. W. MCLAREN, F. J. GONZALEZ, and C. R. WOLF, (1990a), Nucleic Acids Res. **18**, 189.
MILES, J. S., J. E. MOSS, R. R. MEEHAN, and C. R. WOLF, (1990b), Genomics **7**, 445–448.
MIZUKAMI, Y., Y. FUJII-KURIYAMA, and M. MURAMATSU, (1983a), Biochemistry **22**, 1223–1229.
MIZUKAMI, Y., K. SOGAWA, Y. SUWA, M. MURAMATSU, and Y. FUJII-KURIYAMA, (1983b), Proc. Natl. Acad. Sci. USA **80**, 3958–3962.
MOLOWA, D. and G. CIMIS, (1991), Fed. Proc. **5**, A1165.
MONTMINY, M. R., K. A. SEVARINO, J. A. WAGNER, G. MANDEL, and R. H. GOODMAN, (1986), Proc. Natl. Acad. Sci. USA **83**, 6682–6686.
MOORE, C. C. D., S. T. BRENTANO, and W. L. MILLER, (1990), Mol. Cell. Biol. **10**, 6013–6023.
MOORE, C. C. D., D. W. HUM, and W. L. MILLER, (1991), Fed. Proc. **5**, A1165.
MOREL, Y., J. ANDRÉ, B. URING-LAMBERT, G. HAUPTMANN, H. BÉTUEL, M. TOSSI, G. F. MAGUELONE, M. DAVID, J. BERTRAND, and W. L. MILLER, (1989), J. Clin. Invest. **83**, 527–536.
MORISHIMA, N., H. YOSHIOKA, Y. HIGASHI, K. SOGAWA, and Y. FUJII-KURIYAMA, (1987), Biochemistry **26**, 8279–8285.
MORNET, E., J. DUPONT, A. VITEK, and P. C. WHITE, (1989), J. Biol. Chem. **264**, 20961–20967.
MORNET, E., P. CRÉTÉ, F. KUTTENN, M. -C. RAUX-DEMAY, J. BOUÉ, P. C. WHITE, and A. BOUÉ, (1991), Am. J. Hum. Genet. **48**, 79–88.
MOROHASHI, K., K. SOGAWA, T. OMURA, and Y. FUJII-KURIYAMA, (1987), J. Biochem. **101**, 879–887.
MOROHASHI, K. and T. OMURA, (1990), J. Biochem. **108**, 1050–1056.
MOUW, A. R., D. A. RICE, J. C. MEADE, S. C. CHUA, P. C. WHITE, B. P. SCHIMMER, and K. L. PARKER, (1989), J. Biol. Chem. **264**, 1305–1309.
MUERHOFF, A. S., K. J. GRIFFIN and E. F. JOHNSON, (1991), Fed. Proc. **5**, A1162.
NEBERT, D. W. and F. J. GONZALEZ, (1987), Ann. Rev. Biochem. **56**, 945–993.
NEBERT, D. W., M. ADESNIK, M. J. COON, R. W. ESTABROOK, F. J. GONZALEZ, F. P. GUENGERICH, I. C. GUNSALUS, E. F. JOHNSON, B. KEMPER, W. LEVIN, I. R. PHILLIPS, R. SATO, and M. R. WATERMAN, (1987), DNA **6**, 1–11.
NEBERT, D. W. and F. J. GONZALEZ, (1989), in: Frontiers in Biotransformation, (K. RUCKPAUL and H. REIN, eds.), Akademie-Verlag, Berlin.
NEBERT, D. W. and J. E. JONES, (1989), Int. J. Biochem. **21**, 243–252.
NEBERT, D. W., D. R. NELSON, and R. FEYEREISEN, (1989), Xenobiotica **19**, 1149–1160.
NEBERT, D. W., D. R. NELSON, M. ADESNIK, M. J. COON, R. W. ESTABROOK, F. J. GONZALEZ, F. P. GUENGERICH, I. C. GUNSALUS, E. F. JOHNSON, B. KEMPER, W. LEVIN, I. R. PHILLIPS, R. SATO, and M. R. WATERMAN, (1989), DNA **8**, 1–13.

NEBERT, D. W., D. R. NELSON, M. J. COON, R. W. ESTABROOK, R. FEYEREISEN, Y. FUJII-KURIYAMA, F. J. GONZALEZ, F. P. GUENGERICH, I. C. GUNSALUS, E. F. JOHNSON, J. C. LOPER, R. SATO, M. R. WATERMAN, and D. J. WAXMAN, (1991), DNA Cell Biol. **10**, 1–14.
NEF, P., T. M. LAREBEE, K. KAGIMOTO, and U. A. MEYER, (1990), J. Biol. Chem. **265**, 2903–2907.
NELSON, D. R. and H. W. STROBEL, (1987), Molec. Biol. Evol. **4**, 572–593.
NEUHOLD, L. A., Y. SHIRAYOSHI, K. OZATO, J. E. JONES, and D. W. NEBERT, (1989), Mol. Cell. Biol. **9**, 2378–2386.
OKEY, A. B., (1990), Pharmac. Ther. **45**, 241–298.
OKINO, S. T., U. R. PENDURTHI, and R. H. TUKEY, (1992), J. Biol. Chem., **267**, 6991–6998.
OONK, R. B., K. L. PARKER, J. L. GIBSON, and J. S. RICHARDS, (1990), J. Biol. Chem. **265**, 22392–22401.
OWERBACH, D., Y. M. CRAWFORD, and M. B. DRAZNIN, (1990), Mol. Endocrinol. **4**, 125–131.
PADMANABAN, G., P. N. RANGARAJAN, V. VENKATEWAR, M. VENKATESWARA RAO, P. UPADHYA, (1990), in: Biological Oxidations, (C. C. REDDY, G. A. HAMILTON, and K. M. MADYASTHA, eds.), Vol. **1** Academic Press, Inc. San Diego, 39–51.
PARKER, K. L., B. P. SCHIMMER, D. D. CHAPLIN, and J. G. SEIDMAN, (1986), J. Biol. Chem. **261**, 15353–15355.
PENDURTHI, U. R., J. G. LAMB, N. NGUYEN, E. F. JOHNSON, and R. H. TUKEY, (1990), J. Biol. Chem. **265**, 14662–14668.
PICADO-LEONARD, J. and W. L. MILLER, (1987), DNA **6**, 439–448.
POLAND, A., E. GLOVER, and A. S. KENDE, (1976), J. Biol. Chem. **251**, 4936–4946.
POLAND, A. and J. C. KNUTSON, (1982), Ann. Rev. Pharmacol. Toxiol. **22**, 517–554.
QUATTROCHI, L. C., U. R. PENDURTHI, S. T. OKINO, C. POTENZA, and R. H. TUKEY, (1986), Proc. Natl. Acad. Sci. USA **83**, 6731–6735.
QUATTROCHI, L. C. and R. H. TUKEY, (1989), Mol. Pharmacol. **36**, 66–71.
RANGARAJAN, P. N. and G. PADMANABAN, (1989), Proc. Natl. Acad. Sci. **86**, 3963–3967.
RAO, M. V., P. N. RANGARAJAN, and G. PADMANABAN, (1990), J. Biol. Chem. **265**, 5617–5622.
RAY CHAUDHARI, B., D. W. NEBERT, and A. PUGA, (1990), Mol. Endocrinol. **4**, 1773–1781.
RICE, D. A., L. D. AITKEN, G. R. VANDENBARK, A. R. MOUW, A. FRANKLIN, B. P. SCHIMMER, and K. L. PARKER, (1989), J. Biol. Chem. **264**, 14011–14015.
RICE, D. A., M. S. KIRKMAN, L. D. AITKEN, A. R. MOUW, B. P. SCHIMMER, and K. L. PARKER, (1990a), J. Biol. Chem. **265**, 11713–11720.
RICE, D. A., M. S. KRONENBERG, A. R. MOUW, L. D. AITKEN, A. FRANKLIN, B. P. SCHIMMER, and K. L. PARKER, (1990b), J. Biol. Chem. **265**, 8052–8058.
RODRIQUES, N. R., I. DUNHAM, C. Y. YU, M. C. CARROLL, R. R. PORTER, and R. D. CAMPBELL, (1987), EMBO J. **6**, 1653–1661.
RUMSBY, G., M. C. CARROLL, R. R. PORTER, and R. D. CAMPBELL, (1986), J. Med. Genet. **23**, 204–209.
RUMSBY, G., A. H. FIELDER, W. M. HAGUE, and J. W. HONOUR, (1988), J. Med. Genet. **25**, 596–599.
SAATCIOGLU, F., D. J. PERRY, D. S. PASCO, and J. B. FAGAN, (1990a), J. Biol. Chem. **265**, 9251–9258.
SAATCIOGLU, F., D. J. PERRY, D. S. PASCO, and J. B. FAGAN, (1990b), Mol. Cell. Biol. **10**, 6408–6416.
SCHUETZ, E. G. and P. S. GUZELIAN, (1984), J. Biol. Chem. **259**, 2007–2012.
SHEN, E. S. and J. P. WHITLOCK, JR., (1989), J. Biol. Chem. **264**, 17754–17758.
SHERRATT, A. J., D. E. BANET, M. W. LINDER, and R. A. PROUGH, (1989), J. Pharmacol. Exp. Ther. **249**, 667–672.
SIMMONS, D. L., P. MCQUIDDY, and C. B. KASPER, (1987), J. Biol. Chem. **262**, 326–332.

Simpson, E. R., J. Lund, R. Ahlgren, and M. R. Waterman, (1990), Mol. Cell Endocrinol. **70**, C25–C28.
Sinclair, P. R., E. G. Schuetz, W. J. Bement, S. A. Haugen, J. F. Sinclair, B. K. May, D. Li, and P. S. Guzelian, (1990), Arch. Biochem. Biophys. **282**, 386–392.
Skoda, R. C., F. J. Gonzalez, A. Demierre, and U. A. Meyer, (1988), Proc. Natl. Acad. Sci. USA **85**, 5240–5243.
Sladek, F. M., W. Zhong, E. Lai, and J. E. Darnell, Jr., (1990), Genes Dev. **4**, 2353–2365.
Sogawa, K., O. Gotoh, K. Kawajiri, and Y. Fujii-Kuriyama, (1984), Proc. Natl. Acad. Sci. USA **81**, 5066–5070.
Sogawa, K., O. Gotoh, K. Kawajiri, T. Harada, and Y. Fujii-Kuriyama, (1985), J. Biol. Chem. **260**, 5026–5032.
Sogawa, K., A. Fujisawa-Sehara, M. Yamane, and Y. Fujii-Kuriyama, (1986), Proc. Natl. Acad. Sci. USA **83** 8044–8048.
Sogawa, K., H. Handa, A. Fujisawa-Sehara, T. Hiromasa, M. Yamane, and Y. Fujii-Kuriyama, (1989), Eur. J. Biochem. **181**, 539–544.
Song, B. -J., H. V. Gelboin, S. -S. Park, S. C. Yang and F. J. Gonzalez, (1986), J. Biol. Chem. **261**, 16689–16697.
Speiser, P. W., B. Dupont, B. Rubenstein, A. Piazza, A. Kastelan, and M. I. New, (1985), Am. J. Hum. Gen. **37**, 650–667.
Speiser, P. W., M. I. New, and P. C. White, (1988), New Eng. J. Med. **1**, 19–23.
Srivastava, G., M. J. Bawden, A. J. Hansen, and B. K. May, (1989), Eur. J. Biochem. **178**, 689–692.
Srivastava, G., A. J. Hansen, M. J. Bawden, and B. K. May, (1990), Mol. Pharmacol. **38**, 486–493.
Sugimura, H., N. E. Caporaso, G. L. Shaw, R. V. Modali, F. J. Gonzalez, R. N. Hoover, J. H. Resau, B. F. Trump, A. Weston, and C. C. Harris, (1990), Carcinogenesis, **11**, 1527–1530.
Suwa, Y., Y. Mizukami, K. Sogawa, and Y. Fujii-Kuriyama, (1985), J. Biol. Chem. **260**, 7980–7984.
Szyf, M., B. P. Schimmer, and J. G. Seidman, (1989), Proc. Natl. Acad. Sci. **86**, 6853–6857.
Toda, K., M. Terashima, T. Kawamoto, H. Sumimota, Y. Yokoyama, I. Kuribayashi, Y. Mitsuuchi, T. Maeda, Y. Yamamoto, Y. Sagara, H. Ikeda, and Y. Shizuta, (1990), Eur. J. Biochem. **193**, 559–565.
Tusie-Luna, M. -T., P. W. Speiser, M. Dumic, M. I. New, and P. C. White, (1991), Molec. Endocrin. **5**, 685–692.
Ueno, T. and F. J. Gonzalez, (1990a), Mol. Cell. Biol. **10**, 4495–4505.
Ueno, T. and F. J. Gonzalez, (1990b), Nucleic Acids Res. **18**, 4623–4624.
Umeno, M., O. W. McBride, C. S. Yang, H. V. Gelboin, and F. J. Gonzalez, (1988a), Biochemistry **27**, 9006–9013.
Umeno, M., S. Morris, E. Matsunaga, H. V. Gelboin, and F. J. Gonzalez, (1988b), Nucleic Acid Res. **16**, 6249.
Umeno, M., B. J. Song, C. Kozak, H. V. Gelboin, and F. J. Gonzalez, (1988c), J. Biol. Chem. **263**, 4956–4962.
Wen, L. P., R. T. Ruettinger, and A. J. Fulco, (1989), J. Biol. Chem. **264**, 10996–11003.
Wen, L. P., N. Koeiman, and J. P. Whitlock Jr., (1990), Proc. Natl. Acad. Sci. **87**, 8545–8549.
Werkmeister, J. W., M. I. New, B. Dupont, and P. C. White, (1986), Am. J. Hum. Genet. **39**, 461–463.
White, P. C., M. I. New, and B. Dupont, (1986), Proc. Natl. Acad. Sci. USA **83**, 5111–5115.
Whitlock, J. P. Jr., (1987), Pharmac. Rev. **39**, 147–161.
Whitlock, J. P. Jr., (1989), Trends Pharmacol. Sci. **10**, 285–288.

WHITLOCK, J. P. JR., M. S. DENISON, J. M. FISHER, and E. S. SHEN, (1989), Mol. Biol. Med. **6**, 169–178.
WHITLOCK, J. P. JR., (1990), Annu. Rev. Pharmac. Toxicol. **30**, 251–277.
WILHELMSSON, A., S. CUTHILL, M. DENIS, A. C. WIKSTROM, J. A. GUSTAFSSON, and L. POELLINGER, (1990), EMBO. J. **9**, 69–76.
WONG, G., T. ITAKURA, K. KAWAJIRI, L. SKOW, and M. NEGISHI, (1989), J. Biol. Chem. **264**, 2902–2927.
XANTHOPOULOS, K. G., V. R. PREZIOSO, W. S. CHEN, F. M. SLADEK, R. CORTESE, and J. E. DARNELL, JR., (1991), Proc. Natl. Acad. Sci. USA **88**, 3807–3811.
XU, M. -Q., S. D. KATHE, H. GOODRICH-BLAIR, S. A. NIERZWICKI-BAUER, and D. A. SHUB, (1990), Science **250**, 1566–1570.
YAMAZOE, Y., M. SHIMADA, N. MURAYAMA, and R. KATO, (1987), J. Biol. Chem. **262**, 7423–7428.
YANAGIDA, A., K. SOGAWA, K. I. YASUMOTO, and Y. FUJII-KURIYMA, (1990), Mol. Cell. Biol. **10**, 1470–1475.
YANASE, T., M. KAGIMOTO, N. MATSUI, E. R. SIMPSON, and M. R. WATERMAN, (1988), Mol. Cell. Endocrinol. **59**, 249–253.
YANASE, T., M. KAGIMOTO, S. SUZUKI, K. HASHIBA, E. R. SIMPSON, and M. R. WATERMAN, (1989), J. Biol. Chem. **264**, 18076–18082.
YANASE, T., D. Sanders, A. SHIBATA, N. MATSUI, E. R. SIMPSON, and M. R. WATERMAN, (1990), J. Clin. Endocrinol. Metab. **70**, 1325–1329.
YOKOTANI, N., K. SOGAWA, S. MATSUBARA, O. GOTOH, E. KUSUNOSE, M. KUSUNOSE, and Y. FUJII-KURIYAMA, (1990), Eur. J. Biochem. **187**, 23–29.
YOSHIOKA, H., M. LANG, G. WONG, and M. NEGISHI, (1990), J. Biol. Chem. **265**, 14612–14617.
YOUNGBLOOD, G. L., C. SARTORIUS, B. A. TAYLOR, and A. H. PAYNE, (1991), Genomics, **10**, 270–275.
ZAPHIROPOULOUS, P. G., W. R. FOLK, and M. J. COON, (1986), Biochem. Biophys. Res. Commun. **134**, 499–505.
ZAPHIROPOULOUS, P. G., S. WESTIN, A. STRÖM, A. MODE, and J. A. GUSTAFSSON, (1990), DNA Cell Biol. **9**, 49–56.
ZHAO, J., J. K. LEIGHTON, and B. KEMPER, (1987), Biochem. Biophys. Res. Commun. **146**, 224–231.
ZHAO, J., G. CHAN, S. GOVIND, P. BELL, and B. KEMPER, (1990), DNA Cell Biol. **9**, 37–48.

Chapter 2

Topology and Biogenesis of Microsomal Cytochrome P-450s

M. Sakaguchi and T. Omura

1. Introduction

In eukaryotic cells, different forms of cytochrome P-450 (P-450) are distributed between the endoplasmic reticulum (ER) and mitochondria. Both microsomal and mitochondrial forms of P-450s are integral membrane proteins which are tightly bound to the membranes by hydrophobic interaction.

Microsomal P-450s are major integral membrane proteins of the ER membrane. They form an enzyme system in the membrane with NADPH-cytochrome P-450 reductase and cytochrome b_5 and catalyze monooxygenation reactions. The integral membrane protein nature of the P-450s are important both for efficient electron transfer from the reductase components and for efficient catalysis of the metabolism of lipopohilic substrates. Microsomal P-450s are synthesized on the membrane-bound ribosomes and inserted into the ER membrane during the synthesis. Their final topology in the membrane is achieved only co-translationally. Thus, the membrane topology of the P-450s is determined at the step of their biosynthesis. In this section, we review the recent progress of the study on the membrane topology and biogenesis of microsomal P-450s.

2. Membrane topology of microsomal P-450s

The topology of microsomal P-450s in the membrane has been probed by (1) site specific antibodies, (2) protease digestions, (3) chemical modifications, and (4) statistical analysis of their primary structures.

(1) *Use of antibodies in studying the topology of P-450s in the membrane*

Matsuura et al. (1978, 1979, 1983) showed that the antibodies against the whole molecule of rat P-450(PB) (P-450 2B1) reacted with the cytoplasmic side of the ER membrane but not with the luminal side. Thus, the major antigenic determinants of the P-450 molecule were localized on the cytoplasmic side of the membrane, where the antibodies could react with the epitopes. To determine the localization of various segments of the P-450 molecule DeLemos-Chiarandini et al. (1987) raised site specific antibodies to the synthetic peptides corresponding to fifteen different portions of rat liver P-450 PB-4 (P-450 2B1) and used the antibodies for immunoelectron microscopy of the rough microsomal membranes. It was demonstrated that all the epitopes except the amino-terminal hydrophobic segment (1st to 20th amino acid residues) were located on the cytoplasmic surface of the membrane. The antibodies against the amino-terminal sequence consisting of 31 amino acid residues reacted with the solubilized and purified P-450 but very poorly with the P-450 in the membrane, whereas the antibodies against the segment from 24th to 38th residues gave relatively strong signal of colloidal gold on the cytoplasmic surface of the rough microsomes. These observations suggested that only the amino-terminal segment shorter than

23 amino-acid residues long was in the membrane and that the following carboxy-terminal portions were exposed on the cytoplasmic surface of the membrane.

(2) *Protease digestion of purified and membrane-bound P-450s*

Although microsomal proteins (KREIBICH et al., 1983), including cytochrome b5 (ITO and SATO, 1968) and NADPH-cytochrome P-450 reductase (OMURA et al., 1967; ITO and SATO, 1969) were dissected into membrane binding hydrophobic domains and catalytically active hydrophilic domains, microsomal P-450s could not be dissected into functional domains by limited proteolysis. As an alternative strategy for probing the membrane topology of P-450s with proteases, extensive protease treatment of the intact microsomal membranes or the reconstituted proteoliposomes containing purified P-450s was carried out. In these studies, the peptides released from or remaining in the membrane were analyzed.

VERGÈRES et al. (1989) digested purified P-450 2B1 embedded in liposomes with trypsin and analyzed the peptide fragment remained bound to the liposomes. In the experimental strategy, it was postulated that the transmembrane hydrophobic segment(s) participating in the membrane anchoring should be recovered in the liposome fraction. The amino acid sequencing of the peptide fragment showed that only the amino-terminal segment (1st to 21st amino acid residues) was recovered in the liposomes.

BROWN and BLACK (1989) treated intact rabbit liver microsomal membrane, which were enriched in P-450 isozyme 2 (P-450 2B4), with trypsin and analyzed the peptides released from the membrane to the soluble fraction. The P-450 showed nine hydrophobic stretches in the molecule when examined by hydropathy profile analysis (HEINEMANN and OZOLS, 1982, 1983; TARR et al., 1983). All the hydrophilic segments between the seven hydrophobic stretches after the 50th amino acid residue were accessible to the externally added trypsin, indicating that all of these hydrophilic segments were located on the cytoplasmic surface of the membrane. Thereby, they concluded that the seven hydrophobic stretches do not span in the membrane.

These two observations suggested that only the amino-terminal hydrophobic segment was in the transmembrane orientation and participated in the membrane anchoring. However, they could not exclude the possibility that some other portions of the P-450 molecule interacted with the membrane in a loop or an amphipathic α-helical configurations.

(3) *Chemical modifications of membrane bound P-450s*

The membrane sideness of the amino-terminal methionine residue of microsomal P-450s was examined by labeling it with FITC (fluorescein isothiocyanate), a membrane impermeable reagent which selectively reacts with the α-amino-group under certain conditions (BERNHARDT et al., 1983). VERGÈRES et al. (1989) reported that the amino-terminal methionine residue of P-450 2B1 molecule integrated into liposomes was not accessible to FITC, whereas that of the purified P-450 in the solution was readily labeled with the reagent. Thereby, they concluded that only the amino-terminal 21 amino-acids long peptide spans the membrane and that the amino-terminal methionine was in the lumen of the proteoliposomes.

The FITC labeling experiments were also carried out using intact liver microsomal membranes from phenobarbital treated rats or rabbits. Two different observations were reported. One is that the amino-terminal α-amino group of P-450 2B1 was not accessible to FITC in the liver microsomal membrane from phenobarbital treated rats and became accessible to the reagent in the presence of a detergent (VERGÈRES et al., 1991). Another is that P-450 LM2 (P-450 2B4) in the intact rabbit microsomal membranes could be labeled with FITC (BERNHARDT et al., 1988). The reason for the discrepancy between these two observations is not clear. Strict control of the experimental conditions would be required for getting a definite conclusion from the chemical modification experiments.

(4) *Statistical analysis of the primary sequences of P-450s*

Primary structures of a large number of P-450s have been reported. It is likely that the conformations of membrane-bound P-450s are similar to each other as well as to that of a bacterial soluble P-450 (P-450cam), whose three dimensional structure has been elucidated (POULOS et al., 1985, 1987). NELSON and STROBEL (1988) have examined the geometry of the hydrophobic segments of various P-450 in the membrane by statistical analysis of the primary sequences of membrane-bound P-450s and P-450cam. Upon arguments based on the known membrane sideness of several peptide segments and the location and orientation of the heme moiety relative to the membrane, they eliminated nine hydrophobic segments out of 11 which were postulated and predicted as potential transmembrane segments by the multiple alignment, and proposed that the hydrophobic segments except those in the amino-terminal portion do not span the membrane.

NELSON and STROBEL (1989) predicted the secondary structures of membrane-bound P-450s and aligned them with that of P-450cam, and showed that the three-dimensional structures of membrane-bound P-450s were highly similar to that of P-450cam, except that the counterpart of the amino-terminal hydrophobic segment of the membrane-bound P-450s was not found in the soluble P-450. TRETIAKOV et al. (1989) predicted the membrane associated α-helical segment in P-450s by using the algorithm suitable for estimation of membrane binding property and showed that the amino-terminal hydrophobic α-helical segment is the only one in the transmembrane orientation. OUZOUNIS and MELVIN (1991) carried out an equally weighted multialignment of seven eukaryotic P-450 families and denoted profile patterns of primary and predicted secondary structures which are sufficiently conserved. They showed that these conserved regions corresponded to the structural elements of α-helix-rich domains of P-450cam. These lines of prediction suggested that structural elements of the microsomal P-450s exist in the soluble bacterial P-450 and that multiple transmembrane α-helices seem unlikely.

Taken together, we can conclude that only the amino-terminal hydrophobic segment of microsomal P-450s span the membrane and the following portion of the molecules is exposed on the cytoplasmic surface of the ER.

On the other hand, it has been reported that the freedom of the P-450 molecule in the microsomal membrane is restricted (KAWATO et al., 1982; ETTER et al., 1991). This observation might be explained by assuming the following. First, microsomal P-450s do not possess a flexible "hinge" segment between the membrane anchoring signal-anchor sequence and the following cytoplasmic portion. Second, some hydrophobic segments might contribute to fix the P-450 molecule on the membrane in a loop or an amphipathic α-helical configuration.

3. Cotranslational insertion of P-450s into ER membrane

(1) *Signal sequence-mediated membrane insertion of microsomal P-450s*

As well known, purified P-450s are not spontaneously incorporated into preformed liposomes, while some other components of the microsomal electron transfer system, such as cytochrome b_5 and NADH-cytochrome b_5 reductase are readily incorporated into the liposomes. Efficient incorporation of solubilized P-450s into liposomes can only be achieved by the detergent dialysis method.

From in vivo experiments using whole animals, it was shown that the newly synthesized P-450 was primarily detected in the rough ER membrane and then appeared in the smooth ER membrane (NEGISHI et al., 1976; FUJII-KURIYAMA et al., 1979). The messenger RNAs coding for P-450s are enriched in the bound polysomes (BAR-NUN et al., 1980) indicating that they are synthesized on the rough ER membrane such as secretory proteins. Upon in vitro synthesis of P-450 in the presence of rough microsomal membrane, the newly synthesized molecules were incorporated into the membrane, while they could no longer be inserted into the membrane after the completion of the translation. The P-450 (P-450 2B1) molecule incorporated into the membrane in vitro was indistinguishable from those synthesized in vivo in their susceptibility to detergent solubilization (BAR-NUN et al., 1980). Taken together, the P-450 molecules are inserted into the membrane cotranslationally.

According to the signal hypothesis (BLOBEL and DOBBERSTEIN, 1975) P-450s were expected to possess a signal sequence, which targets the ribosomes synthesizing the nascent peptides with the signal sequence to ER membrane and mediates bound ribosome formation. HAUGEN et al. (1977) first pointed out that the amino acid sequence of the amino-terminal portion of rabbit liver PB-inducible P-450 (P-450 2B1) is similar to that of the signal sequence of secretory proteins. BAR-NUN et al. (1980) demonstrated that the P-450 2B1 molecules synthesized in vitro showed the same molecular weight as the purified P-450 and possessed the same amino-terminal sequence. These observations indicated that microsomal P-450s possess no cleavable signal sequence at their amino-terminus. On the other hand SAKAGUCHI et al. (1984) showed that the insertion of the nascent $P\text{-}450_1$

(P-450 2B4) molecule into the rough microsomal membrane depended on signal recognition particle (SRP; WALTER et al., 1981) and the peptide chain elongation was inhibited by SRP in the absence of the membrane. These results demonstrated that mature P-450 molecules possess a sequence equivalent with the signal sequence.

(2) *Topogenic sequences of microsomal P-450s*

The uncleavable signal sequence was searched for in P-450 molecules by using several chimeric proteins and in vitro translation system. Amino terminal portions including the highly hydrophobic segments were fused with the mature portions of secretory proteins, such as interleukin 2, growth hormone, and parathyroid hormone (Fig. 1). These chimeric proteins synthesized in the presence of rough microsomal membranes were inserted into the membranes and acquired resistancy to the alkali extraction, and the secretory protein portions of the chimeric proteins remained on the cytoplasmic side of the membrane (SAKAGUCHI et al., 1987; SZCZESNA-SKORUPA et al., 1988; MONIER et al., 1988). Furthermore, the insertion of a chimeric protein containing the amino terminal 39 residues of P-450 1A1 was inserted into the membrane depending on SRP (SAKAGUCHI et al., 1987). Figure 1 shows the amino-terminal segments of microsomal P-450s used in the in vitro membrane insertion experiments. Thus the internal signal sequences could be localized in the short amino-terminal portion of the P-450s, and interestingly it functioned not only as an insertion signal sequence but also as a stop-transfer sequence, leaving the immediately following portion of the protein molecules on the cytoplasmic surface of the membrane. SAKAGUCHI et al., (1992) named the amino-terminal signal sequence "type I signal-anchor sequence (SAI)" as described in the following section.

```
                          10         20         29
P-450 1A1:   MVSDFGLPTF ISATELLLAS AVFCLVFWV  -    (IL2)

                          10         20    25
P-450 2C2:   MDLVVVLGLC LSCLLLLSLW KQSHG      -    (PTH)

                          10         20
P-450 2B1:   MEPSILLLLA LLVGFLLLLV   -   (GH)

                          10         20
      IL2:   MYSMQLASCV TLTLVLLVNS   -
```

Fig. 1. Amino acid sequences of the amino terminal segments of microsomal P-450s used in the in vitro membrane insertion experiments. Proteins used as the reporters are indicated in the parentheses; IL2, interleukin2 (SAKAGUCHI et al., 1987); PTH, parathyroid hormone (SZCZESNA-SKORUPA et al., 1988); GH, growth hormone (MONIER et al., 1988). Bottom line indicates the authentic signal sequence of mouse IL2 (KASHIMA et al., 1985). Acidic amino acid residues are indicated by boxes. Hydrophobic stretches are underlined.

It should be noted that normal signal sequences mediate the translocation of the following portion of the protein molecule across the membrane leaving its amino-terminus on the cytoplasmic surface (SHAW et al., 1988). In contrast to the normal function of the signal sequence, MONIER et al. (1988) demonstrated that the amino-terminal portion of opsin, including a N-glycosylation site which was fused at the amino-terminus of P-450 2B1 was glycosylated, when it was synthesized in the presence of rough microsomal membrane. Since N-glycosylation occurs only in the luminal side of the microsomal membrane, the amino-terminus of the P-450 molecule was translocated across the membrane. Furthermore, purified human aromatase (P-450 19) possessed a high mannose type sugar chain which could be removed by endoglycosidase H digestion (HARADA, N., personal communication). It is, therefore, highly likely that the amino terminal portions of microsomal P-450s are located in the lumen of the membrane and that the orientation of the signal anchor sequence of microsomal P-450s in the membrane is opposite to normal signal sequences (see Fig. 2).

In the in vitro assay system SAKAGUCHI et al. (1987) also demonstrated that a chimeric protein containing the segment from 99th to 514th residues of P-450 1A1 was not integrated into the microsomal membrane. MONIER et al. (1988) showed that P-450 2B1 with the deletion of the 2nd to 42nd amino acid residues including the signal-anchor sequence was no longer inserted into the membrane. These observations indicated that other portions of P-450 molecules than the amino-terminal segments cannot function as a signal sequence for the membrane insertion.

SZCZESNA-SKORUPA et al. (1989) gave an interesting experiment that P-450 2C2 whose amino-terminal negative charge was exchanged with positive charges was completely translocated across the membrane and glycosylated in the lumen. This indicated that there was no segment which could interrupt the peptide translocation (so called "stop transfer sequence") in the P-450 molecule. KUROIWA et al. (1990, 1991) demonstrated using the in vitro translation system that the

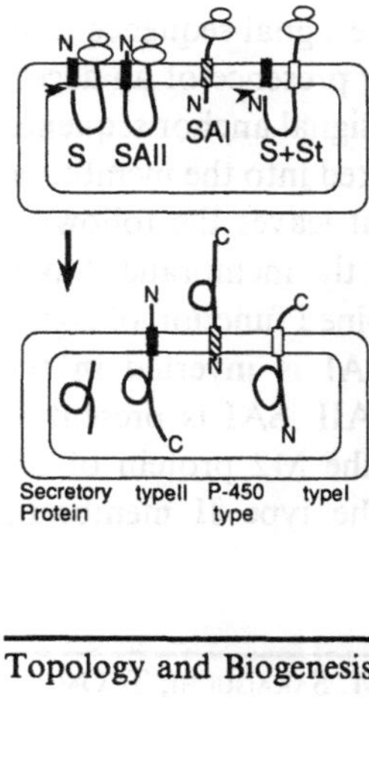

Fig. 2. Orientations of the topogenic sequences of simple transmembrane proteins. The peptide segments are arbitrarily represented by lines. Four types of topogenic sequences, signal, type I signal-anchor, type II signal-anchor, and stop-transfer sequences are indicated by S, SAI, SAII, and St, respectively. N and C denote amino- and carboxyl-terminus of the proteins, respectively. Arrowheads indicate the cleavage by signal peptidase. The upper figure shows the proteins during translation by ribosomes, and the lower figure indicates the topology of the proteins after the completion of the translation.

function of the stop-transfer sequences were primarly determined by the hydrophobicity of the sequence. In contrast, it was also reported that one segment located between 167 and 185 residues of P-450b (P-450 2B1) could function as a stop transfer sequence (MONIER et al., 1988). This seems to suggest that there is a hydrophobic segment in some species of P-450 which can interact with the membrane besides the signal-anchor segment. As indicated in the previous section, however, there is little possibility that the segment spans the membrane.

Considering the topogenic function of the amino-terminal segment described above, it seems likely that the amino-terminus of P-450 is located in the luminal side of the membrane and the P-450 molecule shows the orientation similar to that of type I membrane protein ($N_{lumen}/C_{cytoplasm}$) as shown in Figure 2. Thus, the membrane insertion signal sequence of P-450 designated as type I signal-anchor sequence (SAI) is different from the signal anchor sequences of type II membrane proteins ($N_{cytoplasm}/C_{lumen}$) (SAKAGUCHI et al., 1992).

4. Structural requirements for the topogenic function of the P-450-type signal-anchor sequence

Topologies of the integral membrane proteins cotranslationally inserted into ER membrane are determined by so-called topogenic sequences in their primary structures (SABATINI et al., 1982). The topogenic sequences can be functionally classified into the following four categories (Fig. 2). Signal sequences (S) initiates the peptide translocation across or peptide insertion into ER membrane. It is cleaved by the signal peptidase in the lumen. Stop-transfer sequence (St) interrupts the peptide chain translocation and anchors the protein to the membrane. The orientation of the type I membrane protein molecule is achieved by the sequential actions of S and St. Signal-anchor sequence of type II membrane protein (SAII) is an uncleavable signal sequence which mediates the translocation of the following portion of the peptide chain and anchors the protein to the membrane. Since the orientation of SAII in the membrane is the same with signal sequence (S), it can be categorized as a variant of the signal sequence. The functional difference between SAII and S is the absence or presence of an accessible site to the signal peptidase (SHAW et al., 1988). Type I signal anchor sequence (SAI) is that of P-450-type membrane proteins. SAI is inserted into the membrane and anchors the protein in an SRP-dependent fashion but leaves the following portion of the protein molecule on the cytoplasmic side of the membrane (SAKAGUCHI et al., 1987; HULL et al., 1988). It possesses the combined function of signal sequence and stop-transfer sequence. In this context, SAI is inverted in the orientation in the membrane as compared with S and SAII. SAI is present in P-450 type membrane proteins; microsomal P-450s and the M2 protein of the influenza A virus (HULL et al., 1988). SAII exists in the type II membrane proteins, such as transferrin receptor (Zerial et al., 1986), invariant chain of class II histocompatibility antigen (Lipp and Dobberstein, 1986), and asialoglycoprotein receptor (Spiess and Lodish, 1986). Available evidence suggested that these two types of signal-anchor sequence are inserted into the membrane by the participation of similar components of the protein translocation machinery for secretory proteins (High et al., 1991 a, b; Thrift et al., 1991).

Structural features of the amino-terminal portions of P-450s which maintain the unique topogenic function of SAI were investigated as follows. Various mutations were introduced into the chimeric proteins described in the previous section, in which the amino-terminal portions of several P-450s were fused to the secretory proteins, and their topology in the membrane was examined in the in vitro expression system. When the negatively charged amino acid residues, which are present in the amino-terminus of almost all microsomal P-450s, were changed to positively charged residues, the mutated proteins were cotranslationally translocated across the membrane, indicating that the modified amino-terminal segments functioned as a normal signal sequence (S) or SAII (Szczesna-Skorupa et al., 1988, 1989; Sato et al., 1990). The number of the positive charges required for the functional conversion was dependent on the hydrophobicity of the following segment (Sato et al., 1990). Such functional conversion between the topogenic sequences has been reported with other eukaryotic membrane proteins (Haeuptle et al., 1989; Parks and Lamb, 1991; Beltzer et al., 1991). Sato et al. (1990) also demonstrated that the functional conversion from SAI to S was achieved by partial deletions of the hydrophobic segment even when the negatively charged

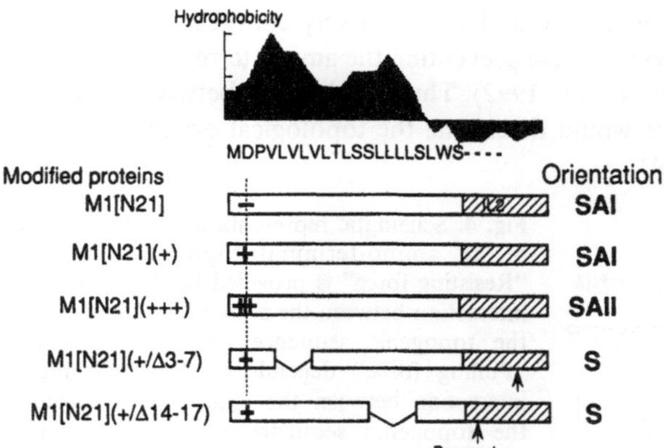

Fig. 3. Effects of the charge replacements and the partial deletions of the hydrophobic segment on the topogenic function of the signal-anchor sequence of P-450M1 (P-450 2C11). The original signal-anchor segment (1st to 20th amino acid residues), which is shown under the hydropathy profile in the figure, and the modified sequences were assayed for their topogenic functions in vitro by using mature interleukin 2 portion as the reporter (Sato et al., 1990). Signal peptidase processing sites are shown by arrows.

amino acid residues were retained. These observations indicated that the SAI of P-450s could be converted to S or SAII by the exchange of the negative charges at amino-terminus with positive charges or by shortening the hydrophobic segment. Figure 3 shows the typical results of in vitro insertion assay of the mutated chimeric proteins.

Recently, SAKAGUCHI et al. (1992) systematically examined the structural requirements for membrane insertion and membrane translocation signals by using artificially constructed model sequences. Model sequences containing 7–10 leucine residues functioned as normal signal sequences irrespective of their amino-terminal charges. When the hydrophobic segments were longer than 12 leucine residues, however, the functions of the sequences were dependent on the charges at the amino-terminus.

Taken together, we can conclude that the functions of the amino-terminal topogenic sequence are determined by the balance between the length of the hydrophobic segment and the amino-terminal charges. This conclusion is in accord with the general structural features of microsomal P-450s; the net charge of the amino-terminal portions of almost all P-450s are negative (HARTMANN et al., 1989) and the amino-terminal hydrophobic segments are much longer than the hydrophobic segments (h regions; VON HEIJNE, 1985) of secretion signal sequences (GOTOH and FUJII-KURIYAMA, 1989).

We propose that the capacity of a polypeptide segment to function as a eukaryotic secretion signal sequence is primarily determined by the hydrophobic segment (h region) which can loop into the protein translocation machinery in the membrane, and that longer hydrophobic segments would tend to pull the amino-terminal portion into the membrane and the positively charged residues at the amino-terminus would provide a force preventing the amino-terminus from entering the membrane (SAKAGUCHI et al., 1992). Thus, the balance between the pulling force and the resisting force would determine the topological orientation of the topogenic sequences (Fig. 4).

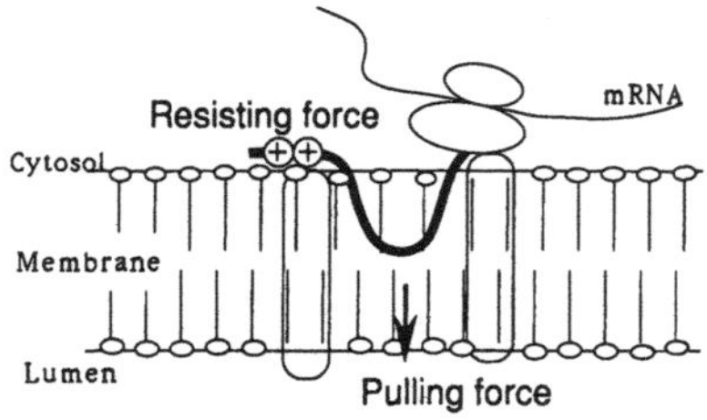

Fig. 4. Schematic representation of the function of an amino-terminal topogenic sequence. "Resisting force" is provided by the electrostatic interaction between the amino-terminal charge of the topogenic sequence and the membrane. "Pulling force" depends on the hydrophobic interaction between the hydrophobic stretch of the topogenic sequence and the membrane interior.

5. Membrane insertion and folding of P-450s

The nascent polypeptides of microsomal P-450s elongating on rbosomes bind to ER membrane by the amino terminal signal-anchor sequence and fold on the membrane. Thus, it is likely that the cotranslational interaction of the nascent peptides with the membrane is important for correct folding of P-450s. CLARK and WATERMAN (1991) expressed the native and amino-terminal truncated P-$450_{17\alpha}$ (P-450 17) in COS 1 cells. The modified P-450, in which an amino terminal stretch from 2nd to 17th amino acid residues corresponding to the signal-anchor sequence was deleted, was recovered in the membrane fraction from the COS cells, but it showed no 450 nm spectrum of the carbon monooxide compound and had no enzymatic activity. This suggested that the membrane insertion of the signal-anchor segment is necessary for the formation of the native conformation of the P-450. This observation also suggested that the modified P-450 molecules incorrectly associated with the membrane by some other hydrophobic segments. Similar results have been obtained using a yeast expression system (YABUSAKI et al., 1987). A modified P-450 1A1 lacking the amino-terminal segment as well as the full length molecule was found to be associated with the yeast microsomal membrane. However, the modified P-450 showed very low enzymatic activity. MONIER et al. (1988) reported that when P-450b (P-450 2B1) was expressed in HeLa cells, the intact molecules gave an ER-type reticular distribution pattern upon the observation of the cells by immuno-fluorescence microscopy, whereas the mutated P-450b lacking the amino-terminal 2nd to 42nd amino acid residues was distributed diffusely throughout the cytoplasm. It was thus suggested that the amino-terminal sequence was important for the localization of the P-450 molecules to the ER membrane. It is clear that the signal-anchor sequence is essential for the formation of the native conformation of P-450s and their microsomal localization in eukaryotic cells.

The mechanism of the incorporation of heme moiety, which is closely related to the folding of P-450 molecules, has not been clarified. It is not known whether the heme incorporation occurs co- or post-translationally and whether it is mediated by some factors or not.

In contrast to the results obtained with mammalian cells, several different observations were reported using the bacterial expression system. LARSON et al. (1991a, b) expressed the native P-450 2E1 and its modified form lacking the amino-terminal 3rd to 29th residues in *E. coli* cells. Both of them were recovered to the membrane fraction and remained associated with the membrane even after the alkali extraction. The modified form solubilized with detergents from the membrane showed 450 nm spectrum of the carbon monooxide compound and normal enzymatic activity. Furthermore, its reactivity with NADPH-cytochrome P-450 reductase on the reconstituted liposomes was identical with that of the full size molecule. These results suggested that the mutated P-450 molecule could be anchored to the membrane by some other portions than the amino-terminal segment. LI and CHIANG (1991) reported that the truncated P-450c7 (P-450 7) molecule, in which the amino terminal 2nd to 24th residues were deleted, was mainly recovered in the soluble fraction when expressed in *E. coli* cells. The purified soluble P-450 showed enzyme activity and a reduced-CO difference spectrum of P-450. It is likely that membrane association of the truncated

P-450s depends on the P-450 species. These observations with *E. coli* expression systems seem to suggest that the signal-anchor sequence is not essential for correct folding of the expressed P-450 molecules in the bacterial cell.

The mechanism of protein incorporation into bacterial membranes could be much different from that into the ER membrane in mammalian cells. P-450s are inserted into the ER membrane only cotranslationally and the insertion is mediated by SRP. However, different protein insertion processes have been observed in other organisms. The alkane inducible P-450 of *Candida maltosa* can be post-translationally inserted into the membrane isolated from the microorganism, whereas it was not integrated into the dog pancreas rough microsomal membrane post-translationally (WIEDMANN et al., 1988). The majority of the proteins with signal sequences can be inserted into the membrane of *E. coli* post-translationally (WICKNER and LODISH, 1985). SRP homologue has not been found in the bacterial cells. Possibly, alternative factors, such as GroE proteins and other chaperonins in the cytosol, would prevent the abnormal folding and mediate the post-translational insertion of the proteins into the membrane.

Taken together, in mammalian cells, cotranslational insertion of P-450 molecules into microsomal membrane by the amino terminal signal-anchor sequence triggers the correct folding and localization of the molecules on the membrane. On the other hand, the correct conformation of P-450 molecules can be formed in bacterial cells even when the segment is deleted.

6. Turnover of microsomal P-450s

The ER membrane proteins are continuously turning over in the cell and the composition of ER depends on a steady-state equilibrium between the synthesis and degradation of the individual components (OMURA, 1980). The turnover rates of microsomal membrane proteins, including P-450s, were investigated by tracing the decay of the radioactivities of P-450 proteins and heme moieties. The half-lives of P-450(PB) (P-450 2B1), P-450(MC) (P-450 1A2), NADPH-cytochrome P-450 reductase, and cytochrome b_5 in the livers of untreated rats were 25–30, 15, 35–50, and 50–80 hour, respectively (SADANO et al., 1983). The heme of P-450s is apparently dissociable and turns over faster than the protein moiety. The half-lives of microsomal P-450s depend on the molecular species and are shorter than those of other electron-transfer components of the ER membrane (SADANO and OMURA, 1983; PARKINSON et al., 1983; SHIRAKI and GUENGERICH, 1984).

Microsomal P-450s synthesized on the membrane-bound polysomes are localized in the ER membrane in eukaryotic cells and not found in other membranes in the exocytosis-endocytosis pathway (YAMAMOTO et al., 1985; BRANDS et al., 1985). YAMAMOTO et al. (1985) demonstrated that P-450(PB) (P-450 2B1) was not detected on the membrane of Golgi complex by immunoelectron microscopy, even when the P-450 was maximally induced by phenobarbital treatment of the rats. They concluded that microsomal P-450s are excluded from the vesicles in the process of the bulk flow en route to Golgi membranes. The different turnover rates of microsomal membrane proteins including P-450s suggest that some specific degradation mechanism is involved in the turnover of microsomal P-450s and regulates their intracellular concentrations, although MASAKI et al. (1987) reported that P-450(PB) (P-450 2B1) and NADPH-cytochrome P-450 reductase were detected in the auto-phagosomes prepared from leupeptin-treated rat liver, suggesting that these proteins were segregated to be degraded by the autophagic pathway. Recent progress of the study on ER protein degradation system (BONIFACINO and LIPPINCOTT-SCHWARTZ, 1991) could shed light on the specific degradation of P-450 molecules in ER.

7. References

BAR-NUN, S., KREIBICH, G., ADESNIK, M., ALTERMAN, L., NEGISHI, M., and SABATINI, D. D., (1980), Proc. Natl. Acad. Sci. USA **77**, 965–969.

BELTZER, J. P., FIEDLER, K., FUHRER, C., GEFFEN, I., HANDSCHIN, C., WESSELS, H. P., and SPIESS, M., (1991), J. Biol. Chem. **266**, 973–978.

BERNHARDT, R., NGOC DAO, N. T., STIEL, H., SCHWARZE, W., FRIEDRICH, J., JÄNIG, G. -R., and RUCKPAUL, K., (1983), Biochim. Biophys. Acta **745**, 140–148.

BERNHARDT, R., KRAFT, R., and RUCKPAUL, K., (1988), Biochem. Int. **17**, 1143–1150.

BLOBEL, G. and DOBBERSTEIN, B. (1975), J. Cell Biol. **67**, 835–851.

BONIFACINO, J. S. and LIPPINCOTT-SCHWARTZ, (1991), Current Opinion in Cell Biol. **3**, 592–600.

BRANDS, R., SNIDER, M. D., HINO, Y., PARK, S. S., GELBOIN, H. V., and ROTHMAN, J. E., (1985), J. Cell Biol. **101**, 1724–1732.

BROWN, C. A. and BLACK, S. D., (1989), J. Biol. Chem. **264**, 4442–4449.

CLARK, B. J. and WATERMAN, M. R., (1991), J. Biol. Chem. **266**, 5898–5904.

DELEMOS-CHIARANDINI, C., FREY, A. B., SABATINI, D. D., and KREIBICH, G., (1987), J. Cell Biol. **104**, 209–219.

ETTER, H. -U., RICHTER, C., OHTA, Y., EINTERHALTER, K. H., SASABE, H., and KAWATO, S., (1991), J. Biol. Chem. **266**, 18600–18605.

FUJII-KURIYAMA, Y., NEGISHI, M., MIKAWA, R., and TASHIRO, Y., (1979), J. Cell Biol. **81**, 510–519.

GOTOH, O. and FUJII-KURIYAMA, Y., (1989), in: Frontiers in Biotransformation, (K. RUCKPAUL and H. REIN, eds.), Akademie-Verlag, Berlin, pp. 195–245.

HAEUPTLE, M. -T., FLINT, N., GOUGH, N. M., and DOBBERSTEIN, B., (1989), J. Cell Biol. **108**, 1227–1236.

HARTMANN, E., RAPOPORT, T. A., and LODISH, H. F., (1989), Proc. Natl. Acad. Sci. USA **86**, 5786–5790.

HAUGEN, D. A., ARMES, L. G., YASUNOBU, K. T., and COON, M. J., (1977), Biochem. Biophys. Res. Commun. **77**, 967–973.

HEINEMANN, F. S. and OZOLS, J., (1982), J. Biol. Chem. **257**, 14988–14999.

HEINEMANN, F. S. and OZOLS, J., (1983), J. Biol. Chem. **258**, 4195–4201.

HIGH, S., FLINT, N., and DOBBERSTEIN, B., (1991a), J. Cell Biol. **113**, 25–34.

HIGH, S., GÖRLICH, D., WIEDMANN, M., RAPOPORT, T. A., and DOBBERSTEIN, B., (1991b), J. Cell Biol. **113**, 35-44.

Hull, J. D., Gilmore, R., and Lamb, R. A., (1988), J. Cell Biol. **106**, 1489–1498.
Ito, A. and Sato, R., (1968), J. Biol. Chem. **243**, 4922–4923.
Ito, A. and Sato, R., (1969), J. Cell Biol. **40**, 179–189.
Kashima, N., Nishi-Takaoka, C., Fujita, T., Taki, S., Yamada, G., Hamuro, J., and Taniguchi, T., (1985), Nature **313**, 402–404.
Kawato, S., Gut, J., Cherry, R. J., Winterhalter, K. H., and Richter, C., (1982), J. Biol. Chem. **257**, 7023–7029.
Kreibich, G., Sabatini, D. D., and Adesnik, M., (1983), Methods Enzymol. **96**, 530–542.
Kuroiwa, T., Sakaguchi, M., Mihara, K., and Omura, T., (1990), J. Biochem. **108**, 829–834.
Kuroiwa, T., Sakaguchi, M., Mihara, K., and Omura, T., (1991), J. Biol. Chem. **266**, 9251–9255.
Larson, J. R., Coon, M. J., and Porter, T. D., (1991a), J. Biol. Chem. **266**, 7321–7324.
Larson, J. R., Coon, M. J., and Porter, T. D., (1991b), Proc. Natl. Acad. Sci. USA **88**, 9141–9145.
Li, Y. C. and Chiang, J. Y. L., (1991), J. Biol. Chem. **266**, 19186–19191.
Lipp, J. and Dobberstein, B., (1986), J. Cell Biol. **102**, 2169–2175.
Masaki, R., Yamamoto, A., and Tashiro, Y., (1987), J. Cell Biol. **104**, 1207–1215.
Matsuura, S., Fujii-Kuriyama, Y., and Tashiro, Y., (1978), J. Cell Biol. **78**, 503–519.
Matsuura, S., Fujii-Kuriyama, Y., and Tashiro, Y., (1979), J. Cell Sci. **36**, 413–435.
Matsuura, S., Masuda, R., Sakai, O., and Tashiro, Y., (1983), Cell Struct. Funct. **8**, 1–9.
Monier, S., Luc, P. V., Kreibich, G., Sabatini, D. D., and Adesnik, M., (1988), J. Cell Biol. **107**, 457–470.
Negishi, M., Fujii-Kuriyama, Y., Tashiro, Y., and Imai, Y., (1976), Biochem. Biophys. Res. Commun. **71**, 1153–1160.
Nelson, D. R. and Strobel, H. W., (1988), J. Biol. Chem. **263**, 6038–6050.
Nelson, D. R. and Strobel, H. W., (1989), Biochemistry **28**, 656–660.
Omura, T., Siekevitz, P., and Palade, G. E. (1967), J. Biol. Chem. **242**, 2389–2396.
Omura, T., (1980), Pharmac. Ther. **8**, 489–499.
Ouzounis, C. A. and Melvin, W. T., (1991), Eur. J. Biochem. **198**, 307–315.
Parkinson, A., Thomas, P. E., Ryan, D. E., and Levin, W., (1983), Arch. Biochem. Biophys. **255**, 216–236.
Parks, G. D. and Lamb, R. A., (1991), Cell **64**, 777–787.
Poulos, T. L., Finzel, B. C., Gunsaulus, I. C., Wagner, G. C., and Kraut, J., (1985), J. Biol. Chem. **260**, 16122–16130.
Poulos, T. L., Finzel, B. C., and Howard, A. J., (1987), J. Mol. Biol. **195**, 687–700.
Sabatini, D. D., Kreibich, G., Morimoto, T., and Adesnik, M., (1982), J. Cell Biol. **92**, 1–22.
Sadano, H. and Omura, T., (1983), J. Biochem. **93**, 1375–1383.
Sakaguchi, M., Mihara, K., and Sato, R., (1984), Proc. Natl. Acad. Sci. USA **81**, 3361–3364.
Sakaguchi, M., Mihara, K., and Sato, R., (1987), EMBO J. **6**, 2425–2431.
Sakaguchi, M., Tomiyoshi, R., Kuroiwa, T., Mihara, K., and Omura, T., (1992), Proc. Natl. Acad. Sci. USA **89**, 16–19.
Sato, T., Sakaguchi, M., Mihara, K., and Omura, T., (1990), EMBO J. **9**, 2391–2397.
Shaw, A. S., Rottier, P. J. M., and Rose, J. K., (1988), Proc. Natl. Acad. Sci. USA **85**, 7592–7596.
Shiraki, H. and Guengerich, F. P., (1984), Arch. Biochem. Biophys. **235**, 86–96.
Spiess, M. and Lodish, H. F., (1986), Cell **44**, 177–185.
Szczesna-Skorupa, E., Browne, N., Mead, D., and Kemper, B., (1988), Proc. Natl. Acad. Sci. USA **85**, 738–742.
Szczesna-Skorupa, E. and Kemper, B., (1989), J. Cell Biol. **108**, 1237–1243.

Tarr, G. E., Black, S. D., Fujita, V. S., and Coon, M. J., (1983), Proc. Natl. Acad. Sci. USA **80**, 6552–6556.
Thrift, R. N., Andrews, D. W., Walter, P., and Johnson, A. E., (1991), J. Cell Biol. **112**, 809–821.
Tretiakov, V. E., Degtyarenko, K. N., Uvarov, V. U., and Archakov, A. I., (1989), Arch. Biochem. Biophys. **274**, 429–439.
Vergères, G., Winterhalter, K. H., and Richter, C., (1989), Biochemistry **28**, 3650–3655.
Vergères, G., Winterhalter, K. H., and Richter, C., (1991), Biochim. Biophys. Acta **1063**, 235–241.
von Heijne, G., (1985), J. Mol. Biol. **184**, 99–105.
Walter, P., Ibrahimi, I., and Blobel, G., (1981), J. Cell Biol. **91**, 545–550.
Wickner, W. T. and Lodish, H. F., (1985), Science **230**, 400–406.
Wiedmann, M., Wiedmann, B., Voigt, S., Wachter, E., Müller, H. -G., and Rapoport, T., (1988), EMBO J. **7**, 1763–1768.
Yabusaki, Y., Murakami, H., Sakaki, T., Shibata, M., and Ohkawa, H., (1988), DNA **7**, 701–711.
Yamamoto, A., Masaki, R., and Tashiro, Y., (1985), J. Cell Biol. **101**, 1733–1740.
Zerial, M., Melancon, P., Schneider, C., and Garoff, H., (1986), EMBO J. **5**, 1543–1550.

Chapter 3

Inhibitors of Cytochrome P450 and Possibilities for their therapeutic Application

M. A. Correia and P. R. Ortiz de Montellano

1. Introduction

The cytochrome P450 monooxygenases are a large and diverse class of hemoproteins whose biological functions can be broadly subdivided into two general categories. One category is that of enzymes that catalyze the synthesis or catabolism of specific endogenous substrates. This category includes enzymes like lanosterol 14α-demethylase, which catalyzes a key transformation in the biosynthesis of cholesterol, and aromatase, which promotes the last step in the biosynthesis of estradiol (JEFCOATE, 1986). The other category is that of cytochrome P450 enzymes that primarily oxidize (or occasionally reduce) xenobiotics to more polar, and sometimes more toxic, metabolites (YANG and LU, 1987; NELSON and HARVISON, 1987). The distinction between these two categories of cytochrome P450 enzymes is not absolute in that members of the first class accept some xenobiotic substrates and members of the second class some endogenous substrates. In general, however, the biosynthetic forms of cytochrome P450 are much more substrate-specific, and often catalyze more complex transformations, than the enzymes primarily involved in xenobiotic metabolism.

Cytochrome P450 enzymes can also be subdivided according to whether they obtain the reducing equivalents required for catalysis from the flavoprotein NADPH-cytochrome P450 reductase or from an iron-sulfur protein such as adrenodoxin or putidaredoxin (PETERSON and PROUGH, 1986). In mammals, the iron-sulfur-dependent enzymes are located in the mitochondria and are involved in biosynthetic processes (JEFCOATE, 1986). All of the xenobiotic-metabolizing enzymes are supported by NADPH-cytochrome P450 reductase and are primarily located in the endoplasmic reticulum (DISTLERATH and GUENGERICH, 1987). This segregation of function and mechanism is not absolute, however, as exemplified by lanosterol 14α-demethylase, a microsomal enzyme involved in cholesterol biosynthesis (YOSHIDA and AOYAMA, 1984; TRZASKOS et al., 1986a).

The inhibition of cytochrome P450 enzymes is of clinical interest for both therapeutic (prospective) and toxicological (retrospective) reasons. Most of the potential clinical uses of cytochrome P450 inhibitors involve inhibition of the synthesis or catabolism of endogenous substrates. As a case in point, the inhibition of aromatase is currently under intensive study as an approach to the treatment of estrogen-dependent breast cancer (BRODIE et al., 1987; COLE and ROBINSON, 1990). On the other hand, the secondary inhibition of cytochrome P450 enzymes, sometimes due to the low isozyme specificity of cytochrome P450 inhibitors, is an undesirable property of many lipophilic drugs (e.g. PASANEN et al., 1988; AYUB and LEVELL, 1989). The intentional inhibition of xenobiotic-metabolizing enzymes, in principle, provides a mechanism for altering the pharmacokinetic properties or toxicity of drugs or harmful environmental agents. In practical terms, however, the inhibition of drug metabolizing enzymes has primarily been of clinical concern as a side effect of drug therapy.

2. The catalytic chemistry of cytochrome P450 relevant to enzyme inhibition

2.1. Basic principles of catalysis

The catalytic cycle of cytochrome P450 is initiated by binding of the substrate (Ortiz de Montellano, 1986). This event alters the redox potential of the heme group and results in reduction of the heme group from the ferric to the ferrous state by an electron provided by cytochrome P450 reductase or an iron-sulfur redox partner. Oxygen binds to the ferrous hemoprotein, as it does to ferrohemoglobin, to give an unstable intermediate that can be visualized as a ferrous dioxygen [$Fe^{II} \cdot O_2$] or ferric superoxide [$Fe^{III} \cdot O_2 \cdot^-$] complex. Transfer of a second electron to this complex by one of the redox partners formally reduces the complex to the ferrous superoxide [$Fe^{II} \cdot O_2 \cdot^-$] or ferric peroxy dianion [$Fe^{III} \cdot O_2^{2-}$] state. Although this species may be involved in specialized cytochrome P450 chemistry (vide infra), it is generally thought to undergo scission of the oxygen–oxygen bond to give a molecule of water and the final oxidative species (Ortiz de Montellano, 1986). The oxidative species has not been directly observed, so its identity rests on inferences made from the chemistry of the system and analogies with well-characterized hemoprotein and metalloporphyrin models. These inferences suggest that the oxidative species is a ferryl oxygen ($Fe^{IV} = O$) coupled to a porphyrin or protein radical (Ortiz de Montellano, 1986). The two oxidation equivalents represented by the $Fe^{IV} = O$ and the porphyrin or protein radical are required for the oxygen transfer reaction, a net two-electron oxidation of the substrate that returns the enzyme to the resting ferric state. It is now clear, however, that the two oxidation equivalents can, at least in some instances, come into play in discrete steps rather than in a concerted two-electron oxidation.

2.2. Carbon hydroxylation

Carbon hydroxylation is perhaps the most characteristic of the cytochrome P450-catalyzed reactions. The enzyme readily inserts oxygen into the C–H bonds of all types of tetravalent carbon atoms, including the terminal methyl groups of alkyl chains (Hardwick et al., 1987), but rarely if ever inserts oxygen into the much stronger vinyl or aromatic C–H bonds. A substantial body of evidence now supports a reaction sequence that involves abstraction of the hydrogen by the ferryl oxygen followed by rebound collapse of the resulting iron-bound hydroxyl radical equivalent with the carbon radical (Ortiz de Montellano, 1986):

$$P^{\cdot}Fe^{IV}{=}O + C\text{-}H \longrightarrow PFe^{IV}\text{-}OH + C^{\cdot} \longrightarrow PFe^{III} + C\text{-}OH$$

The evidence includes large isotope effects (Hjelmeland et al., 1977; Groves et al., 1978), inversion of stereochemistry (GROVES et al., 1978), addition of the hydroxyl group to a different carbon than the one the hydrogen was removed from due to delocalization of the radical over a conjugated framework (GROVES and SUBRAMANIAN, 1984), and rearrangement of the carbon radical skeleton between the hydrogen abstraction and oxygen transfer steps (ORTIZ DE MONTELLANO and STEARNS, 1987).

2.3. Heteroatom oxidation

The catalytic oxidation of heteroatoms by hemoproteins is mediated not only by cytochrome P450 but also by the peroxidases. The products of the reactions differ, however, in that the peroxidases simply remove an electron from the heteroatom to give a radical or radical cation whereas the cytochrome P450 enzymes add an oxygen atom to the heteroatom or to the carbon adjacent to the heteroatom. The cytochrome P450-catalyzed reactions nevertheless appear to involve initial electron abstraction, much as the peroxidase reactions (ORTIZ DE MONTELLANO, 1986). The heteroatom radical then collapses with the iron-bound oxygen or, after deprotonation of the adjacent carbon, with the resulting carbon radical. This yields the heteroatom oxide or the α-hydroxylated product that subsequently eliminates the heteroatom:

$$P^{\cdot}Fe^{IV}{=}O + \ddot{X} \longrightarrow PFe^{IV}\text{-}O + \dot{X}^{+} \longrightarrow PFe^{III} + \overset{+}{X}\text{-}O^{-}$$

$$P^{\cdot}Fe^{IV}{=}O + \ddot{X}\text{-}C\text{-}H \longrightarrow PFe^{IV}\text{-}O + \dot{X}^{+}\text{-}C\text{-}H \longrightarrow$$

$$PFe^{IV} + \ddot{X}\text{-}C^{\cdot} + H^{+} \longrightarrow PFe^{III} + \ddot{X}\text{-}C\text{-}OH$$

The evidence for initial electron abstraction from the heteroatom includes low isotope effects for dealkylation reactions (FOSTER, 1985), the correlation of oxidation potentials with substrate oxidation rates (GALLIANI et al., 1986; MACDONALD et al., 1989), and the transfer of substrate-derived carbon radicals to the heme group in the oxidation of certain heterocyclic compounds (AUGUSTO et al., 1982; LEE et al., 1988).

2.4. Olefin and acetylene oxidation

The mechanisms involved in the oxidation of the different types of non-aromatic π-bonds remain less well defined than the mechanisms for the oxidation of C–H bonds and heteroatoms (ORTIZ DE MONTELLANO, 1986). The normal metabolites of the cytochrome P450-catalyzed oxidation of olefins are epoxides (ORTIZ DE MONTELLANO, 1985). In some instances, however, oxygen transfer to one carbon of a π-bond oxidizes it to a carbonyl group with concomitant migration of one of its substituents to the adjacent vinylic carbon (Ortiz de Montellano, 1985; Liebler and Guengerich, 1983). In the case of acetylenes, the only metabolites that have been identified are the corresponding acetic acids derived from the initially formed ketene products by such a rearrangement (see Figure) (Ortiz de Montellano, 1985, 1986):

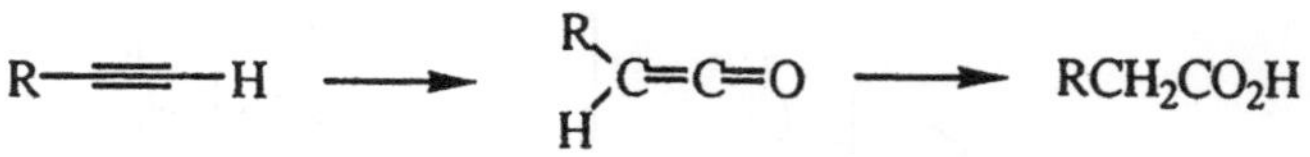

A number of proposals have been made concerning the intermediate(s) involved in transfer of the ferryl oxygen to the π-bond (Ortiz de Montellano, 1986). These include electron abstraction to give a radical cation species (Ortiz de Montellano et al., 1982; Traylor and Xu, 1988), addition to give a radical or cationic intermediate (Ortiz de Montellano et al., 1982; Traylor and Xu, 1988), formation of a metallaoxetane complex (Collman et al., 1984, 1990), and rate limiting formation of a charge-transfer complex (Groves and Watanabe, 1986; Garrison et al., 1989). A distinction cannot yet be made for the biological reactions between these alternatives, particularly since one intermediate may dominate in one reaction and a second in another. Bruice has suggested on the basis of model studies that charge transfer complex formation is the key to the reaction, although the other species may be involved in individual reaction outcomes or with particular substrates (Garrison et al., 1989). The oxidation of acetylenes is unlikely to involve epoxidation of the triple bond to the highly disfavored unsaturated epoxide (Ortiz de Montellano, 1985). Indeed, evidence has been obtained in the oxidation of phenylacetylene by cytochrome P450IIB1 that the acetylenic hydrogen migrates to the internal carbon as the oxygen is added to the terminal carbon to give the ketene metabolite (Komives and Ortiz de Montellano, 1987). Addition of the oxygen to the internal carbon, on the other hand, results in inactivation of the enzyme (vide infra).

2.5. Carbon–carbon bond cleaving reactions

The cytochrome P450 reactions that involve cleavage of carbon–carbon bonds are particularly important in the area of sterol biosynthesis, and consequently are among the reactions of greatest interest as targets for the development of clinically useful inhibitors. The transformations of interest in this context are the 14α-demethylation of lanosterol in the hepatic biosynthesis of cholesterol, cleavage of the side chain of cholesterol by cytochrome $P450_{SCC}$ (P450XIA1) to give pregnenolone, lysis by cytochrome $P450_{17\alpha}$ (P450XVIIA1) of the C_{17}–C_{20}

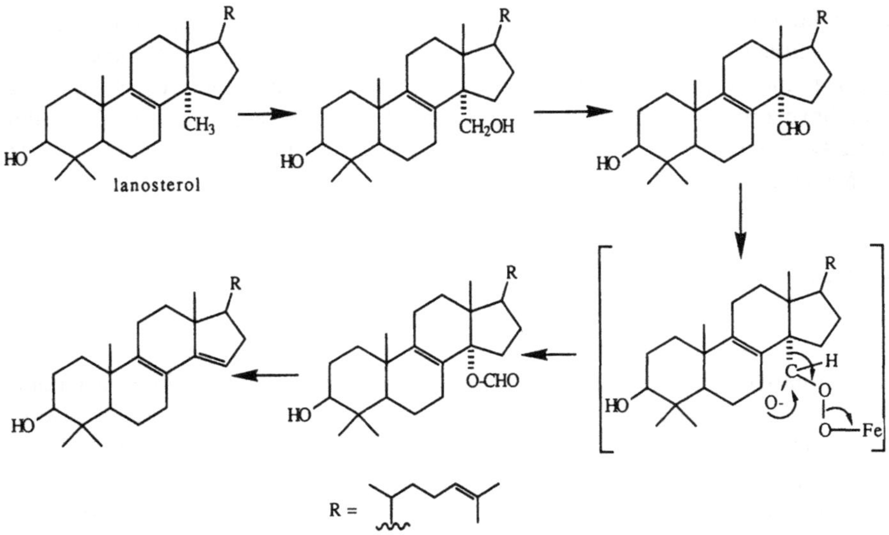

Fig. 1. The mechanism proposed for lanosterol 14α-demethylase.

carbon bond of progesterone to give androstenedione, and 19-demethylation/aromatization of androstenedione by aromatase (P450XIXA1) to give estradiol.

Lanosterol demethylation is catalyzed by a microsomal cytochrome P450 enzyme that has not yet been purified or cloned from human tissue but has been isolated from rat liver and *Saccharomyces cerevisiae* (Yoshida and Aoyama, 1984; Trzaskos et al., 1986a). The *Saccharomyces cerevisiae* enzyme has been cloned and its amino acid sequence determined (Kalb et al., 1987). The enzyme catalyzes three successive two-electron oxidation steps. The first two steps are thought to be conventional hydroxylation reactions of the 14α-methyl group that produce, respectively, the 14α-hydroxymethyl and 14α-carboxaldehyde derivatives of lanosterol (Fig. 1) (Trzaskos et al., 1986b). The final, and unconventional, step of the sequence involves oxidative removal of the 14α-aldehyde moiety as formic acid with concomitant formation of the 14,15-desaturated sterol (Fischer et al., 1991). Evidence has been provided for the formation of an additional intermediate in the conversion of the aldehyde to the $\Delta^{14,15}$-saturated sterol (Fischer et al., 1991). The reported unstable intermediate is the formyloxy species obtained by insertion of oxygen between the carbon of the aldehyde and C-14 of the sterol framework. The mechanism of the oxygen insertion into the carbon–carbon bond is unclear but may proceed by a conventional Baeyer-Villiger mechanism in which addition of an iron-bound "hydroperoxide" to the carbonyl group is followed by rearrangement to the formyloxy product) (Fig. 1). Reduction of the $\Delta^{14,15}$-double bond, migration of the $\Delta^{8,9}$-double bond to the $\Delta^{4,5}$-position, removal of the two 4-methyl groups by a non-P450 oxidase, and saturation of the $\Delta^{24,25}$-double bond produces cholesterol.

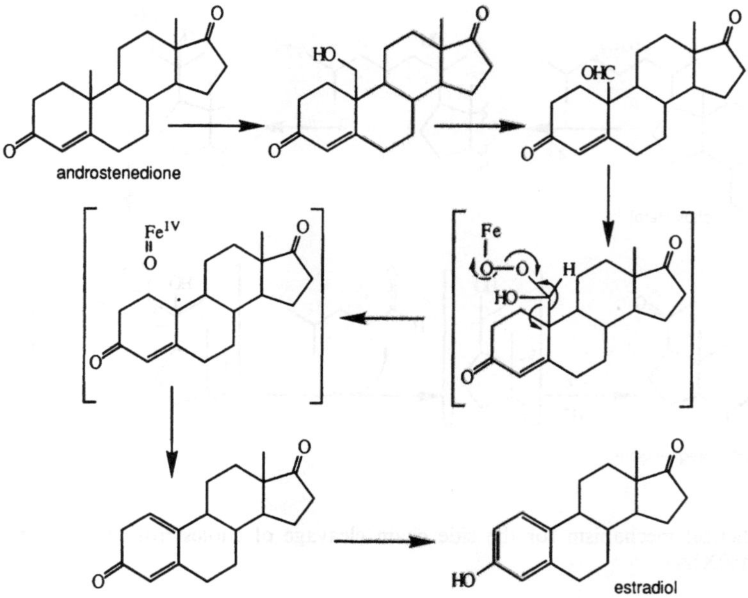

Fig. 2. One of the possible mechanisms for the aromatization of androstenedione by aromatase.

The reaction catalyzed by aromatase is very similar to that catalyzed by lanosterol 14α-demethylase in that a methyl group is removed with concomitant introduction of a double bond. The first two steps of the reaction again appear to be conventional hydroxylations of the 19-methyl of androstenedione to produce the 19-carboxaldehyde derivative (Cole and Robinson, 1990). As with lanosterol 14α-demethylase, the aldehyde is lost in the third catalytic step as formic acid with concomitant introduction of a double bond into the sterol framework (Fig. 2) (Cole and Robinson, 1990). In the case of aromatase, introduction of a $\Delta^{1,10}$-double bond into the already unsaturated ($\Delta^{4,5}$) sterol A-ring produces the doubly unsaturated ketone that readily tautomerizes to the phenol of estradiol. The details of the carbon–carbon bond cleavage step remain to be elucidated and may or may not be similar to those of the lanosterol 14α-demethylase reaction. Model studies concur with the suggestion (Akhtar et al., 1982) that peroxide addition to the aldehyde can cause deformylation/aromatization (Cole and Robinson, 1988; Watanabe and Ishimura, 1989; Cole et al., 1990), but an oxyformyl derivative analogous to that reported for 14-demethylation has not been detected. It is possible that the formyl derivative is too reactive to be detected by the procedures used to date, but it is also possible that the reaction proceeds, as suggested by Stevenson et al. (1988) and shown in Figure 2, by carbon–carbon bond homolysis and subsequent electron transfer.

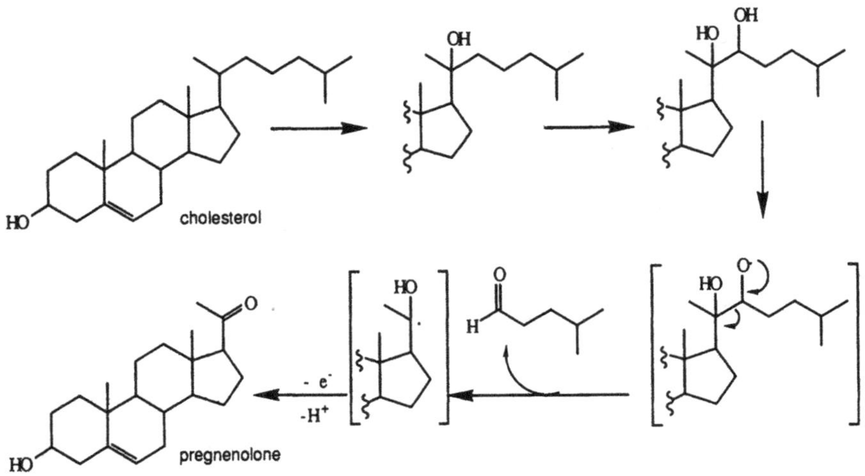

Fig. 3. Hypothetical mechanism for the side chain cleavage of cholesterol catalyzed by cytochrome P450XIA1.

Cholesterol side-chain cleavage requires scission of an internal carbon–carbon bond rather than removal of a methyl group and therefore must proceed by an intermediate other than an aldehyde. The first two oxidative steps are conventional, stereospecific introductions of hydroxyl groups at C-20 and C-22 of cholesterol to give a vicinal diol (Fig. 3) (Burstein et al., 1975; Byon and Gut, 1980). Carbon–carbon bond cleavage takes place in the third, still poorly understood, oxidative step. Model studies confirm the ability of the ferryl species to catalyze cleavage of the carbon–carbon bond between two vicinal hydroxyl groups (Murray and Sligar, 1985; Okamoto et al., 1988). The cleavage may involve a concerted oxidation in which one of the hydroxyl groups first adds to the ferryl oxygen or a mechanism in which one of the hydroxyl groups is oxidized to an alkoxy radical as shown in Figure 3. Subsequent cleavage of the carbon bond then produces pregnenolone and 4-methylpentanal from 20,21-dihydroxycholesterol.

Cleavage of the C_{17}–C_{20} carbon–carbon bond of progesterone by cytochrome P450XVIIA1, unlike the other carbon–carbon bond cleavage reactions, involves

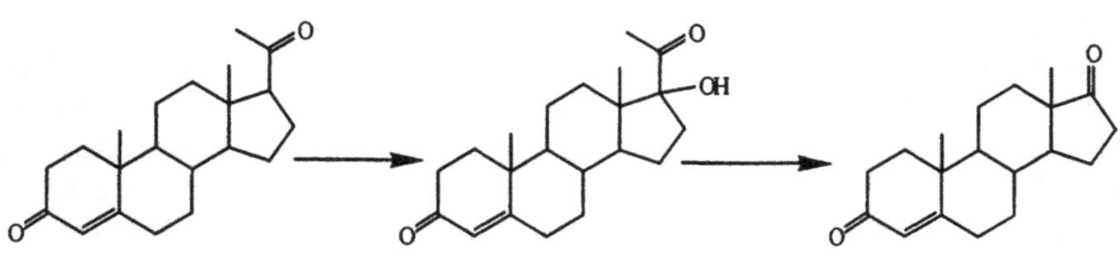

Fig. 4. One of two probable mechanisms for the conversion of progesterone to androstenedione by cytochrome P450XVIIA1.

two rather than three oxidative steps and yields androst-4-en-3,17-dione and acetic acid as products (Fig. 4). The difference in the number of catalytic steps is insignificant, however, because one of the carbon atoms involved in the $C_{17}-C_{20}$ cleavage reaction is already at the ketone oxidation state. The first of the two oxidative steps catalyzed by the enzyme is a conventional C-17 hydroxylation (NAKAJIN et al., 1984; ZUBER et al., 1986). $^{18}O_2$-studies indicate that the 17α-hydroxyl derives from molecular oxygen, and that one atom of molecular oxygen is incorporated into the acetic acid but not into the 17-keto function of the sterol during the carbon–carbon bond cleavage step (MILLER et al., 1991). This can be rationalized by either of the two mechanisms suggested for the analogous cholesterol side chain cleavage reaction (Fig. 3). Addition of the ferryl dioxide species to the ketone followed by homolysis of the dioxygen bond would lead to the observed products, as would oxidation of the 17α-hydroxyl to the alkoxy radical followed by bond scission and transfer of the ferryl oxygen to the carbonyl radical to generate the acetate moiety (MILLER et al., 1991).

3. The clinically-relevant inhibition of cytochrome P450 enzymes

Cytochrome P450 enzymes can be inhibited both reversibly and irreversibly. Reversible inhibition usually depends on non-covalent interactions of the inhibitors with the active site amino acid residues and/or the heme group. Irreversible inhibition, on the other hand, requires modification of the protein or the heme group by the inhibitor. Agents that are intrinsically reactive and directly modify protein residues are too non-specific and potentially toxic to be of clinical utility and are not discussed here. The potentially most useful irreversible inhibitors are those in which a chemically inert functionality is converted by the catalytic action of the target enzyme into a chemically reactive moiety that reacts with the heme group or an active site residue. Ideally, the activated species reacts efficiently within the catalytic site in which it is formed and does not diffuse out into the medium to react with other proteins. A number of approaches to the construction of such mechanism-based, or "suicide" inhibitors, are available, but only those mechanisms that have been exploited in the design of clinically interesting mechanism-based inhibitors are discussed here. The interested reader is referred to a number of recent reviews for more general discussions of the range of options available for the design of reversible and irreversible inhibitors of cytochrome P450 enzymes (TESTA and JENNER, 1981; ORTIZ DE MONTELLANO and REICH, 1986; MURRAY, 1987; ORTIZ DE MONTELLANO, 1988; COVEY, 1988).

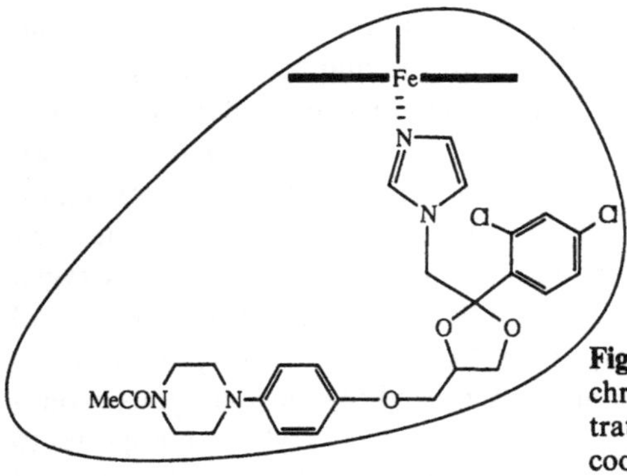

Fig. 5. Binding of ketoconazole to a cytochrome P450 active site schematically illustrating both lipophilic binding and coordination to the heme iron.

3.1. Competitive inhibition

The simplest mechanism for the inhibition of cytochrome P450 enzymes is competition between the inhibitor and the substrate for occupancy of the active site. Competitive inhibition is effective when the inhibitor binding affinity is significantly greater than that of the substrate. Most of the useful competitive inhibitors are compounds that are lipophilic, and thus bind well in the lipophilic active sites of cytochrome P450 enzymes, and in addition contain a nitrogen or other heteroatom that coordinates to the heme iron atom (Testa and Jenner, 1981; Ortiz de Montellano and Reich, 1986; Murray, 1987). The synergism between hydrophobic interactions and metal coordination enhances binding of the inhibitor to the enzyme and prevents its catalytic turnover by occupying the iron site to which the oxygen normally binds. Isozyme specificity is conveyed to such inhibitors by the tightness of the fit of the inhibitor to the active site of the target cytochrome P450 and the degree to which the same structure is excluded by the active sites of other cytochrome P450 enzymes. An additional requirement is that the coordinating atom must be appropriately located relative to the iron atom when the inhibitor is bound in the active site. Coordination of the heteroatom is readily detected in most instances by the formation of a spectrum with the Soret band shifted to approximately 430 nm or, in the difference spectrum, by the formation of a Type II spectrum with a broad trough between 390 and 410 nm and a peak at 425 to 435 nm (Jefcoate, 1978; Schenkman et al., 1981). These spectroscopic changes are observed, for example, in the inhibition of sterol 14α-demethylation by ketoconazole (Fig. 5) (Yoshida and Aoyama, 1987) and the inhibition of sterol side chain cleavage by 22-amino-23,24-bisnor-5-cholen-3β-ol (Fig. 6) (Sheets and Vickery, 1982). Indeed, coordination of the 22-amino group of the sterol to the heme iron has been used to argue that the 22-position of cholesterol is located at the appropriate distance from the iron for oxidation by the ferryl oxygen (Sheets and Vickery, 1982).

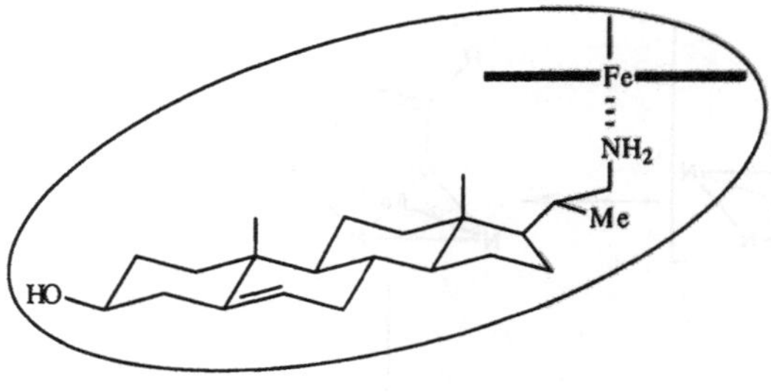

Fig. 6. Schematic binding of an aminosterol to the active site of cytochrome P450XIA1.

3.2. Mechanism-based inactivating agents

Mechanism-based inactivating agents differ from competitive inhibitors in that they not only compete with substrates for binding to the active site but, in addition, are catalytically oxidized to reactive species that covalently modify and thereby inactivate the enzyme (ORTIZ DE MONTELLANO and REICH, 1986). Mechanism-based inhibitors generally inactivate the enzyme by adding to either the prosthetic heme group or the protein, although other mechanisms for covalent modification of the hemoprotein, such as amino acid oxidation, are possible.

3.2.1. Heme modification

A variety of functional groups have been found to result in modification of the prosthetic group when oxidized by a cytochrome P450 enzyme. They include acetylenes and olefins (ORTIZ DE MONTELLANO and REICH, 1986; ORTIZ DE MONTELLANO, 1988), 4-alkyldihydropyridines (AUGUSTO et al., 1982; DE MATTEIS et al., 1982), 1-aminobenzotriazoles (ORTIZ DE MONTELLANO and MATHEWS, 1981; ORTIZ DE MONTELLANO et al., 1984), and alkylhydrazines (ORTIZ DE MONTELLANO et al., 1983). Each of these structures is oxidized by cytochrome P450 to a reactive species that reacts with the nitrogens of the prosthetic heme group (ORTIZ DE MONTELLANO and REICH, 1986; ORTIZ DE MONTELLANO, 1988). In the case of olefins and acetylenes, the isolated adducts indicate that heme alkylation involves addition of the ferryl oxygen to the internal carbon of the π-bond and the porphyrin nitrogen to the terminal carbon. The products isolated are protoporphyrin IX derivatives with an N-(2-hydroxyalkyl) or N-(2-oxoalkyl) function, respectively. These alkylation reactions do not result from reaction of the heme with the epoxide (olefin) or putative unsaturated epoxide (acetylene) metabolites. The mechanism of the alkylation event is still unclear, but it appears to involve a transient species formed by reaction of the ferryl oxygen with the π-bond. In general, acetylenes are more effective than olefins for the inactivation of cytochrome P450 because the oxidation of fewer acetylene molecules is required

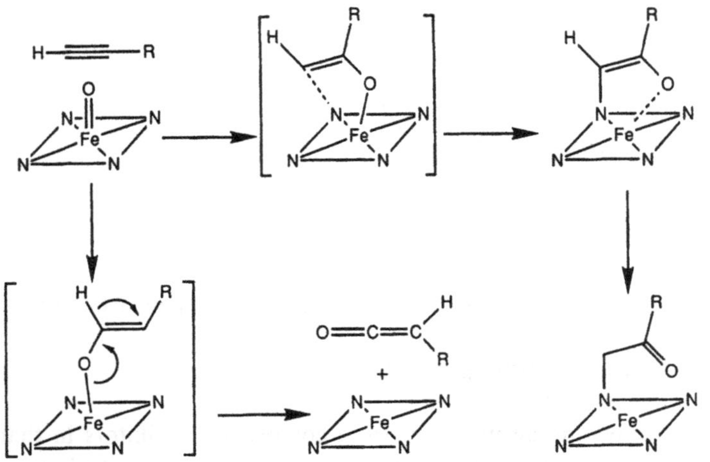

Fig. 7. Proposed mechanism for the oxidation of terminal acetylenes to ketenes and heme N-alkylating species.

before the enzyme is inactivated. In the case of terminal acetylenes, delivery of the oxygen to the terminal carbon produces a ketene that hydrates to the acid or, in appropriate circumstances, may react with a protein residue (CAJACOB et al., 1988; HAMMONS et al., 1989). Delivery of the oxygen to the internal carbon, on the other hand, results in alkylation of the prosthetic heme group (Fig. 7) (KUNZE et al., 1982). Mechanisms have also been proposed to explain N-alkylation of the heme during the catalytic turnover of the 4-alkyldihydropyridines and 1-aminobenzotriazoles. In the case of the dihydropyridines, one-electron oxidation is proposed to be followed by extrusion of the 4-alkyl group as a radical that adds to the heme (AUGUSTO et al., 1982). Oxidation of 1-aminobenzotriazole, on the other hand, produces benzyne or a related species that alkylates two of the nitrogens of the heme group (ORTIZ DE MONTELLANO and MATHEWS, 1981; ORTIZ DE MONTELLANO et al., 1984).

3.2.2. Protein modification

The best documented example of the inactivation of cytochrome P450 through protein modification by a mechanism-based inhibitor is that of chloramphenicol (HALPERT and NEAL, 1980). The key functionality of chloramphenicol, a dihalomethyl group, is oxidized to a species that acylates a lysine residue (Fig. 8) (HALPERT, 1981). This acylation event interferes with the transfer of electrons from cytochrome P450 reductase to the heme group of the cytochrome P450 and thereby prevents catalytic turnover of the enzyme (HALPERT et al., 1985). The key evidence for this is the finding that the enzyme oxidizes substrates

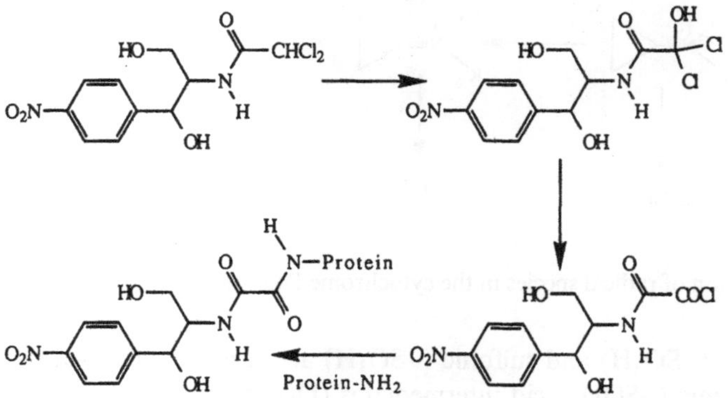

Fig. 8. Activation of chloramphenicol to a protein alkylating agent by the catalytic action of cytochrome P450.

normally if the need for reducing equivalents is circumvented by providing it with iodosobenzene as the oxidizing agent. The dihalomethyl function has been exploited in a number of other inactivating agents (HALPERT et al., 1986; HALPERT et al., 1989; STEVENS et al., 1991), including 21,21-dichloroprogesterone, which inactivates the bovine adrenal enzyme (P450XVII) that catalyzes the C-17 hydroxylation and C-17/C-20 scission of progesterone (STEVENS et al., 1991).

The oxidation of nitrogen and sulfur functions can result in modification of the protein. Among the nitrogen and sulfur compounds that inactivate cytochrome P450 enzymes when oxidized are cyclopropylamines (HANZLIK and TULLMAN, 1982; MACDONALD et al., 1982; TULLMAN and HANZLIK, 1984), compounds with a carbon–sulfur or phosphorus–sulfur double bond (e.g., HALPERT and NEAL, 1981; OLDHAM, 1989), and alkyl thiols (DECKER et al., 1989; OLDHAM, 1989). The cyclopropylamine and sulfhydril functions have been employed in mechanism-based inhibitors targeted at clinically relevant cytochrome P450 enzymes. The cyclopropylamine group is thought to inactivate cytochrome P450 after it is oxidized by the enzyme to the cyclopropylamine radical cation which rapidly rearranges to the ring opened carbon radical (Fig. 9) (HANZLIK and TULLMAN, 1982; MACDONALD et al., 1982; TULLMAN and HANZLIK, 1984). The exact mechanism of the reaction between the cyclopropylamine radical cation, or the carbon radical derived from it, and the protein is unclear but radiolabeling experiments show that the cyclopropylamine is covalently bound to the protein.

The inactivation of adrenal steroidogenic and hepatic cytochrome P450 enzymes by spironolactone is accompanied by loss of the heme group and covalent binding of the drug to the protein. These reactions depend on initial hydrolytic unmasking of the 7α-thiol function (MENARD and MARTIN, 1974; MENARD et al., 1979a, 1979b). Oxidation of the 7α-thiol group appears to be critical for enzyme inactivation (Fig. 10). The thiol group is converted in this oxidative

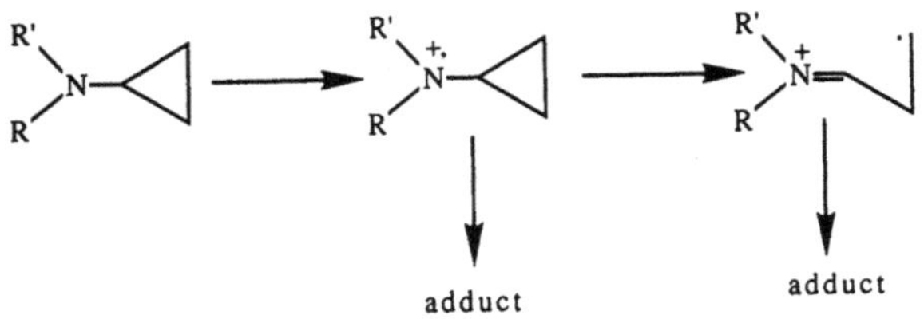

Fig. 9. Oxidative formation of radical species in the cytochrome P450-catalyzed oxidation of cyclopropylamines.

process to the sulfinic ($-SO_2H$) and sulfonic ($-SO_3H$) acids, presumably via the thiol radical and sulfenic ($-SOH$) acid intermediates (DECKER et al., 1989). The species responsible for the inactivation reaction, and the nature of the covalent link between the thiosterol and the protein, remain unidentified. The most likely candidates for the reactive species, however, are the sulfenic acid or its thiol

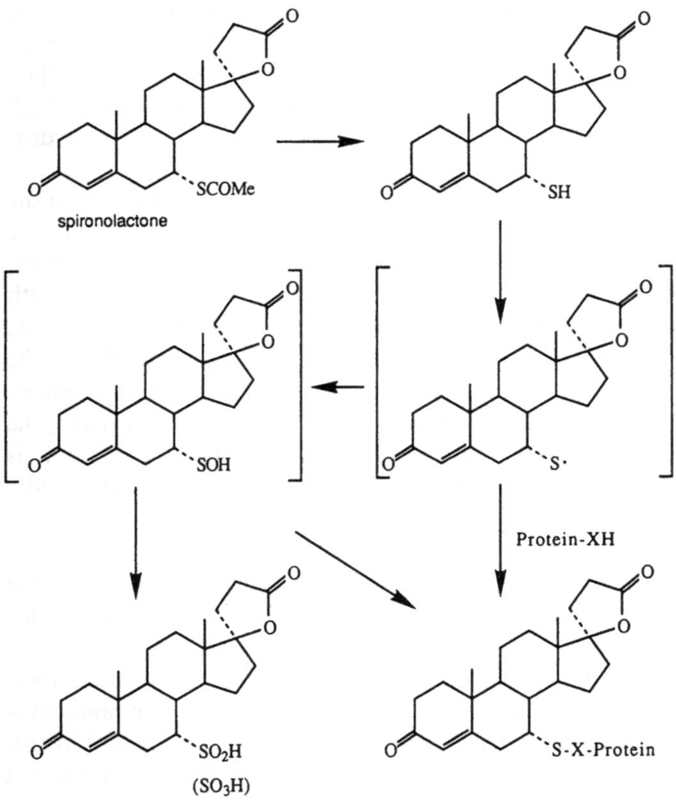

Fig. 10. Oxidation by cytochrome P450 of spironolactone to reactive intermediates that may be responsible for inactivation of the enzyme.

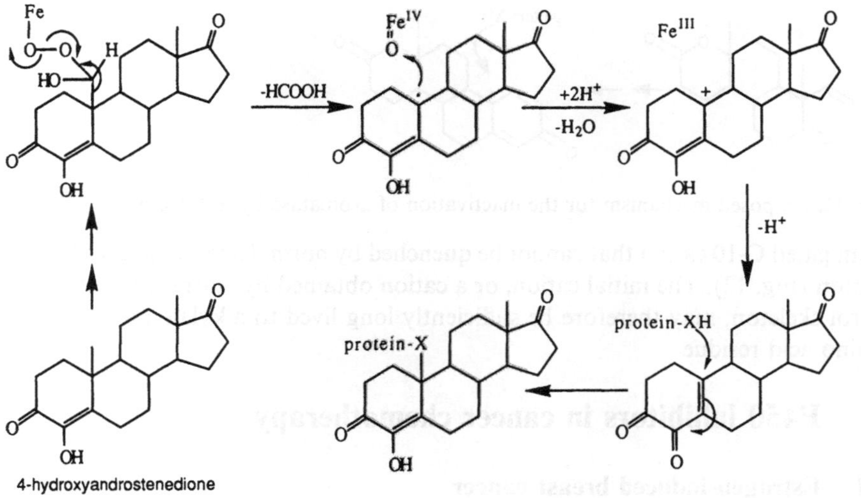

Fig. 11. Proposed mechanism for the inactivation of aromatase by 4-hydroxyandrostenedione.

radical precursor. The thiol radical would be expected to react with aromatic residues whereas the sulfenic acid would react with nucleophiles such as cysteine thiol groups. It is also possible that the thiol radical reacts with oxygen to form the thiol peroxy radical (RS–O–O·) which causes oxidative degradation of the heme group.

Of all the mechanism-based cytochrome P450 inhibitors, the one whose mechanism of action is probably least well understood is 4-hydroxyandrostenedione, a potent irreversible inhibitor of aromatase (Brodie et al., 1981; Covey and Hood, 1982; Covey, 1988). One possible but speculative mechanism (Fig. 11) involves diversion of the normal catalytic process by electron transfer to the ferryl complex to give a sterol cation. Proton loss from this intermediate yields a highly electrophilic, unsaturated vicinal diketone. Addition of a protein nucleophile to the carbonyl-activated $\Delta^{4,10}\pi$-bond then results in covalent binding of the sterol to the protein. This mechanism is consistent with the observation that the 4-amino (Di Salle et al., 1990), but not the 4-methoxy (Marsh, 1985), 4-chloro (Marsh et al., 1985), or 4-thioalkyl or -thioaryl (Abul-Hajj, 1986) analogues cause irreversible inactivation. It is also consistent with the fact that inactivation only occurs if the steroid bears the 19-methyl that is normally removed by the enzyme (Covey and Hood, 1982), and with the suggested involvement of a radical intermediate in the deformylation reaction (Cole and Robinson, 1988; Watanabe and Ishimura, 1989; Cole et al., 1990). A related mechanism can be used to explain the inactivation of aromatase by sterols with a $\Delta^{1,2}$ double bond such as testolactone (Covey and Hood, 1982). In this case, electron transfer yields a

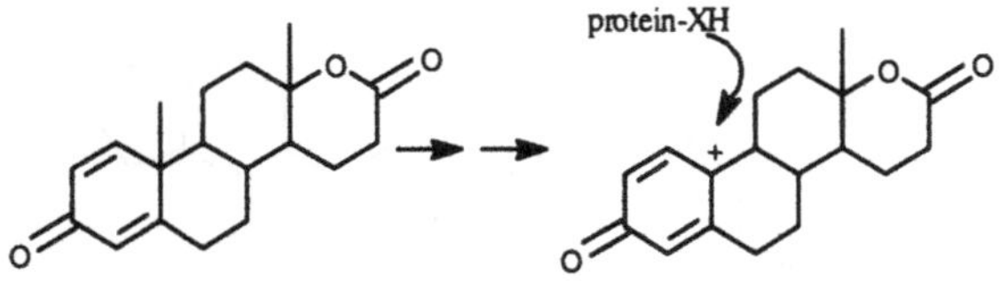

Fig. 12. Proposed mechanism for the inactivation of aromatase by testolactone.

conjugated C-10 cation that cannot be quenched by normal elimination of the C-1 proton (Fig. 12). The initial cation, or a cation obtained by rearrangement of the sterol skeleton, may therefore be sufficiently long lived to alkylate an active site amino acid residue.

4. P450 inhibitors in cancer chemotherapy

4.1. Estrogen-induced breast cancer

It is now quite obvious that human cancers of the breast and prostate gland are frequently endocrine-dependent, and therapeutic strategies therefore are designed to counteract, if not abort, their promotion by sex steroid hormones. For instance, the recognition that in 75% of postmenopausal and 60% of premenopausal patients, breast cancers are estrogen-dependent (McGuire, 1980), has provided the impetus for estrogen-lowering approaches. While ablative surgical procedures such as oophorectomy in females and orchidectomy in males are always an option, albeit invasive, they are associated with high morbidity and mortality. Fortunately, the current chemotherapeutic arsenal includes several less traumatic alternatives for lowering physiological levels of sex steroids with reasonable degrees of success (Brodie and Santen, 1986; Cavalli, 1986; Coombes et al., 1987; Osborne, 1988; Brodie et al., 1988; Banting et al., 1989; Lonning et al., 1989a; Goldhirsch, 1990; Miller, 1990). These include luteinizing hormone releasing hormone (LHRH)-agonists (e.g. Zoladex^{R}) which reduce circulating sex steroids by interfering with the hypothalamic-pituitary-gonad-sex-steroid target tissue axis and are effective in premenopausal patients, as well as antiestrogens (e.g. Tamoxifen) and inhibitors of P450-dependent estrogen biosynthesis in postmenopausal patients.

4.1.1. Role of estrogens in breast cancer

Stimulation of certain breast cancer growth by estrogens is characteristic of the intrinsic effects of estrogens in their target tissues (Jordan, 1986). Thus estrogens may induce (i) cell replication of the epithelial cells of breast tumors in common with those of the vagina and uterus, (ii) expression of the progesterone receptors in the uterus as well as some breast tumors, and/or (iii) enhanced prolactin secretion by the pituitary gland (JORDAN, 1986; and references therein). Such estrogen-dependent effects are mediated through specific estrogen receptors in the target tissues. Accordingly, detection of estrogen receptors in malignant cells is a valuable clinical reporter of the estrogen sensitivity of a given tumor and allows the unequivocal phenotyping of the tumors as estrogen-receptor positive or negative. However, since breast tumors may be heterogenous in cellular character, not all constitutive malignant cells may be amenable to such phenotyping. While, predictably, estrogen-receptor negative tumors would not be expected to respond to estrogen deprivation, not all estrogen-receptor positive tumors are responsive either (JORDAN, 1986). Tumor cell heterogeneity may be one reason for such refractoriness. The other is the apparent absence of the progesterone receptor, an additional requirement for the effectiveness of "endocrine therapy" (JORDAN, 1986). In one mechanistic model of breast tumor development (JORDAN, 1986), the two receptors appear to mediate a catenation of estrogen-regulated events intricately linked with cell growth factors and cell replication and differentiation. Thus, breast tumor cells capable of expressing these dual receptors and thus inherently capable of estrogen-regulated cell replication, proliferation and differentiation, apparently exhibit a relatively high responsiveness to such therapy, whereas estrogen-receptor positive but progesterone-receptor negative cells (which can replicate in response to growth factors but not differentiate) appear to be relatively resistant (JORDAN, 1986; and references therein; DICKSON et al., 1990). Since most estrogen-sensitive tumors eventually lose such responsiveness on progression, it is obvious that endocrine therapy will only be beneficial as long as the tumors remain "estrogen-responsive", and have not progressed to the irrevocable "estrogen-independent" state.

A well established clinical approach in this endocrine therapy is the successful use of non-steroidal anti-estrogens such as tamoxifen (Nolvadex^{R}) in the treatment of early breast cancer (Early Breast Cancer Trialists' Collaborative Group: Results of tamoxifen trials, 1985–1990; FURR and JORDAN, 1984). However, the drug is tumoristatic rather than tumoricidal, and it suffers the common limitation of a competitive inhibitor, in that its action is reversible and adequately high therapeutic concentrations need to be continuously maintained over prolonged periods of time for effective therapy. Furthermore, reflex increases in steroidogenesis in premenopausal patients, or resistance and side-effects in postmenopausal patients, also develop after tamoxifen therapy and may limit its efficacy (Early Breast Cancer Trialists' Collaborative Group: Results of tamoxifen trials, 1985–1990; MANNI and PEARSON, 1980; BRODIE et al., 1990; MILLER, 1990).

The failure of tamoxifen therapy in these cases on the one hand, and the overall need to lower cellular estrogen availability on the other, have fueled the continuous search for better and more specific estrogen-lowering chemotherapeutic agents. To a large extent this search has focused on inhibitors of estrogen biosynthesis, specifically those capable of potent inhibition of P450 XIXA1, *aroma-*

tase, the enzyme responsible for the aromatization of androgens to estrogens (Fig. 2) (BRODIE et al., 1986, 1988, 1990; SANTEN, 1986; COVEY, 1988; OSBORNE, 1988; BANTING et al., 1989; WAUVE and JANSSEN, 1989; LONNING et al., 1989a, 1990; COLE and ROBINSON, 1990; MILLER, 1990). Such aromatization although ubiquitous, occurs predominantly in the ovary in premenopausal patients but largely peripherally (adrenals, fat, muscle, liver, brain) in postmenopausal subjects (LONGCOPE, 1987; LONNING et al., 1989a, 1990; MILLER, 1990). Aromatase activity is also reportedly increased in mammary fat surrounding the tumor (O'NEILL et al., 1988).

Estradiol levels apparently are also elevated in breast cancer tissues (VAN LANDEGHEM et al., 1985). Breast tumor cells not only contain aromatase, but also the enzymes sulfatase and 17β-hydroxysteroid dehydrogenase, capable of sequentially desulfating estrone sulfate (taken up from the plasma) and converting it to estradiol, the physiologically active estrogen (MANNI and SANTEN, 1988; MILLER and O'NEILL, 1990). Although in postmenopausal women, the latter route is quantitatively the predominant source of estrogens in breast cancer tissues, nevertheless the generation of estrone sulfate itself also is critically dependent on the aromatization of androstenedione to estrone in peripheral tissues (MILLER and O'NEILL, 1990; MILLER, 1990). Additionally, breast cancer tissues also are known to avidly sequester plasma estrogens (MCNEIL et al., 1986; JAMES et al., 1989; MILLER, 1990), and this uptake may be quantitatively more important than their ability to biosynthesize estrogens.

4.1.2. Aromatase inhibitors

The peripheral localization of aromatase in postmenopausal patients in organs/sites not easily accessible for surgical removal, makes the quest for pharmacological intervention through inhibition of the enzyme doubly attractive. Furthermore, in principle, pharmacological intervention is superior to surgical endocrine gland ablation since it is (i) reversible, (ii) specific to estrogen biosynthesis, thereby circumventing the side effects manifested on deprivation of other desirable non-estrogenic steroids and (iii) effective in a broader estrogenic target tissue range (BRODIE et al., 1990; MILLER, 1990). Agents exploited for their potential to inhibit estrogen biosynthesis include reversible (competitive) inhibitors and irreversible (mechanism-based or suicide) inactivators of aromatase (BRANDT et al., 1988; COVEY, 1988; OSBORNE, 1988; BRODIE et al., 1988, 1990; BANTING et al., 1989; VAN WAUVE and JANSSEN, 1989; LONNING et al., 1989a, 1990; ROBINSON, 1990; MILLER, 1990). However, given the relatively poor turnover rates of aromatase and its high affinity for androstenedione (Km $\approx$ 30nM), effective therapy with competitive inhibitors requires high dosages, often in the toxic range. Mechanism-based inactivators on the other hand, offer overall better therapeutic prospects as low dosage regimens apparently suffice.

4.1.2.1. Reversible non-steroidal inhibitors

Aminoglutethimide (AG) **and analogs** (Table 1) are widely used clinically in the treatment of estrogen-responsive breast tumors in postmenopausal women, and have been clearly useful as the second line of therapy in tamoxifen resistant patients (MANNI and SANTEN, 1988; and references therein). AG blocks peripheral tissue aromatase activity with a potency superior to that of testolactone, but somewhat inferior to that of *4-hydroxyandrostenedione* (4-OHA; see below; BRODIE et al., 1988; MANNI and SANTEN, 1988). The D-isomer is the more potent inhibitor (GRAVES and SALHANICK, 1979). This blockade of P450 aromatase entails coordination of one of its nitrogens with the iron of the P450 prosthetic heme, as indicated by the Type II binding spectrum (Section 3.1.). Because of this feature P450-dependent cholesterol side-chain-cleavage enzyme [P450scc (XIA1)], P450XXIA2 (21-hydroxylase) and P450XIB1 (11β-hydroxylase), key enzymes in cortisol production, are also vulnerable to AG-inhibition (DEXTER et al., 1967; COHEN and FOA, 1969). Consequently, supplementation of AG-treated patients with hydrocortisone is usually required (MANNI and SANTEN, 1988). AG-mediated inhibition of these steroidogenic enzymes, along with its relatively potent inhibition of P450-dependent 18-hydroxylation, also results in significantly reduced production of the mineralocorticoid aldosterone (SANTEN and BRODIE, 1982; MANNI and SANTEN, 1988). AG also produces clinical signs of hypothyroidism, possibly because of its noted ability to inhibit the in vitro synthesis of both thyroxine and triiodothyronine (RALLISON et al., 1964). Thus, clearly, AG is not a specific aromatase inhibitor and can compromise other vital steroid-dependent processes.

When constantly infused in patients at the standard dose of 1000 mg, it reduces peripheral aromatase activity by 95–98%, and lowers plasma and urinary estrogen to levels comparable to those achieved with surgical adrenalectomy (SANTEN et al., 1978, 1981). Furthermore, at mean plasma concentrations ($\approx 30\,\mu M$) attained with such a daily dosage regimen, AG also markedly ($\approx 80\%$) inhibits the in vitro aromatase activity of breast cancer tissue homogenates (SANTEN et al., 1982). However, AG apparently is ineffective in reducing estrogen biosynthesis in premenopausal patients with advanced breast cancer, and its use is therefore restricted largely to postmenopausal patients with breast cancers (HARRIS et al., 1982). In these, it appears to have a successful response rate of $\approx 32\%$ and is often as effective as tamoxifen (MANNI and SANTEN, 1988; and references therein). Moreover, it apparently is superior to tamoxifen in patients with bone as the predominant site of disease (SANTEN et al., 1981; HARVEY et al., 1984). Surprisingly, given the independent modes of action of each drug, combined therapy with AG and tamoxifen fails to appreciably synergize the anti-tumor effects (SMITH et al., 1982).

Approximately 30–40% of the AG-treated patients display toxic side effects such as lethargy, dizziness, skin rash, orthostatic hypotension, ataxia, drug fever,

Table 1. Competitive inhibitors of aromatase: non-steroids

	Trivial/Chemical Name	Reference
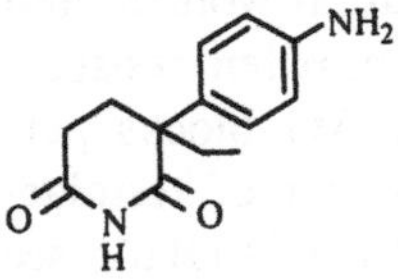	Aminoglutethimide (AG)	GOWER, 1974 SHAW et al., 1988
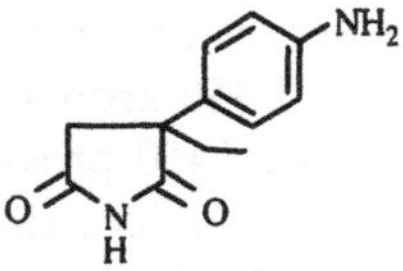	3-(4-aminophenyl)-3-ethyl-pyrrolidine-2,5-dione	DALY et al., 1986
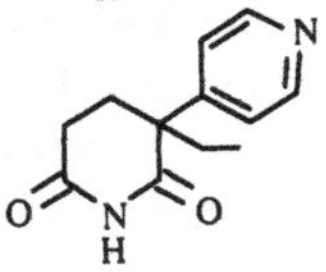	Pyridylaminoglutethimide (PG)	FOSTER et al., 1985
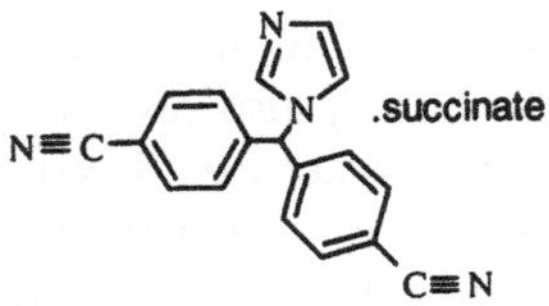	Bis-(p-cyanophenyl)-imidazo-1-yl-methane hemi-succinate (CGS 18320B)	BHATNAGAR et al., 1990
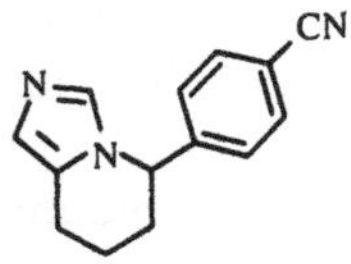	{4-(5,6,7,8-terahydro-imidazo[1,5α]-pyridin-5-yl) benzonitrile}.HCl (CGS 16949A)	STEELE et al., 1987 BHATNAGAR et al., 1990
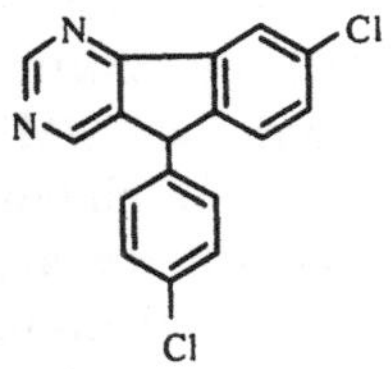	8-chloro-5-(4-chlorophenyl)-5H-indeno[1,2-d]pyrimidine (LY-113174)	HIRSCH et al., 1987
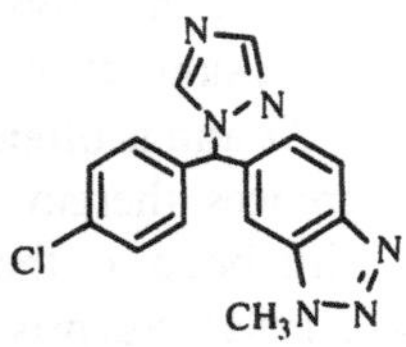	6-[(4-chlorophenyl)1H 1,2,4-triazol-1-yl)methyl]-1-methyl-1H-benzotriazole (R-76713)	WOUTERS et al., 1989a

hypothyroidism, and hematological disorders such as thrombocytopenia, leukopenia, and pancytopenia as well as a low incidence of bone marrow toxicity (SANTEN et al., 1982; MANNI and SANTEN, 1988). In the majority, these toxic symptoms subside with time and only in $<5\%$ is the discontinuation of AG necessary. However, discontinuation due to side effects is much more frequent in older women (>65yr). Some of these side effects (nausea, drowsiness and ataxia) appear to be dose-related rather than idiosyncratic and approaches to reduce AG dosage without compromising its therapeutic effectiveness are being explored.

The inhibition of adrenal P450s other than aromatase has led to the search for structural AG analogs with relatively selective aromatase inhibitory potential. Modified AG congeners that retain the glutarimide moiety have been synthesized and indeed are more selective aromatase inhibitors (FOSTER et al., 1983, 1986a–c; HARTMANN and BATZL, 1986; LEUNG et al., 1987; SEAGO et al., 1988; LAUGHTON et al., 1990). Of these *pyridoglutethimide [PG, 3-ethyl-3-(4-pyridyl) piperidine-2,6-dione]* has been licensed for Phase II trials as a potential anticancer agent against estrogen-dependent breast cancer. The drug, albeit slightly less potent than AG, is a relatively selective human aromatase inhibitor without any effect on P450scc (XIA1), thereby obviating the need for corticosterone supplementation (LEUNG et al., 1987). In addition, its long half-life and lack of appreciable CNS activity in mice makes it a promising substitute for AG (FOSTER et al., 1985).

The glutarimide moiety of AG is apparently not absolutely essential for anti-aromatase activity, since its replacement with a cyclohexyl group yields a compound that inhibits aromatase activity much more potently than that of P450scc (XIA1) (KELLIS and VICKERY, 1984; NICHOLLS et al., 1986; BANTING et al., 1989). Similarly, substitution of the glutarimide group of AG with *3-(4-aminophenyl)-3-alkyl-pyrrolidine-2,5-diones*, also yields relatively selective aromatase inhibitors (DALY, 1985; NICHOLLS et al., 1986). For instance, *3-(4-aminophenyl)-3-ethyl-pyrrolidine-2,5-dione* was as active as AG in inhibiting estrogen biosynthesis but had no appreciable effect on P450scc (XIA1) (BANTING et al., 1989; and references therein). Replacement of the glutarimide of AG with substituted or unsubstituted 3-azabicyclo [3.1.0]hexane-2,4-dione as in *1-(4-aminophenyl)-3-pentyl-3-*azabicyclo[3.1.0]hexane-2,4-dione and *1-(4-aminophenyl)-3-butyl-3-azabicyclo-[3.1.0]hexane-2,4-dione* also produced relatively potent and selective aromatase inhibitors, with no appreciable anti-P450XXIA activity (ROWLANDS et al., 1988). These glutarimide-lacking analogs indeed represent substantial design improvements, since the glutarimide ring of AG is incriminated in its side effects (ataxia, skin rashes) (SALHANICK, 1982). Elongation of the 3-ethyl moiety of AG also appears to endow the analog with more potent and relatively selective inhibition of aromatase activity, with the 3-isopentyl analog exhibiting a 100-fold greater potency than AG (LEUNG et al., 1987). Optimal aromatase inhibition is also achieved with 1-alkyl analogs of AG, with the 1-n-propyl analog exhibiting the highest specificity for aromatase with minimal antiglucocorticoid activity (BANTING et al., 1989; and references therein). Similarly, introduction of an N(1)-octyl function into PG or modification of its 3-ethyl to the 3-octyl group, affords 13–30-fold more potent anti-aromatase analogs than the parent drug (LAUGHTON et al., 1990a).

CGS 16949 {4-(5,6,7,8-tetrahydroimidazo[1,5α]-pyridin-5-yl)benzonitrile monochloride} and analogs: Of the vast array of imidazole drugs (vide infra), CGS 16949 and its congener CGS 18320 *[bis-(p-cyano-phenyl)-imidazo-1-yl-methane hemisuccinate]* are the most promising breast cancer therapeutic agents. They are relatively 700-fold more potent than AG as non-steroidal competitive inhibitors of human placental aromatase (STEELE et al., 1987; BHATNAGAR et al., 1990). Furthermore, they are much more selective inhibitors of aromatase than of P450scc, P450XXI and P450XIB (BHATNAGAR et al., 1989, 1990). However, both in vitro and in vivo, CGS16949 appears to inhibit aldosterone production (18-hydroxylase activity) in rats, at concentrations 100-fold lower than those of AG (LAMBERTS et al., 1989; DEMERS et al., 1990). This particular feature is, however, much less pronounced with CGS 18320 (BHATNAGAR et al., 1990). Experimentally in rats, when three indices (ovarian estrogen content, androstenedione-induced uterine hypertrophy and serum estradiol and LH-concentrations) of the in vivo efficacy of each aromatase inhibitor were examined, both CGS16949 and CGS 18320 were found to be $\approx$ 300- to $>$1000-fold more potent than AG, with CGS 18320 nominally superior to the other two in exhibiting a longer duration of action as well as higher potency (BHATNAGAR et al., 1990). More importantly, both CGS16949 and CGS 18320 were found to cause marked and dose-dependent regression of 7,12-dimethylbenzanthracene (DMBA)-induced mammary tumors in rats (SCHIEWECK et al., 1988; BHATNAGAR et al., 1990). Further CGS16949 and CGS 18320 at ED_{50} of 0.1–0.5 mg/kg/day and 0.05 mg/kg/day, respectively, almost totally suppressed the appearance of new tumors (BHATNAGAR et al., 1990). This is to be contrasted with the corresponding marginal effects of AG (100 mg/kg/day, the maximally tolerated dose) (SCHIEWECK et al., 1988; BHATNAGAR et al., 1990). CGS16949 also inhibited human mammary tumor cell growth in vitro (BHATNAGAR et al., 1990). These experimentally promising findings, together with its relatively well-tolerated profile in preclinical studies, have launched the Phase I clinical trials of CGS16949 in postmenopausal patients with advanced breast cancer (SANTEN et al., 1989; LIPTON et al., 1990; STEIN et al., 1990). Preliminary findings indicate that CGS16949 is a potent inhibitor of estrogen biosynthesis in postmenopausal women, and that it is extremely well tolerated even in patients who responded adversely to AG. Minor side effects occurred but were mild and transient. Furthermore, at the maximal effective dose of 2 mg/kg/day, no significant alterations in cortisol, androstenedione, testosterone or aldosterone were observed, indicating that at this dose the drug functions as a relatively selective aromatase inhibitor (LIPTON et al., 1990; DEMERS et al., 1990). Moreover, although aldosterone was significantly lowered in another study, no clinically apparent cases of adrenocortical insufficiency were encountered (STEIN et al., 1990).

Collectively, these preliminary clinical findings are encouraging and indicate that CGS16949 suppresses plasma estradiol in humans to the same extent as either AG or 4-OHA. Furthermore, given the limited sample size and the nature of the patient sample (unresponsive to other aromatase inhibitors or with estrogen-receptor negative tumors), the overall 16% response rate versus ≈ 30–35% for AG or 4-OHA, may be considerably underestimated.

Bis-chlorophenyl-pyrimidine analogs: The recognition that the infertility in adult male rats exposed neonatally to the agricultural fungicide *fenarimol [α-(2-chlorophenyl)-α-(4-chlorophenyl)-5-pyrimidine-methanol]* was due to its CNS aromatase inhibition (HIRSCH et al., 1987), spurred the development of two agents *LY-56110 [α, α-bis (4-chlorophenyl)-5-methylpyrimidine]* and *LY-113174 [8-chloro-5-(4-chlorophenyl)-5H-indeno(1,2D)pyrimidine]* as prospective inhibitors of human aromatase (TAYLOR et al., 1987; HIRSCH et al., 1987). Although, LY-56110 is indeed a very potent orally active aromatase inhibitor ($IC_{50} \approx 29\,nM$) in rats, and produced an impressive 80% regression of DMBA-induced mammary tumors, it had to be abandoned due to its unusually long half-life, inhibition of hepatic mixed function oxidase (MFO) activity on acute administration, and inducibility of hepatic MFOs on prolonged exposure (LINDSTROM and WHITAKER, 1987). This led to the rational design and development of LY-113174, which retained all the desirable features of its progenitor, had an acceptable biological-half life and was devoid of the undesirable MFO-inducing potential (HIRSCH et al., 1987). Both compounds yield a spectrum indicative of nitrogen coordination to the P450 heme with rat ovarian microsomes (LINDSTROM and WHITAKER, 1987). More recently, diaryl-3-pyridine methanes and diaryl-3-pyridinylcarbinols have also been synthesized and found to be respectable aromatase inhibitors (JONES et al., 1990).

R 76713 (6[(4-chlorophenyl)(1H-1,2,4-triazol-1-yl)methyl]-1-methyl-1H-benzotriazole: This compound is a new, relatively selective and very potent inhibitor for human placental aromatase, with almost the ideal inhibitory profile for an aromatase inhibitor (WOUTERS et al., 1989a,b; 1990a,b). R76713 is a racemate, with the (+) *enantiomer (R 83842)* exhibiting the lower IC_{50} values for human placental aromatase activity, and therefore the most potent of the two (VANDEN BOSSCHE et al., 1990a). In this system, R 83842 is 1029-fold more potent than AG, with R76713, its (-) enantiomer R83839, and 4-OHA displaying 548-, 32- and 34-fold greater potencies than AG (VANDEN BOSSCHE et al., 1990a). Furthermore, no appreciable inhibition of other steroidogenic enzymes or liver microsomal P450s was observed unless concentrations exceeded 1000-times the IC_{50} values for aromatase inhibition (VANDEN BOSSCHE et al., 1990a; WOUTERS et al., 1990b).

This promising in vitro inhibitory potential is confirmed by in vivo studies in animals, to date reported only with the racemic R76713 (WOUTERS et al., 1989b). R76713 at 1 mg/kg, acutely lowers plasma estradiol levels for over 16 h in female rats, whereas chronically this dose produces plasma estradiol reductions comparable to those attained with ovariectomy. It also is a potent inhibitor of peripheral aromatization in male cynomolgus monkeys, without any detectable effects on plasma testostrone, corticosterone, aldosterone, progesterone or 11-deoxycorticosterone, indicating that R76713 treatment does not affect steroidogenic enzymes other than those involved in estrogen formation (WOUTERS et al., 1989a,b). More importantly, R76713 treatment is comparable to ovariectomy in causing complete regression of DMBA-induced or dimethylnitrosourea (NMU)-induced mammary tumors in rats (DE COSTER et al., 1990a,b). This agent also appears to be as useful as ovariectomy in causing the regression of experimental endometrial auto-transplants (DE COSTER et al., 1990a,b). Successful reports of efficient plasma estradiol lowering by R76713 in preclinical studies in human volunteers are highly encouraging, and together with the absolute lack of any detectable clinical toxicity, either physiological or hematological (WOUTERS et al., 1989b; DE COSTER et al., 1990b), underscore its clinical portent as an excellent aromatase inhibitor, particularly if its even more selective (+) enantiomer, R 83842, is developed.

4.1.2.2. Reversible/irreversible steroidal inhibitors

Testolactone (Teslac^R, D-homo-17α-oxaandrosta-1,4-diene-3,17-dione): This agent has been employed since 1960 in the palliative treatment of human breast cancer, although its mechanism of action as a weak irreversible inhibitor of aromatase was revealed only in 1979 (SEGALOFF et al., 1962; GOLDENBERG et al., 1973; BARONE et al., 1979). The compound (see Tables 2, 3) is inferior to AG or 4-OHA in its clinical effectiveness, producing an objective response of $\approx$ 14.5% versus $\approx$ 33% for the others (VOLK et al., 1974; MANNI and SANTEN, 1988). Such inferiority apparently is explained by its relatively weak binding to, and consequently slow inactivation of, aromatase (see Fig. 12 for a possible mechanism).
Atamestane (1-methylandrosta-1,4-diene-3,17-dione, SH-489): The introduction of a C_1-methyl into 1,4-androstadiene-3,17-dione, apparently enhances its affinity (Ki $\approx$ 2 nM vs Ki of 29 nM for 4-OHA) for the human placental aromatase while slowing its inactivation of the enzyme [$t_{1/2}$ (min), 45.3 for SH-489 vs 2.1 for 4-OHA], thereby reducing the production of aromatized estrogenic products (HENDERSON et al., 1986a–c; DI SALLE et al., 1989, 1990). In female rats, it lowered serum estrogen levels and produced 24% regression of DMBA-induced mammary tumors (NISHINO et al., 1989). It has no intrinsic androgenic or antigonadotropic activity in rats (HENDERSON et al., 1986a–c). The compound is currently being evaluated in Phase I clinical trials by Schering for possible therapy of estrogen-dependent conditions (breast cancer and benign prostatic hypertrophy). Other 1,2-methylene-substituted congeners are also presently being evaluated.
Org-30958 [19-(ethyldithio)androst-4-ene-3,17-dione]: This compound is a 19-substituted analog of androst-4-ene-3,17-dione steroid inhibitors currently in Phase I clinical trial for the treatment of estrogen-dependent breast cancers. The ethyldithio substitution apparently renders the steroid more stable extracellularly than the free thiol Org-30365 (19-mercapto-androst-4-ene-3,17-dione), which is a

Table 2. Competitive inhibitors of aromatase: steroids

	Substituents	Reference
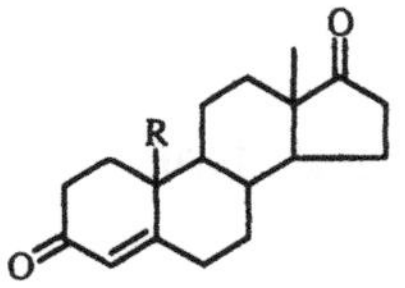	R =  R = β-OOH	SHIH et al., 1987 CHILDERS et al., 1987a, b KELLIS et al., 1987 COVEY et al., 1984
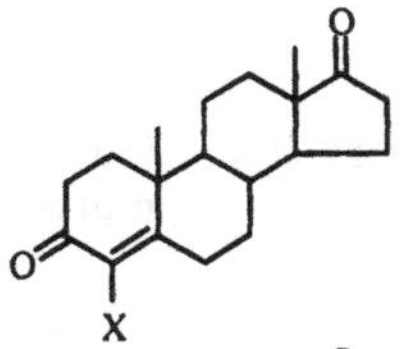	X = SC_6H_4-oCH_3 = SC_6H_5 = $SH(CH_3)_2$ X = F	ABUL-HAJJ, 1989, 1990 ROWLANDS et al., 1986
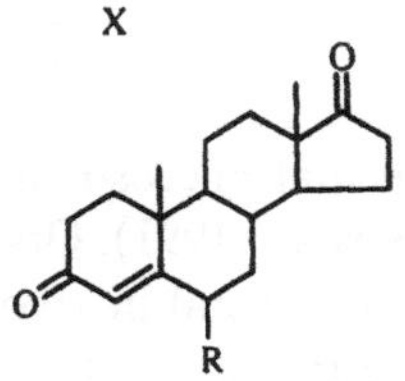	R = C=O = OH = α-OOH = β-OOH	TAN et al., 1983
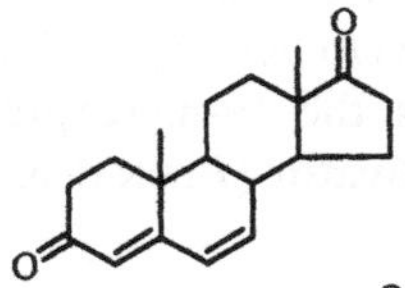		COVEY and HOOD, 1982
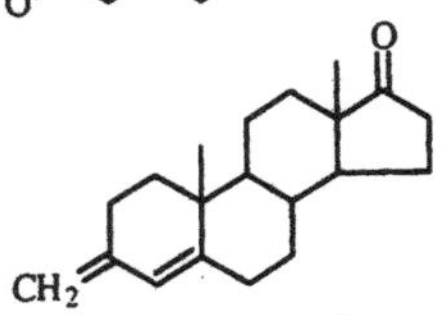		MIYARI and FISHMAN, 1986
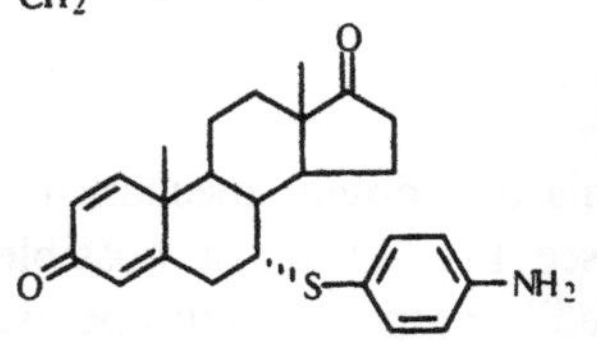	Δ^4 (7α-APTA) $\Delta^{1,4}$(7α-APTAD)	BRUEGGEMEIER et al., 1987 1990
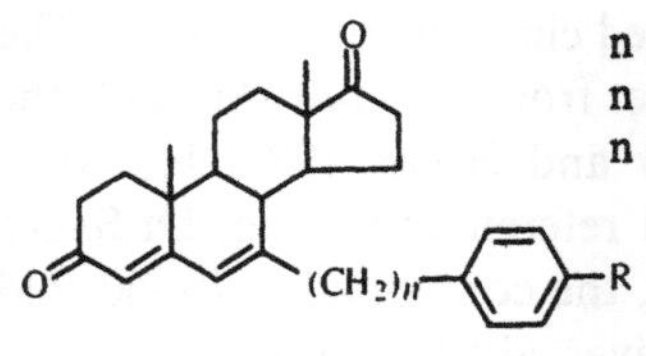	n = 1, R = H n = 1, R = NO_2 n = 1, R = NH_2	LI and BRUEGGEMEIER, 1990

Table 2. Continued

	Substituents	Reference
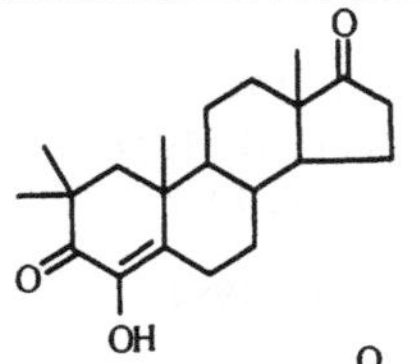		FURTH et al., 1990
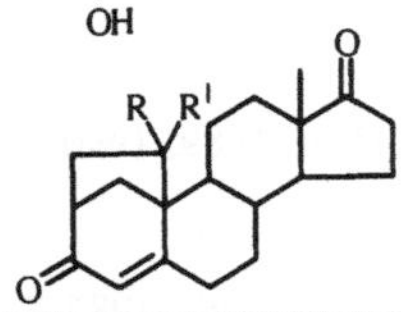		BURKHART et al., 1991 (in press)

16- or 8-fold more potent irreversible inhibitor of human placental microsomal aromatase in vitro than 4-OHA or SH489, respectively (GEELEN et al., 1991). Org 30958 is also 8-fold more active than either 4-OHA or SH-489 in several in vivo assays of aromatase inhibitory potential, effectively lowering estrogen levels in several animal models. It is a remarkable prodrug since its in vivo activity is due to the intracellular reduction of its disulfide which releases the 19-mercapto analog Org-30365, a potent mechanism-based aromatase inactivator (BEDNARSKI and NELSON, 1989).

4.1.2.3. Irreversible steroidal inhibitors

4.1.2.3.1. 4-substituted steroids

4-Hydroxyandrostenedione (4-OHA; CGP-32349): This is a potent mechanism-based inactivator of human placental aromatase (see Fig. 11 for a possible mechanism), whose potential as a "rationally designed" prospective therapeutic agent has materialized, since it is now being evaluated clinically (cf. Table 3). The Ki values for human placental aromatase range from 10–170 nM and the $t_{1/2} \approx 1.23$ to 2.9 min (BRODIE et al., 1981; COVEY and HOOD, 1982; JOHNSTON and METCALF, 1984; BRODIE and WING, 1987; and references therein; DI SALLE et al., 1990). On the basis of such inhibition alone, the compound is 40–60 fold more potent than AG (DI SALLE et al., 1990). However, although it is as active as AG in reducing ovarian estrogen secretion, it is much more effective than AG in causing regression of DMBA- or NMU-induced mammary tumors (BRODIE et al., 1979). In contrast to AG, it also is highly effective in lowering plasma estradiol levels and in suppressing LH levels in rats with DMBA-induced mammary

Table 3. Irreversible inhibitors of aromatase: steroids

	Substituents	References
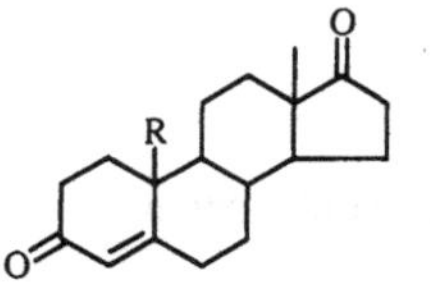	R = $CH_2C{\equiv}CH$ R = $CH_2C{\equiv}C{-}CH_3$ R = $CH{=}C{=}CH_2$ R = $CHOH{-}C{\equiv}CH$	Covey et al., 1981 Johnston et al., 1987
	R = SH	Bednarski and Nelson, 1989
	R = CH_2SH R = $CH_2SSCH_2CH_3$	Geelen et al., 1991
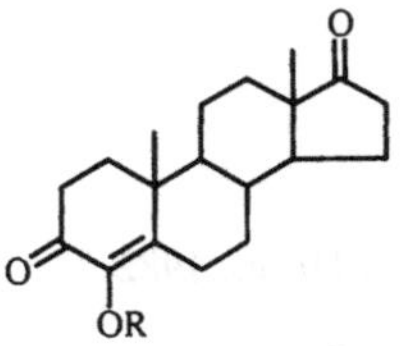	R = H R = $COCH_3$	Brodie et al., 1983
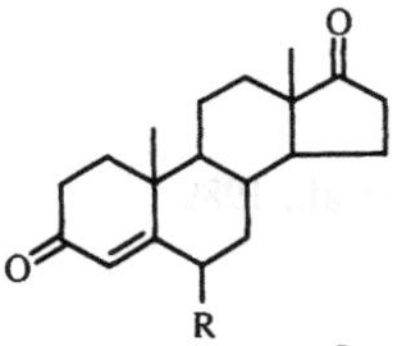	R = β-Br	Osawa et al., 1987
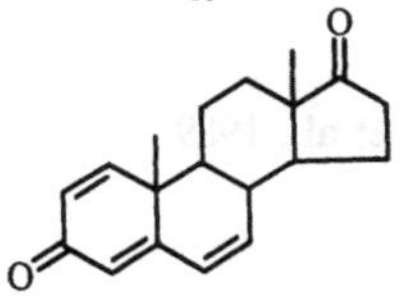 ATD		Henderson et al., 1986 Covey and Hood, 1981
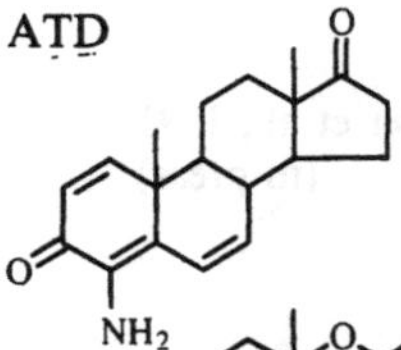	$\Delta^{1,4,6}$ (FC24928) $\Delta^{4,6}$ $\Delta^{1,4}$ Δ^{4}	Di Salle et al., 1990a
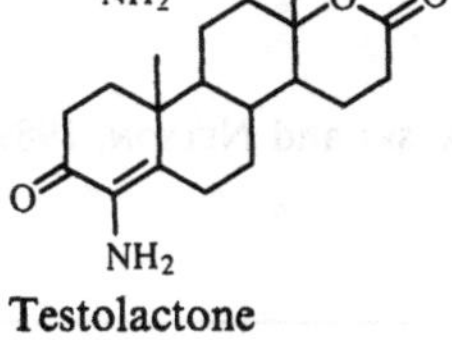 Testolactone		Barone et al., 1979
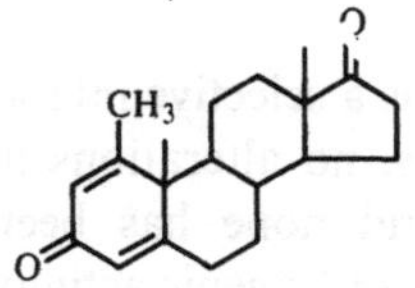 SH-489		Henderson et al., 1986 Di Salle et al., 1990b

Table 3. Irreversible inhibitors of aromatase: steroids

	Substituents	References
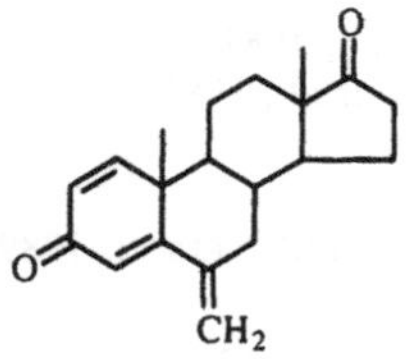 FCE-24304		Di Salle et al., 1990b
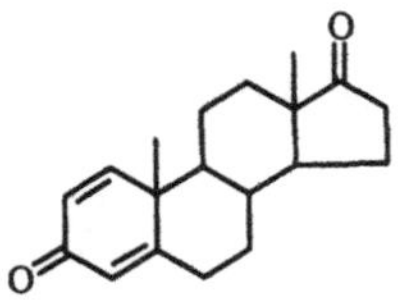		Covey and Hood, 1982
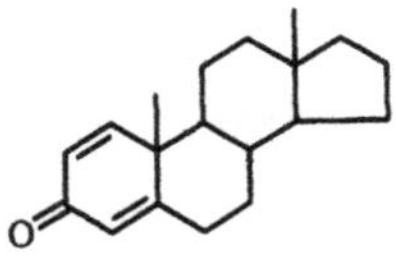		Sherwin et al., 1989
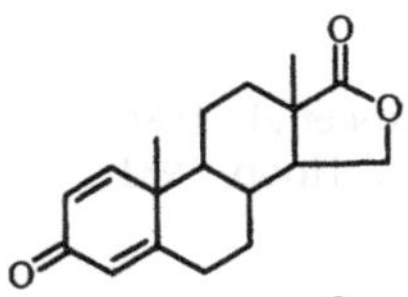		Sherwin et al., 1989
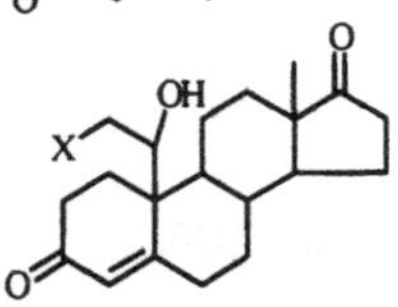	X = Br(19R) = Br(19R) = Cl = I	Burkhart et al., 1991 (in press)
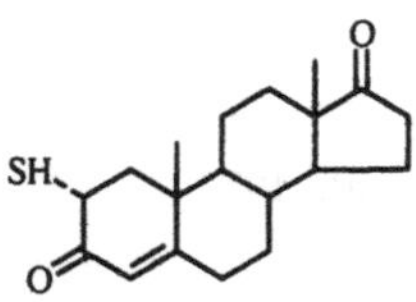		Bednarski and Nelson, 1989

tumors (Brodie et al., 1990). Furthermore, because the drug is a selective inhibitor of aromatase and does not affect other adrenal enzymes, no alterations in serum glucocorticoid or mineralocorticoid are expected and none has been encountered (Brodie et al., 1983). The drug does exhibit weak androgenic activity (≈ 1% of testosterone) in experimental animals, a manifestation yet to be uncovered in humans (Manni and Santen, 1988).

This collectively promising profile stimulated several clinical trials of the drug.

Reports of Phase II studies in postmenopausal patients with breast cancer are highly encouraging with a 23–27% objective response and substantial (19–27%) stabilization of the disease (BRODIE et al., 1988, 1990; HOFFKEN et al., 1990). Relief of bone pain (associated with metastases) was also a frequently noted benefit of 4-OHA. The drug is particularly well tolerated, and effective in patients who have relapsed from tamoxifen or from whom AG had to be withdrawn due to side effects. Minor problems with the higher intramuscular doses were sterile abscesses at the injection site in some patients and hot flushes in others (BRODIE et al., 1988, 1990).

Although oral dosing should circumvent such side effects, it was initially discounted because the relatively avid first pass glucuronidation of 4-OHA in the liver (GOSS et al., 1986; FOSTER et al., 1986c) was expected to compromise 4-OHA bioavailability. However, recent clinical evaluation in postmenopausal patients with breast cancer indicates that the bioavailability, and more importantly the clinical response, are equivalent on oral and parenteral administration, with 90% of the patients registering free of side-effects (DOWSETT et al., 1987).

The early concerns about the bioavailability of the drug, however, have prompted the design, synthesis and evaluation of other 4-substituted analogs of 4-OHA such as *4-acetoxy-androstene-3,17-dione and 4-hemisuccinate-androstene-3,17-dione*, which are potent inhibitors of human placental aromatase, reduce ovarian estradiol secretion, and cause regression of mammary tumors in rats comparably to 4-OHA (BRODIE et al., 1979). However the introduction of bulky ester groups apparently reduces the potency of these compounds. Further conjugation of the 4-hydroxyandrostene nucleus as in *1,4,6 androstatriene-3,17-dione (ATD)*, results in retention of aromatase inhibitory activity (MARSH et al., 1985), as well as marked tumor regression (≈ 80%) (BRODIE et al., 1982). In further attempts to improve the oral 4-OHA bioavailability, the 4-OH group has been replaced with halogen, thiol, or thioester groups with variable results. Of these, 4-fluoroandrostenedione appears to be the most promising (ROWLANDS et al., 1986).

FCE-24928 (4-aminoandrostan-1,4,6-triene-3,17-dione): Of a series of $\Delta^{1,2}$ and $\Delta^{6,7}$ analogs, FC 24928 represents the most promising candidate because it inactivates human placental aromatase activity as potently as 4-OHA and *FCE-24304 (6-methylene-androstan-1,4-diene-3,17-dione)*, but unlike both 4-OHA and FCE-24304 it has little intrinsic androgenic activity (DI SALLE et al., 1990). Its inactivation rate ($t_{1/2} \approx 4$ min) of human placental aromatase was somewhat faster than that ($t_{1/2} \approx 13.9$ min) of FCE-24304, but comparable to that of 4-OHA ($t_{1/2} \approx 2.1$ min). The compound did not inhibit 5α-reductase or P450scc. In vivo, FC 24928 was considerably more effective than 4-OHA in lowering ovarian aromatase activity in female rats, but had little androgenic activity in immature castrated male rats (DI SALLE et al., 1990). Because of these potentially useful characteristics, Farmitalia Carlo Erba is currently evaluating the compound as a prospective aromatase inhibitor in preclinical trials.

4.1.2.3.2. 6-substituted steroids

FCE-24304 (6-methylene-androsta-4,6-diene-3,17-dione): In parallel studies, the IC_{50} for human placental aromatase was found comparable ($\approx$ 40 nM) to that of 4-OHA (Di Salle et al., 1990). The relative Ki (nM) and $t_{1/2}$ (min) values for the inactivation processes were 26 ± 1.4 and 29.0 ± 7.5, and 13.9 ± 0.7 and 2.1 ± 0.2 for FCE-24304 (cf. Table 3) and 4-OHA, respectively (Di Salle et al., 1989, 1990). Although FCE-24304 was a relatively slow aromatase inactivator, the ED_{50}s for ovarian aromatase activity in PMSG-stimulated rats on subcutaneous and oral administration were 1.8 mg/kg and 3.7 mg/kg for FCE-24304 and 3.1 mg/kg and > 100 mg/kg for 4-OHA, respectively. Thus, clearly, FCE-24304, orally as well as subcutaneously, is relatively a more potent agent in experimental animals (Di Salle et al., 1990). It remains to be determined whether this superiority holds in the clinical tests currently underway. However, in comparison with MDL 18962 and SH-489, FCE-24304 was also much more effective in producing regression of DMBA-induced mammary tumors in rats (Di Salle et al., 1989).

4.1.2.3.3. 10-substituted steroids

10-(2-propynyl)estr-4-ene-3, 17-dione (MDL18,692): Of the "rationally" designed suicide substrates of aromatase, **MDL18,692** (cf. Table 3) is clearly one of the most potent, with Ki values of the order of 3-4 nM, and $t_{1/2}$ of 9 and 6 min for the human and baboon placental aromatases, respectively (Covey et al., 1981; Metcalf et al., 1981; Marcotte and Robinson, 1981; Johnston and Metcalf, 1984; Perel et al., 1986; Johnston, 1987). The compound is also orally active with ED_{50}s of 0.01 mg/kg, iv and 4 mg/kg po for inhibition of peripheral aromatization in baboons (Longcope et al., 1988). Its long-lasting in vivo effects appear to stem from the retention of pharmacological activity by its first pass metabolites (O'Neal Johnston et al., 1990). Preclinical findings in normal human volunteers indicate that *MDL18,692* indeed lowers estrogen levels (O'Neal Johnston, 1990; personal communication). *MDL18,692* is currently in Phase I clinical trials. The compound also inhibits Leydig tumor cell steroidogenesis (Zimniski et al., 1985).

Collectively, these findings have provided an impetus for the development of other high affinity 10-substituted analogs, such as the mechanism-based inhibitor *10-β-mercapto-estr-4-ene-3, 17-dione*, and the competitive inhibitors *(19-R)-10-oxiranyl and 10-thiiranyl-estr-4-ene-3, 17-diones and 10-β- MeSCH$_2$-estr-4-ene-3, 17-dione* (Bernardski and Nelson, 1985, 1989; Childers et al., 1987a,b; Kellis et al., 1987; Shih et al., 1987; Wright et al., 1989). Very recently, C-10, C-2 hydroxyethyl bridged species have been synthesized as stable carbon analogs of the 2β-hydroxylated 19-oxoandrostenedione, a putative intermediate of the aromatization reaction (Fishman and Hahn, 1987), and found to be potent competitive inhibitors of the human placental aromatase (Burkhart et al., 1991). On the other hand, related halohydrin analogs are potent mechanism-based aromatase inhibitors (Burkhart et al., 1991).

4.2. Benign Prostatic Hyperplasia (BPH)

In Caucasian men the incidence of BPH increases with age, such that over 50% of 70-year old men exhibit morphologic evidence of BPH even if otherwise asymptomatic (Birkoff, 1983). Enlargement of the prostate, however, can seriously obstruct the urinary tract in some individuals, necessitating surgical removal for relief. Although, the mortality rate of such surgery is nominally quite low (≈ 1%), the 15–20% associated morbidity rates and the frequently noted side-effects of impotence and incontinence (Grayhack and Sadlowski, 1975) are sufficient inducement for the pursuit of alternative approaches to treatment.

Development of therapeutic approaches for any disease requires an understanding of its essential etiology. The pathogenesis of BPH, however, is unclear, partly due to lack of adequate animal models. The inherent interspecies anatomical/histological differences in the prostate are a well-recognized problem (McNeal, 1981, 1982). Although BPH induced by androstenedione-treatment of castrated dogs has been used as a model (Habenicht et al., 1986, 1989), it differs from human BPH in that the hyperplasia is diffuse, largely epithelial in cell character and somewhat more homogenous in the dog. In man on the other hand, it is much more nodular and heterogenous with prominent stromal rather than epithelial (glandular) recruitment (McNeal, 1981, 1983; Henderson et al., 1986d; Henderson, 1987; Habenicht et al., 1986, 1989). Given the intrinsic differences in biochemical regulation of each cell type, such differences are significant and may have considerable bearing in the evaluation of therapeutic agents (vide infra). They also suggest that the more closely related primates (cynomolgus monkeys and baboons) might serve as better models in spite of some anatomical differences (Karr et al., 1984; Habenicht et al., 1987). Nevertheless, the canine model of BPH has provided useful mechanistic insights which, together with clinically derived circumstantial evidence, tends to implicate sex steroids (androgens and estrogens) in the pathogenesis of human BPH (Krieg and Tunn, 1988; and references therein).

4.2.1. Role of sex steroids in BPH

Abundant clinical and experimental evidence attests to the fact that hypophysectomy or orchiectomy leads to functional decline and atrophy of the prostate, strengthening the widely held notion that prostatic development, structural differentiation and secretory function is regulated by testicular androgens (testosterone and androstenedione) (KRIEG and TUNN, 1988; and references therein). In the prostatic stroma and epithelial tissues, these androgens are rapidly reduced by 5α-reductase, and the 5α-reduced metabolite dihydrotestosterone (DHT), rather than testosterone, appears to principally account for the induction of androgen-receptor mediated prostatic growth, differentiation and function (KRIEG and TUNN, 1988; and references therein). Furthermore, not only is the DHT content usually relatively high in the prostatic cell nuclei, but DHT also accumulates in human BPH tissue nuclei (HUDSON and WHERETT, 1990). Indeed, these observations have contributed to the notion that it might play a permissive role in BPH, and have led to the development of anti-androgens as well as the targeting of 5α-reductase for therapeutic intervention (HENDERSON et al., 1986d, 1987; BARTSCH et al., 1987; HENDERSON, 1987; KRIEG and TUNN, 1988; and references therein). However, anti-androgen therapy by itself apparently is insufficient for successful clinical treatment of the disease (TUNN et al., 1981).

The putative role of estrogens in BPH pathogenesis on the other hand is largely inferential and stems from the observation that estrogens are essential for normal prostatic stroma growth, differentiation and function that, in turn, through production of DHT and/or protein growth factors, are responsible for prostatic epithelial cell development (HENDERSON et al., 1986; HENDERSON, 1987; KARR et al., 1987; BARTSCH et al., 1987; KRIEG and TUNN, 1988; and references therein; OESTERLING et al., 1988). Thus, estrogens appear to regulate normal prostatic growth by stimulating stroma-epithelial interactions (TENNISWOOD, 1986). However, several clinical observations suggest that estrogen-regulation might in fact be heightened in human BPH. For instance, not only does human BPH exhibit a predominant stromal component, with the BPH tissue accumulating estrogens to a higher extent than normal, but such accumulation is most striking in the nuclear fraction of BPH stroma (KRIEG and TUNN, 1988; and references therein). In view of the relatively higher levels and preferential localization of high-affinity estradiol binding estrogen receptors in human BPH stroma, estrogen accumulation might indeed be pathologically significant.

Several ancillary pieces of direct and indirect evidence also reinforce this notion and tend to implicate estrogens in the pathobiology of BPH (HABENICHT et al., 1986; HENDERSON et al., 1986, 1987; HENDERSON, 1987; KARR et al., 1987; KRIEG and TUNN, 1988; and references therein). These include (i) the synergistic role of estrogens in androgen-induced canine prostatic hyperplasia; (ii) the specific estrogen-mediated increase in prostatic nuclear androgen receptors in canine BPH; (iii) the positive correlation between increased prostatic stroma and plasma and urinary estrogens; (iv) the parallelism between the age-dependent increased incidence of human BPH and the increased potential for peripheral aromatization in elderly men; (v) the high sequestration of estrone sulfate, which through the sequential actions of prostatic sulfatase and 17β-hydroxy-steroid dehydrogenase would contribute to the accumulation of estradiol; (vi) the estrogen-mediated alterations in androgen metabolism in BPH tissue slices and (vii) the regressive effect of aromatase inhibitors in experimental PBH. Collectively, these findings strongly support the possibility that estrogens act synergistically with androgens at the cellular level in promoting human BPH.

Although the levels of aromatase are miniscule in the human prostate (BRODIE et al., 1989), and its role in prostatic estrogen synthesis is controversial (STONE et al., 1986; BARTSCH et al., 1987; BRODIE et al., 1989), peripheral aromatization appears to be the primary source of estrogens in aging men. This recognition, together with the collective evidence presented above, has provided a strong rationale for the consideration of aromatase inhibitors as a therapeutic alternative to prostatectomy in BPH (BARTSCH et al., 1987; HENDERSON et al., 1986d, 1987; HENDERSON, 1987; KARR et al., 1987). In principle, an array of aromatase inhibitors (discussed above) should qualify as prospective candidates, but only a few have been evaluated in animal models (with encouraging results) (HENDERSON et al., 1986d, 1987; HENDERSON, 1987; BARTSCH et al., 1987; BRODIE et al., 1989), and only one clinically.

4.2.2. Aromatase inhibitors

As already discussed, testolactone is a weak irreversible inhibitor of aromatase, considerably inferior to either AG or 4-OHA (SEGALOFF et al., 1962; GOLDENBERG et al., 1973; BARONE et al., 1979). Nevertheless, in the single clinical study in 13 BPH patients reported to date (SCHWEIKERT and TUNN, 1987), after 8 wk of treatment, testolactone induced spontaneous micturation with no residual urine in seven patients. In three of the other six, it was only partially successful, since it induced spontaneous micturation but a residual urinary volume of > 100 ml required the continual use of the bladder fistula. In all treated patients, however, testolactone reduced the prostatic volume, with average 26% decreases in the group of seven, while approximately a 15% reduction was observed in the remaining six patients. These findings are encouraging, given the recognized inferiority of testolactone to either AG or 4-OHA in clinical treatment of breast cancer (VAN WAUWE and JANSSEN, 1989), and augur well for the consideration of more potent aromatase inhibitors (HENDERSON et al., 1986d, 1987; HENDERSON, 1987; BARTSCH et al., 1987, 1990; BRODIE et al., 1989).

4.2.3. Other prospective candidates

Aromatase inhibitors evaluated in experimental BPH to date include: 4-OHA, atamestane (SH-489;1-methyl-1,4-androstene-3,17-dione) and FCE-24928 (4-amino-1,4,6-androstatriene-3,17-dione).

4-OHA (20 mg/kg, thrice weekly for 6 or 9 mos) was found to clearly lower

androstenedione-induced serum estradiol levels in castrated male beagles (HABENICHT et al., 1986). It also prevented the estrogen-related marked proliferation and activation of the stroma observed at the lower dose of androstenedione (3 mg/kg), but not at the higher dose of 30 mg/kg required to maintain BPH in dogs. Such failure of 4-OHA at the higher doses of androstenedione is entirely consistent with the first-order kinetics of suicide substrates. It also underscores another inadequacy of the androstenedione-induced canine BPH model for evaluating aromatase inhibitors.
Atamestane (SH-489;1-methyl-1,4-androstene-3,17-dione): This reversible/irreversible aromatase inhibitor has been evaluated both in the cynomolgus monkey and canine BPH model with encouraging results in both. In the monkey model (a closer model of human BPH), subcutaneously administered *SH-489* significantly lowers serum estradiol levels as well as effectively prevents the androstenedione-induced estrogen-related prostatic stromal changes (HABENICHT et al., 1987). In the canine BPH model, on the other hand, *SH-489* by itself significantly lowers serum estradiol levels but fails to lower prostatic weight or reverse androstenedione hypertrophy or hyperplasia. However, when combined with the anti-androgen cyproterone acetate, which produces prostatic atrophy, it completely antagonizes the androstenedione-induced estrogen-related stromal simulation (HABENICHT and EL ETREBY, 1989).
FCE-24928 (4-amino-1,4,6-androstatriene-3,17-dione): Although the evidence that supports consideration of this agent in human BPH has not yet, to our knowledge, been published, Farmitalia Carlo Erba has projected clinical trials for the drug in the 1990s.

Overall, given these limited but encouraging findings with aromatase inhibitors, and the current availability of even more potent and selective agents, the prospects for the use of aromatase inhibitors in the non-invasive treatment of human BPH are excellent and warrant serious consideration. However, given that the canine model is not wholly representative of the human condition for both anatomical and histological reasons (MCNEAL, 1983), such inhibitors will have to be screened in the more reliable, albeit expensive, primate model if drugs with clinical prospect are to be identified.

4.3. Prostate cancer

Second to lung cancer, prostatic cancer is the most frequent cause of cancer death in males in most parts of the industrialized world, accounting for $\approx 10\%$ of all yearly cancer deaths (SILVERBERG and LUVERA, 1986; DENIS, 1988a). As in the case of BPH, prostatic cancer is also age-dependent, with $> 75\%$ being diagnosed between the ages of 55 and 75, frequently after it has progressed to the metastatic form.

Prostate cancer is essentially a glandular disease and may present in any one of its morphologically distinct developmental disease stages, ranging from focal to locally extensive non-metastatic, to hormone-dependent metastatic, and finally to the hormone-independent metastatic form that is refractory to most treatments and ultimately fatal (SCHROEDER, 1988; CHEVALIER et al., 1990). Although the cause of prostate cancer is unknown, the growth, differentiation and function of the prostate is dependent on functioning testes and therefore on androgens, specifically DHT (HAVLIN and TRUMP, 1988; TURKES and GRIFFITHS, 1989). More importantly, this dependence is retained by the prostatic cell through progressive malignant transformations until the advanced clinically irreversible hormone-independent metastatic stage is reached. Thus, in addition to prostatectomy, treatments of patients at less advanced stages have been aimed at reducing androgen levels and/or action, and include bilateral orchidectomy, hypophysectomy, adrenalectomy, 5α-reductase inhibitors, LHRH antagonists and anti-androgens (TURKES and GRIFFITHS, 1989). Thus, it is not surprising that androgen withdrawal leads to objective response in 50% of the patients, while $\approx$ 80% exhibit subjective improvement (DENIS, 1988a).

Reponses to testicular androgen deprivation are not always complete and are frequently transient because, after orchidectomy, the adrenals become the primary source of androgens, and sufficient androgens are produced to promote the growth of the hormone-dependent metastatic prostate cancer cell (HAVLIN and TRUMP, 1988; TURKES and GRIFFITHS, 1989; MARTINI et al., 1990). In patients with relapsed prostate cancer, adrenal ablation is therefore advocated as a second-line therapeutic measure (DENIS, 1988b). However, the high morbidity and the short survival (6–12 months) associated with adrenalectomy, has prompted the consideration and use of inhibitors of androgen biosynthesis (BLANKESTEIN and BAKKER, 1985; DENIS, 1988b; HAVLIN and TRUMP, 1988; TURKES and GRIFFITHS, 1989). Since many of these are also aromatase inhibitors (HAVLIN and TRUMP, 1988; TURKES and GRIFFITHS, 1989; VAN WAUVE and JANSSEN, 1989), and since estrogens are well recognized prostatic growth stimulators (BELANGER et al., 1990), the dual benefit of combined androgen and estrogen deprivation if any, remains to be determined. In this context, it is noteworthy that very recently the aromatase inhibitor 4-OHA was found to palliate patients with advanced hormone resistant prostatic cancer by an as yet to be elucidated mechanism of action (SHEARER et al., 1990).

4.3.1. Inhibitors of androgenic steroid biosynthesis

As already discussed, DHT, the 5α-reduced testosterone, is the biologically active androgen in the prostatic tissue, and may be generated from several other androgenic precursors (androstenedione, DHEA) (HAVLIN and TRUMP, 1988; TURKES and GRIFFITHS, 1989; and references therein; BARTSCH et al., 1990). Although 5α-reductase is a key enzyme targeted in the rational design of inhibitors of steroid synthesis, enzymes such as the *17α-hydroxylase/C_{17-20} lyase* and other steroidogenic enzymes that generate androgenic precursors (Fig. 4) are obviously also of therapeutic importance (HAVLIN and TRUMP, 1988; TURKES and GRIFFITHS, 1989; VAN WAUVE and JANSSEN, 1989; and references therein; MARTINI et al., 1990). Agents targeted against such steroidogenic enzymes include:

Aminoglutethimide (AG): As discussed earlier, AG not only inhibits aromatase and the 18-hydroxylase, but also P450 XIA1 (the cholesterol side chain cleavage enzyme), P450XXIA2 (21-hydroxylase) and P450XIB1 (11β-hydroxylase), key enzymes in cortisol and androgen production (DEXTER et al., 1967; COHEN and FOA, 1969). Thus, it appears to be an excellent inhibitor of adrenal steroidogenesis. Indeed, AG causes significant tumor regression in rats with transplantable Dunning prostatic carcinoma (a commonly accepted model of an androgen-dependent tumor), comparable to that achieved with adrenalectomy alone (SHESSEL et al., 1980). It is also known to affect testicular steroidogenesis in normal men (SANTEN et al., 1974, 1979; SOMOJLIK et al., 1977, 1985). These combined features prompted its clinical trials in patients with advanced prostate cancer (SANFORD et al., 1976; ROSTOM et al., 1982; WORGUL et al., 1983). In earlier studies, AG (500–1000 mg/day) therapy of patients without previous treatment as well as patients who did not respond to estrogen therapy appeared quite promising in terms of both subjective and objective responses, but the high incidence of side effects precluded its recommendation for clinical use (ROBINSON et al., 1974). The report that glucocorticoid supplementation improved tolerance to AG in women with breast cancer (LIPTON and SANTEN, 1974) also renewed interest in such combined therapy for patients with advanced prostate cancer. In major clinical trials in patients with advanced prostate cancer who had failed primary treatment with orchidectomy, estrogens, or both treatments, a mean objective response of 28% (range 0–40%), and a mean subjective response of 38.5% were recorded after daily treatments of AG (1000 mg/day)/hydrocortisone (40 mg/day) (BLOCK, et al., 1984; DRAGO et al., 1984; PONDER et al., 1984; MURRAY and PITT, 1985; HAVLIN and TRUMP, 1988). An overall (subjective/objective) response rate of 45% was also observed with comparable treatment in a fifth study with 37 patients (SOMOJLIK et al., 1985). In that study, AG/hydrocortisone treatment also lowered serum DHT, DHEA-S and testosterone levels by $\approx$ 50% below control levels, thereby attesting to effective suppression of adrenal androgens (SAMOJLIK et al., 1985). Thus, AG, provided it is coadministered with a glucocorticoid, indeed appears to have some beneficial effect in the treatment of advanced prostate carcinoma, even if largely subjective in response character (HAVLIN and TRUMP, 1988). However, the high incidence of side-effects, principally lethargy, skin rashes and hypotension (HAVLIN and TRUMP, 1988; and references therein), limits its usefulness and necessitates the development of alternatives, including analogs such as *pyridoglutethimide*, which to our knowledge has yet to be tried clinically as an alternative.

Other related prospective candidates include the 4-pyridylacetic acid esters, such as *trans-4-methylcyclohexyl 4-pyridylpropanoate*, a relatively potent inhibitor of the rat testicular *17α-hydroxylase/C_{17-20} lyase*, which also inhibits human placental aromatase (McCague et al., 1990; Laughton et al., 1990). On the other hand, the *(1S,2S,3S,5R)-(+)-isopinocampheyl 4-pyridylacetic acid ester*, a relatively selective inhibitor of the lyase, is more potent than ketoconazole (McCague et al., 1990). Similarly, the *4-cyclohexyl-2-methyl-2-(4-pyridyl) propanoate* also appears to inhibit *17α-hydroxylase/C_{17-20} lyase* considerably (8-fold) more selectively than aromatase (McCague et al., 1990). In spite of their in vitro efficacy, their in vivo potential may be compromised not only by cortisol depletion (a common feature of 17α-hydroxylase inhibitors), which can be circumvented by replacement therapy, but also by their hydrolysis by ubiquitous esterases.

Ketoconazole: The potential usefulness of this well known, orally effective antimycotic agent in men with prostate carcinoma was gleaned from its antisteroidogenic side-effects of gynecomastia and/or breast tenderness in patients treated with multiple high doses (400–1200 mg/day) for valley fever or CNS fungal infections (De Felipe et al., 1981; Pont et al., 1982, 1984; Van Tyle, 1984; Sugar et al., 1987). Further scrutiny indeed revealed that after acute administration, the drug rapidly, albeit reversibly, lowered plasma androgen levels in normal volunteers (Santen et al., 1983). This response, however, could be both considerably magnified and prolonged by doubling the drug dose as well as by increasing the dosing frequency. The concomitant accumulation of plasma 17α-hydroxyprogesterone implicated the 17,20 lyase as the potential target, a finding subsequently confirmed in vitro with microsomes from rat testis and bovine adrenal cortex (Vanden Bossche et al., 1987), and with dispersed rat testicular cell cultures (Pont et al., 1982; De Coster, 1984). The drug is reportedly also cytotoxic to prostate carcinoma cells (Eichenberger et al., 1989). These findings confirm that ketoconazole is an effective and potent inhibitor of androgen biosynthesis (Loose et al., 1983) if given in sufficiently high doses, and prompted its exploration in the high dose form for prospective treatment of patients with advanced prostate cancer who had elapsed after orchidectomy or other treatments (Trachtenberg et al., 1983; Trachtenberg, 1984; Tapazoglou et al., 1986; De Coster et al., 1987; Havlin and Trump, 1987).

Clinically, high dose ketoconazole appears to be considerably more effective than AG, as evidenced in a population of 400 patients with advanced prostate cancer, half of which had never been pretreated and the other half of which had elapsed after other hormonal treatments (Amery et al., 1986; De Coster et al., 1986; Tapazoglou et al., 1986; Denis, 1988b). The objective response rates of 83 and 71% in the first group, and 58 and 34% in the relapsed patient group, after 3 and 6 months of high dose ketoconazole, respectively, indicate relatively good therapeutic efficacy for this drug. Subjective responses also were apparently high. Although generally well-tolerated, the side effects of nausea and gastrointestinal intolerance (Sugar et al., 1987) may limit its widespread adoption for this purpose, and has led to the search for better alternatives.

R 75251 [5-[(3-chlorophenyl)(1-imidazol-1-yl)-methyl]-1H-benzimidazole]: This agent is a recently developed imidazole derivative that inhibits both peripheral aromatization as well as testicular androgen biosynthesis (BRUYNSEELS et al., 1990). In rats with transplantable Dunning prostatic carcinoma, R75251 administered in the diet at 80 mg/100 g food/day, failed to lower serum testosterone levels, but comparably reduced the growth rate and the weight of the subcutaneous tumors to the same extent as castration (VAN GINCKEL et al., 1990). Ketoconazole, apparently exhibits similar tumor-reduction potential without lowering testosterone levels (HEYNS et al., 1985; DE COSTER et al., 1986; PONT, 1987; VAN GINCKEL et al., 1990) raising the suspicion that both of these drugs exert direct antitumorigenic activity. The basis for such direct antitumorigenic action is unknown. However, in addition to inhibiting steroidogenesis, these imidazoles also inhibit P450-dependent hydroxylation of retinoic acid (a key regulator of cellular proliferation and differentiation) to its physiologically inactive 4-hydroxylated metabolite (VAN WAUWE et al., 1988, 1990; VAN WAUWE and JANSSEN, 1989). Since retinoic acid is known to exhibit both anticarcinogenic and antitumoral effects, it is postulated that the beneficial effects of both imidazoles might be mediated via cellular conservation of retinoic acid, a possibility currently being investigated (VAN WAUWE et al., 1988; 1990; VAN WAUWE and JANSSEN, 1989). Furthermore, assessment of direct antitumorigenic effects of R75251 might be important clinically. Preclinical studies in male healthy human volunteers (BRUYNSEELS et al., 1990) indicate that while R75251 reduces testicular androgen levels to levels achieved by orchidectomy, it failed to affect adrenal androgen levels, and thus in principle would not qualify as an effective androgen-lowering agent.

4.3.2. Other antiprostatic agents

Several hydroxyperfluoroazobenzenes have been synthesized and evaluated as potential antiprostatic agents (JARMAN et al., 1988, 1990). Among these, *4-hydroxynonafluoroazobenzene* and *2-hydroxynonafluoroazobenzene*, proved to be inhibitors of the rat testicular 17α-hydroxylase/C_{17-20} lyase, comparable to ketoconazole at 100μM concentrations. 2-Hydroxynonafluoroazobenzene in particular was also a relatively potent and selective inhibitor of the C17,20 lyase reaction. A similar preferential inhibitory profile was also noted with *octafluorodibenz[b, f][1,4,5]oxadiazepine* (JARMAN et al., 1990). On the other hand, the successful design of *17β-(cyclopropyl-amino)-androst-5-en-3β-ol (MDL27,302)* as a selective mechanism-based inactivator of human testicular 17α-hydroxylase may instigate the search for other similarly acting structural derivatives (ANGELASTRO et al., 1989).

4.4. P450 inhibitors for treatment of fungal infections

Although the human environment is rife with fungi such as the airborne *Crypto-*

coccus and *Aspergillus* and the soil-denizen *Coccidioides immitis*, and fungi such as Candida are even harboured in the gastrointestinal tract, humans seldom succumb to fungal infections, largely because the body normally is immunocompetent and eminently capable of successfully warding off such attacks (KIRKPATRICK, 1984). However, pathogenic as well as opportunistic [systemic candidiasis, oral candidiasis (thrush)] infections commonly occur (BODEY, 1982, 1984; DREIZEN, 1984). For instance, over 50% of clinical visits to gynecologists apparently are instigated by candidal vulvovaginitis (SOBEL, 1988). Additional concerns include the currently increased incidence of fungal pathogenesis in patient populations that have cancer, are pathologically (leukemia, diabetes, AIDS, ARC), therapeutically (immunosuppressants, intensive chemotherapy) or surgically (bone-marrow or organ transplants) immunocompromised, and/or otherwise physiologically (old age, debilitation) compromised (BODEY, 1982, 1984, MAKSYMIUK et al., 1984; ST. GEORGIEV, 1988; VANDEN BOSSCHE et al., 1990b). The high incidence (58%) of various opportunistic mycoses (thrush, candidal esophagitis and cryptococcal meningitis) among the patients of the ever-growing AIDS epidemic (CHANDLER, 1985; KOVACS et al., 1985) alone argues the need for effective approaches to anti-fungal therapy that are fungicidal rather than merely fungistatic. For full effectiveness, fungistatic therapy requires prolonged treatments which may not always be well tolerated by the host tissues or to which they may develop resistance. This fact together with the high prevalence of additional common fungal pathogenesis (vaginal candidiasis, dermatomycoses) and the dramatic increased incidence of all forms of fungal infections, present a formidable challenge as well as an impetus to the rational development of anti-fungal strategies.

Fungi such as *Saccharomyces cerevisiae, Candida albicans* and *Candida tropicalis* require ergosterol, a vital plasma membrane component, for their normal aerobic proliferation and growth (KALB et al., 1987; VANDEN BOSSCHE, 1989; and references therein). Exogenous supplementation of ergosterol to anaerobic strains of these organisms can in fact adapt them to normal aerobic life (KALB et al., 1987; VANDEN BOSSCHE, 1989). Under normal aerobiosis, the required ergosterol is biosynthesized via oxidative removal of the 14α-methyl group of lanosterol (OHBA et al., 1978) in a key three-step reaction catalyzed by microsomal P450DM (LI) (mechanism depicted in Fig. 1; AYOMA et al., 1984, 1987a), a protein encoded by a single gene (KALB et al., 1986). Impairment of ergosterol synthesis results not only in the accumulation of 14α-methyl sterols and membrane perturbation, but also in the consequent uncoordinated synthesis of chitin, a primary cell wall constituent essential for the morphogenetic transformation of fungi to the invasive pseudomycelial phase (VANDEN BOSSCHE et al., 1986). The absence of chitin in plants and mammals, also makes chitin synthase an additional target for strategic antifungal drug development (KERRIDGE and VANDEN BOSSCHE, 1990). Although inhibitors of ergosterol synthesis primarily are fungistatic, their secondary ability to cause gross cell membrane alterations and leak-

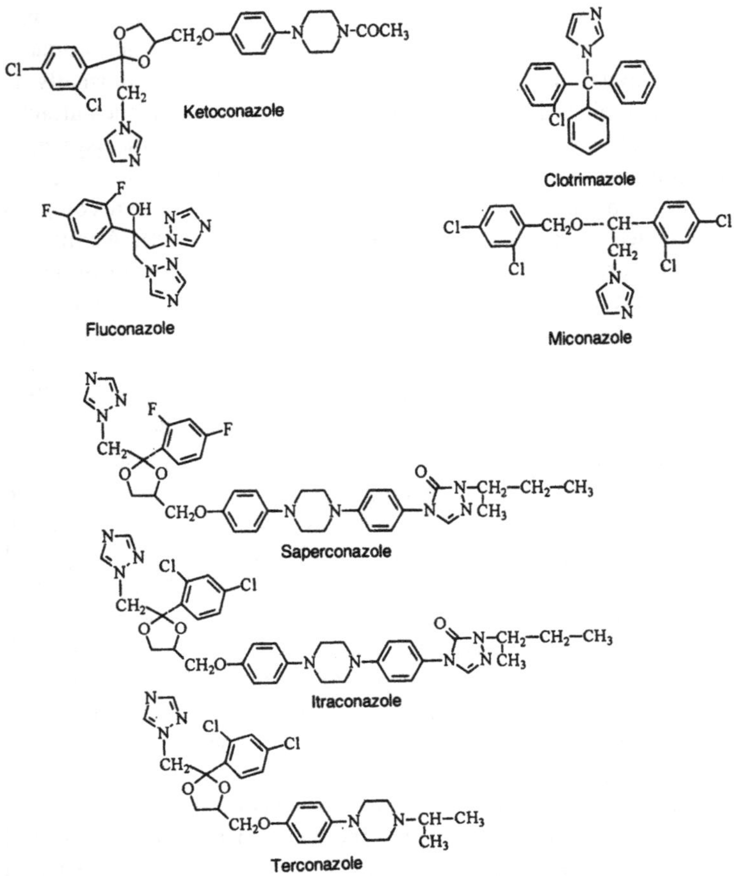

Fig. 13. Structures of some antifungals in clinical use.

age of cell constituents and/or accumulation of toxic substances can be fungicidal (Uno et al., 1982; Polak, 1990). The critical nature of P450-dependent ergosterol synthesis for normal/pathogenic fungal growth and proliferation makes it a focal point for antifungal attack (Yoshida, 1988; Saksena et al., 1989).

4.4.1. 14α-demethylase azole inhibitors

Approaches to antifungal development have largely centered on chemical inhibition of this particular P450-dependent step (Saksena et al., 1989). Specifically designed inhibitors (Fig. 13) include compounds with structural features that (i) increase their lipophilicity and hence their persistence at the P450 substrate binding site, and (ii) contain a function such as an imidazole nitrogen which functionally inhibits the enzyme by complexing with the P450 prosthetic heme iron to give the typical "Type II spectrum", thereby interfering in O_2 binding and activation (Fig. 5). The successful exploitation of the fungistatic potential of pyrimidines such as *Triarimol, Fenarimol, Nuarimol*, pyridine compounds such as *Buthiobate*, and the piperazine *Triforine* in agricultural practice, as well as the clinical efficacy of imidazoles such as the topically active **tioconazole, miconazole, econazole** and **clotrimazole**, reinforced the merit of this strategy (Sisler and Ragsdale, 1983; Vanden Bossche, 1988a; Vanden Bossche et al., 1988a,b). **Miconazole**, for instance, was found quite effective in the treatment of coccidioidomycosis and in disseminated candidiasis (Stevens, 1977, 1983a,b), and when injected intravenously is successful even in the treatment of coccidioidal and/or cryptococcus meningitis (Deresinski et al., 1977). In addition to the fungistatic properties derived from its blockade of the 14α-demethylase, in high enough concentrations miconazole is also believed to be fungicidal, partly because of its ability to elevate intracellular H_2O_2 through its inhibition of both peroxidase and catalase (Vanden Bossche et al., 1982; Polak, 1990), and partly because of its deleterious effects on plasma membrane integrity resulting from its insertion into the membrane lipid bilayer (Brasseur et al., 1983).

However, the extensive and rapid metabolism of agents such as miconazole, and/or their relative water insolubility require multiple intravenous dosing and hence hospitalization which limits their utility to topical application in dermal/intravaginal infections (Graybill, 1990). A significant step in this direction was the introduction in 1978, by Janssen Pharmaceuticals, of **Ketoconazole**, an orally active, relatively broad spectrum, longer acting ($t_{1/2} \approx 1-11$ h) and effective imidazole in several animal models of fungal pathogenesis (Thienpont et al., 1979; Van Cutsem et al., 1983). The drug not only directly inhibits the 14α-demethylase activity but also, through its inhibition of mitochondrial cytochrome c oxidase, is believed to impair respiration and induce anaerobiosis, and thus to synergize the block in ergosterol synthesis (Uno et al., 1982; Polak, 1990). Unlike other imidazoles, it is relatively more polar and does not induce hepatic P450s, but it is highly protein bound (Craven et al., 1983; Graybill, 1990). Although its absorption at normal (low) intragastric pHs is good, its penetration into the cerebrospinal fluid is poor, thus requiring high doses for the treatment of coccidioidal meningitis (Craven et al., 1983; Graybill, 1990). Such high doses revealed its side effects of gastrointestinal intolerance (Sugar et al., 1987; Graybill, 1990), hepatotoxicity, potential for adverse drug–drug interactions (notably with cyclosporin A), oligospermia, impotence, and gynecomastia, the latter three due to its interactions with other steroidogenic P450s (successfully exploited for therapy of prostate cancer, vide supra) (Pont et al., 1984; Britton et al., 1988; Vanden Bossche, 1990b). In spite of its drawbacks and somewhat unacceptable metabolism and elimination, its oral efficacy in blocking ergosterol synthesis and its consequent fungistatic properties, have largely fueled the search for other orally but longer acting azole derivatives.

These pharmacokinetic considerations have led to the development of triazoles such as **terconazole**, which is metabolically more stable, more potent, and longer acting than miconazole or clotrimazole. Its hydrophobic tail is believed to contribute both to its longer persistence at the P450 substrate binding site and, through direct insertion into the fungal membrane phospholipids, to the membrane disruption and fungicidal potential of the drug. Although also a relatively non-toxic drug, its rapid metabolism and elimination limits its clinical use to topical applications in the treatment of vulvovaginal candidiasis (SOBEL, 1988; CAUWENBERGH and VANDEN BOSSCHE, 1988).

Itraconazole is a broad spectrum, orally active, long-acting ($t_{1/2} \approx 19\,h$) triazole with highly fungistatic as well as fungicidal properties against the *Aspergillus* species (VAN CUTSEM et al., 1987; VAN CAUTEREN et al., 1987). It is clearly not only more potent than ketoconazole but also exhibits considerably fewer and relatively minor side effects (GRAYBILL, 1990). In addition to complexing with the P450 prosthetic heme-iron, it appears to bind more tenaciously at the substrate binding site, thereby contributing to its high affinity for the enzyme. Although such interactions contribute to ergosterol depletion and its fungistatic potential, they do not account for its fungicidal properties (POLAK, 1990). It remains to be elucidated whether the latter actually result from its specific orientation in the membrane lipid bilayer. It is the drug of choice for serious fungal infections such as paracoccidioidomycosis and chromoblastomycosis (GRAYBILL, 1990).

Saperconazole, a close analog of itraconazole wherein the two chloro groups are replaced by fluoro groups, is another broad spectrum antifungal agent currently in Phase II studies. It is believed to display an antifungal activity spectrum superior to that of fluconazole (VAN CUTSEM et al., 1988, 1989; RILEY, 1990), as well as a sufficiently better lipophilic/hydrophilic profile than itraconazole (VANDEN BOSSCHE et al., 1990a; RYLEY, 1990).

Fluconazole, another direct result of a systematic series of structure-function studies, is a water soluble, long-acting ($t_{1/2} \approx 25\,h$) and poorly metabolized azole antifungal agent (RICHARDSON et al., 1988). The replacement of the imidazole with the triazole moiety led to its relative selectivity for P450LI versus other steroidogenic P450s. The additional triazole moiety blocks its metabolism and increases its plasma levels while at the same time reducing plasma protein binding, thus maintaining high levels of "free" drug. The difluorophenyl moiety also contributes to its lowered protein binding and higher water solubility (RICHARDSON et al., 1988). The drug is remarkable in that it is both orally and intravenously active, as well as highly clinically effective (93–99% cure-rate), against vaginal candidiasis for up to 6 weeks after a single dose (HUMPHREY et al., 1985; RICHARDSON et al., 1988). More importantly, in addition to its high clinical effectiveness against a host of skin fungal infections, it is also highly efficacious against fungal infections in immunocompromised AIDS, leukemia and cancer patients. Its high CSF/serum concentrations relative to those of ketoconazole and itraconazole are encouraging for its potential use in the treatment of CNS infections (ARNDT et al., 1988; GRAYBILL, 1990). A cure rate of

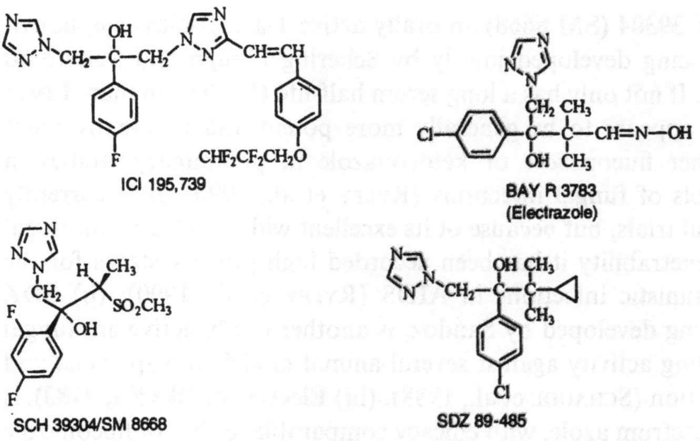

Fig. 14. Structures of anti-fungals in therapeutic development.

> 80% in AIDS patients with otherwise fatal cryptococcal meningitis certainly supports such great promise (GRAYBILL, 1990). Like itraconazole, its toxicity profile is also negligible.

ICI 195739 *[(RS)-2-(2,4-difluorophenyl)-1-(3-[(Z)-4-(2,2,3,3-tetrafluoro-propoxy)styryl]-1,2,4-triazol-yl-3-(1,2,4-triazol-1-yl)propan-2-ol)]*, is a novel 3′-styryl-substituted bistriazolyl tertiary alcohol, and is the most potent, wide spectrum, orally active antifungal agent in animal models of fungal pathogenesis (BOYLE et al., 1988; RYLEY et al., 1988). In common with ketoconazole, itraconazole and fluconazole, it is largely fungistatic rather than fungicidal, and its in vitro activity belies its true in vivo potential (RYLEY et al., 1990). However, in contrast to the other agents, it can lead to > 90% kill of C. parapsilosis and > 99% kill of the representative dermatophyte *Trichophyton rubrum* (RYLEY et al., 1990) at relatively low concentrations, thereby revealing its fungicidal potential against certain pathogenic fungi. In addition, it is also markedly active against *Histoplasma capsusalutum* and is the only azole with demonstrated activity against murine infections of *Blastomyces dermatitidis* (TUCKER et al., 1989; RYLEY et al., 1990). By virtue of its inhibition of sterol-demethylation, it also exhibits very good activity against *Trypanosoma cruzi*, the causative protozoan of Chagas disease prevalent in Latin America, and results in sterilization of chronic trypanosomal infections in mice (RYLEY et al., 1990). It appears to be rather selective for fungal 14α-demethylase, since it only poorly inhibits cholesterol biosynthesis and only interacts weakly with other mammalian steroidogenic P450s (BARRETT-BEE et al., 1990). These characteristics augur well for its clinical prospects as an antifungal agent.

4.4.2. Other 14α-demethylase azole inhibitors and related heterocyclic compounds

Promising new 14α-demethylase inhibitors awaiting in the therapeutic wings (Fig.

14) include: (i) **SCH 39304 (SM 8668)** an orally active 1-azolyl-alkyl-sulphonyl-alkanol derivative, being developed jointly by Schering Plough and Sumitomo (Okuda et al., 1988). It not only has a long serum half-life (57–93 h in man; Lin et al., 1989), but also appears to be generally more potent and frequently more efficacious than either fluconazole or ketoconazole in preliminary studies in several animal models of fungal infections (Ryley et al., 1990). It is currently in Phase II/III clinical trials, but because of its excellent wide spectrum antifungal activity and CSF penetrability it has been accorded high priority status for the treatment of opportunistic infections in AIDS (Ryley et al., 1990); (ii) **SDZ 89-485**, currently being developed by Sandoz, is another orally active antifungal azole, with outstanding activity against several animal models of superficial and systemic fungal infection (Schaude et al., 1988). (iii) **Electrazole (BAY R 3783)**, is yet another broad spectrum azole, with efficacy comparable to that of fluconazole but superior to that of either ketoconazole or itraconazole in several animal models (Plempel and Buchel, 1988; Hector and Yee, 1990). It is notable in that two of its primary metabolites (*BAY U 3624 and BAY U 3625*) are also fungistatic both in vitro and in vivo, and one of the metabolites (BAY U 3625) has a significantly longer half-life (48.3 h) than the parent compound ($t_{1/2}$, 2.1 h), and is considered to be the primary contributor to the anti-fungal activity of BAY R 3783 (Lee et al., 1989). Unfortunately, its prolonged half-life of 7 ± 3 days in human volunteers, was considered unacceptable and further clinical trials are presently suspended (Riley et al., 1990). (iv) **Vibunazole** *[(±)-1-(4-chlorophenoxy),3,3-dimethyl-2-(1,2,4-triazol-1-ylmethyl)butan-2-ol] (BAY N 7133)*, is an orally active broad-spectrum antifungal agent, comparable to ketoconazole in efficacy (Ritter et al., 1988). Its (-) isomer *(BAY 2302)* has considerably higher antifungal activity than the racemate and its pharmacokinetic and safety profiles are currently being evaluated in animal models (Ritter et al., 1988). (v) **Omoconazole** *[2-(1H-Imidazol-1yl)vinyl ether]*, an econazole derivative, is another orally active antifungal imidazole which appears to have the broad spectrum antifungal activity typical of the azole antimycotics, but is considerably safer and longer acting than econazole, miconazole or clotrimazole (Zirngibl et al., 1988). (vi) **Isoxazolidines:** A series of novel substituted *5-phenoxyalkyl-3-phenyl-(1H-imidazol-1-ylmethyl)-2-methylisoxazolidines* have been synthesized and found to be quite active in vitro against a broad spectrum of pathogenic yeasts. The highest potency was displayed by compounds halogenated in either or both phenyl moieties. Those with the 5-(4-halophenoxyalkyl) or 3-(4-fluoro or methoxy) group clearly are the best (Mullen et al., 1988a). Several *3,5-diphenyl-(1H-imidazol-1-ylmethyl)-2-methylisoxazolidine* derivatives also exhibit excellent in vitro antifungal activity against a wide spectrum of pathogenic fungi, with the dichloroanalogs [3-(4-chloro-), 5-(2-chloro-)] and [3-(4-chloro-), 5-(4-chloro-)] proving to be the most active and nominally superior to ketoconazole (De Cory et al., 1988a). Similarly, this team also evaluated a series of *cis-3-(substituted phenyl)-3-(1H-imidazol-1-ylmethyl)-2-methyl-5-{[(substituted phenyl)*

thio (or amino)]methyl} *isoxazolidine derivatives*, with most members of that series exhibiting relatively potent in vitro antifungal activity comparable to that of ketoconazole (DE CORY et al., 1988b). *Cis-3-(4-chlorophenyl)-3-(1H-triazol-1-ylmethyl)-2-methyl-5-(4-chlorophenyl) methylisoxazolidine* and the corresponding *imidazolyl derivatives (PR 988-399 and PR 967-234*, respectively) were found to be comparable to ketoconazole in two different models of rat vaginal candidiasis (BENNET et al., 1988a; MULLEN et al., 1988b). In contrast, the antifungal activity of 3-(substituted phenyl)-3-(1H-1,2,4-triazol-1-ylmethyl)-2-methyl-5-[(substituted phenoxy) methyl]- isoxazolidine derivatives or cis-5-alkoxy-alkyl 3-(substituted phenyl)-3-(1H-imidazol-1-ylmethyl)-2-methyl-5-[(substituted phenoxy) methyl]-isoxazolidine derivatives was relatively weak (BENNET et al., 1988a,b). Several novel *5-phenyl(or phenoxyalkyl)-3-(2-furanyl)-3-(1H-*imidazol-1-ylmethyl)-2-methylisoxazolidine or 5-(phenyl or phenoxy-alkyl)-3-(2-thienyl)-3-(1H-imidazol-1-ylmethyl)-2-methylisoxazolidine derivatives were also found to exhibit potent in vitro antifungal activity comparable to that of ketoconazole (BENNETT et al., 1988b; MULLEN et al., 1988c). Although, in general, the in vitro antifungal profiles of certain isoxazolidine derivatives are comparable to those of ketoconazole, it appears that a few of these agents exhibit a marginally safer profile in preclinical testing in terms of hormonal, CNS and cardiovascular effects (PALMER et al., 1988). These features could be usefully exploited in the development of safer and more efficacious antifungal isoxazolidines. Other antifungal agents of note because of their structural diversity include the *dihydrobenzothiophene azoles (Schering 30566)* and the *pyrido[3,4-e)-1,2,4-triazines* and *related heterocycles* (Lederle), which have antifungal activity comparable to that of miconazole (RANE et al., 1988; REICH et al., 1988).

4.4.3. Aminomethyl-substituted lanosterol derivatives

C-14 aminomethyl and the C-14-imidazolylmethyl derivatives were designed and synthesized with the prospects that introduction of a P450-heme iron complexing function at the site of oxidation of lanosterol would increase the selectivity of the compound for inhibition of ergosterol biosynthesis. Indeed, in vitro, both the 14-aminomethyl- and 14-imidazolylmethyl derivatives are more potent (relative MIC values, μg/ml) than ketoconazole in inhibiting several Candida and dermatophyte infections (COOPER et al., 1989), and at 100μM, inhibit rat liver P450-dependent aminopyrine oxidation to a considerably lesser extent than ketoconazole (COOPER et al., 1988). However, the relative IC_{50} values for inhibition of ^{14}C-mevalonate incorporation into ergosterol reveal that the 14-imidazolyl-methyl derivative is comparable to ketoconazole, whereas the 14-amino-methyl-derivative is inferior (COOPER et al., 1989). Similar structural considerations have led to the synthesis of a series of *14α-methyl-15-aza-D-homosterols*, certain of which exhibit in vitro fungistatic potencies comparable to those of fluconazole, high fungicidal activities, and respectable in vivo efficacies in a murine model of candidadiasis (DOLLE et al., 1990). Similar considerations also have rationalized the synthesis and evaluation of compounds such as *7-oxo-24,25-dihydrolanosterol*, which competitively inhibits P450LI by blocking the ability of its oxyferrocomplex to accept the second electron and thereby effectively uncoupling its catalytic cycle (AOYAMA et al., 1987b). A series of *lanosterol epoxides (7,8-, 8,9-, 8,14-, 14,15-, and 32-oxiranyl-lanostan-3β-ol)*, also synthesized as potential inhibitors of P450LI, are expected to act through coordination of the epoxy-oxygen atom with the prosthetic heme iron (TUCK and ROBINSON, 1988).

4.4.4. Mechanism-based 14α-demethylase inhibitors

More recently, rational design of 14α-demethylase inhibitors has led to synthetic strategies which incorporate a latent oxidizable function at the C-14 position (or in its immediate vicinity), which can be activated to a potentially reactive intermediate and result in suicide inactivation of the enzyme (Section 3.2.). As already discussed under aromatase inhibition, such agents are designed to be highly selective and thus may be expected to be therapeutically potent as well as relatively free of the side-effects associated with the less discriminating P450 inhibitors (i.e. ketoconazole). Although, to our knowledge, no reports of their in vivo antifungal potency and hence of their therapeutic potential are available, several inhibitors have been synthesized and evaluated in vitro. These include lanosterol derivatives wherein the C-14 methyl is replaced by: (a) cyclopropylamino methyl (COOPER et al., 1988), CH_2CHF_2, $CH_2\text{-}C{\equiv}CH$ or CHF_2 (FRYE and ROBINSON, 1988, 1990), or $CHOHCH{=}CH_2$ or CHNOH (KO et al., 1987; MAGOLDA et al., 1987); or (b) propargylic functions [i.e. 32-(ethynyl)lanost-8-en-3β-ol] (ADAMS et al., 1990).

4.5. Potential therapeutic uses of cytochrome P450 inhibitors

4.5.1. Inhibition of retinoic acid metabolism

Vitamin A (retinol) has long been recognized as a critical regulator of various physiological functions such as vision, reproduction, as well as growth, differentiation, maintenance and regeneration of epithelial tissues such as the breast, skin, bladder, trachea, small intestine, lung, testes and prostate (FAVENNEC and CALS, 1988; GANGULY, 1989; DAWSON and OKAMURA, 1990; and references therein). In these tissues, all-trans-retinoic acid (RA), a biologically active metabolite of retinol, appears to play a critical role in the regulation of normal epithelial growth and differentiation (DOWLING and WALD, 1960; DE LUCA, 1985). This notion is reinforced not only by the epithelial growth impairment observed on vitamin A deprivation, but also by the reported ability of natural and synthetic retinoids to suppress the development of chemically induced cancers in those organs (FRAKER et al., 1984; MOON and ITRI, 1984; CHYTYL, 1984; SCHLEICHER et al., 1988; SHERMAN and TRUITT, 1988). However, these retinoid actions appear to be mechanistically versatile. For instance, retinoids not only impair carcinogenic initiation directly by lowering the binding of benzo(a)pyrene to rat liver, stomach and lung DNA, possibly through alteration of its metabolic activation (MCCARTHY et al., 1987), but also synergize the effects of tamoxifen in decreasing the cell proliferation of human mammary carcinoma (FONTANA, 1987) and thus may act as true "anti-promoters". RA is also a potent inhibitor of human and murine melanoma cell proliferation (LOTAN et al., 1981; LOTAN, 1986). Furthermore, in human prostatic tissue, RA is believed to impair the formation of DHT (the active androgenic prostatic cell proliferator) by direct competitive binding to testosterone 5α-reductase, the enzyme that generates it (JUTLEY et al., 1990).

Additionally, RA also appears to play a key role in cell differentiation (DE LUCA, 1985; GANGULY, 1989). Vitamin A deprivation apparently results in keratinization of normally non-keratinizing tissues such as the trachea, cornea, bronchi and urogenital tract, an effect that is reversed by RA supplementation (WOLFBACH and HOWE, 1925; PARNELL and SHERMAN, 1962; GANGULY, 1989; and references therein). Furthermore, the keratinizing metaplasia and hyperkeratosis of the skin induced by retinoid deficiency also reveal that RA may control hyperkeratotic conditions such as psoriasis (WOLFBACH and HOWE, 1925; VAHLQUIST, 1980; JANSSEN et al., 1989). The beneficial effects of RA analogs in other skin diseases such as non-androgenic acne (cystic and conglobata acne) and DARRIER's disease (PECK, 1984; GANGULY, 1989; and references therein; JANSSEN et al., 1989), also implicate a regulatory role for RA in skin pathogenesis.

However, the steady state levels of RA are usually very low, largely because of its extensive and rapid metabolism to polar, biologically inactive, metabolites which apparently compromises its biological efficacy (NAPOLI and MCCORMICK, 1981; SWANSON et al., 1981). Since the major route of its inactivation is conversion to the 4-oxo-product, which entails an initial 4-hydroxylation of the cyclohexenyl moiety (a documented P450 dependent function) (FROLICK et al., 1981; ROBERTS, 1981; LEO and LIEBER, 1985; LEO et al., 1989), inhibition of this reaction could lead to tissue RA conservation. Indeed, the non-specific P450 inhibitor ketoconazole inhibits the hamster liver microsomal P450-dependent 4-hydroxylation of RA at $\geq 0.6\,\mu M$ concentrations (VAN WAUVE et al., 1988), and, when topically applied to the rat epidermis, blocks the subsequent microsomal 4-hydroxylation of RA (VANDEN BOSSCHE et al., 1988c). Furthermore, in F-9 embryonal carcinoma cells, ketoconazole ($10\,\mu M$) not only significantly retards the half life of RA three-fold, but also concomitantly reduces the RA concentration required to produce half-maximal stimulation of cell differentiation from 1 to 0.3 nM (WILLIAMS and NAPOLI, 1987; NAPOLI, 1990). Such modulation of intracellular RA economy through inhibition of RA metabolism is proposed to rationalize the beneficial effects of ketoconazole in patients with scalp psoriasis (FARR et al., 1985). Oral administration of either ketoconazole or R-75 251, in contrast to other P450 inhibitors such as cimetidine, metyrapone, AG, itraconazole and saperconazole, also increases the plasma levels of intravenously injected or endogenously generated RA in rats (VAN WAUVE et al., 1988, 1990). Of the two inhibitors, R-75 251 is the more effective, and its elevation of the plasma RA level is proposed to synergize its beneficial anti-tumorigenic action on the growth of transplantable prostatic adenocarcinoma in rats (VAN GINKEL et al., 1990). More importantly, these findings clearly underscore the likelihood that the biological efficacy of RA may be considerably enhanced through inhibition of its metabolic inactivation.

To date, the P450s participating in the oxidative metabolism of RA have been identified as P450IIC7, P450IIC8, P450IIB1 and possibly P450IA1 (inferred from 3-methylcholanthrene induction of RA 4-hydroxylation on topical application of the carcinogen to the skin of neonatal rats) (LEO et al., 1984, 1989; VANDEN BOSSCHE et al., 1988c). Given the importance of this reaction in the cellular disposal of RA, selective inhibition/inactivation of the participating P450s could provide a viable target in the rational design of therapeutic agents aimed at elevating intracellular RA concentrations. Although such inactivators are presently unknown, the finding that cannabidiol, a mechanism-based inactivator of P450IIIA enzymes (BORNHEIM and CORREIA, 1990) also inactivates the purified mouse liver P450IIC7 ortholog, with consequent marked loss of RA 4-hydroxylase activity (LIU, BORNHEIM and CORREIA, preliminary findings), provides a precedent for the development of novel strategies to prospective mechanism-based inactivators.

4.5.2. Physiological and pathophysiological implications of arachidonic acid metabolism: prospects for therapy with P450 inhibitors

Arachidonic acid (AA) is the well recognized precursor of an array of important biological modulators such as the prostanoids, leukotrienes, lipoxins, and thromboxanes engaged in a myriad of diverse physiological processes (NEEDLEMAN et al., 1986). The generating pathways [i.e. cyclooxygenase (prostaglandin synthase/thromboxane synthase), lipoxygenase] and the intrinsic physiological roles of the individual products have been well characterized (NEEDLEMAN et al., 1986). More recently, the isolation and structural and biological characterization of oxidized AA metabolites from various tissues have led to the identification of yet another physiologically relevant pathway for AA metabolism catalyzed by P450 enzymes (FITZPATRICK and MURPHY, 1989; McGIFF, 1991; HIRT and JACOBSON, 1991; and references therein). Indeed, the in vitro reproducibility of AA metabolism in the presence of NADPH and molecular O_2, and its inhibition by CO, SKF-525A, metyrapone or ketoconazole as well as anti-P450 reductase lgG, are entirely consistent with the participation of one or more P450 forms (CAPDEVILA, 1981a,b; SCHWARTZMANN et al., 1987, 1988). The 2-3 fold stimulation of AA metabolism by kidney microsomes from rabbits treated with the P450-inducers 3-methylcholanthrene and β-naphthoflavone (SCHWARTZMAN et al., 1986), and the more direct documentation of similar catalyses by isolated human and rat liver or kidney P450s (CAPDEVILA et al., 1981a,b; SCHWARTZMAN et al., 1988; FALCK et al., 1990), leave little doubt of P450 participation in AA metabolism. Although in most other tissues, the forms involved in AA oxidation remain to be definitively identified and characterized, the role of the corneal epithelial P450 epoxygenase in AA metabolism has been established through immunoinhibition studies with an anti-liver P450 epoxygenase (SCHWARTZMAN et al., 1987, 1988, 1990).

Specifically, AA metabolites derived from P450-dependent oxidation of AA include (i) the regioisomeric 5,6; 8,9; 11,12, and 14,15 epoxyeicosatrienoic acids (EETs), which on further hydrolysis by epoxide hydrase yield the corresponding diols (DHTs); (ii) the allylic oxidation products 5-, 8-, 9-, 11-, 12- and 15-hydroxyeicosatetraenoic acids (HETEs); and (iii) the ω- and ω-1 alcohols (19- and 20-HETEs). P450-dependent AA oxidations are usually regiospecific and, although not always stereospecific, frequently only one enantiomer exhibits significant biological activity/potency (FITZPATRICK and MURPHY, 1989; KARARA et al., 1989; 1990; CAPDEVILA et al., 1990; MCGIFF, 1991; HIRT and JACOBSON, 1991). Furthermore, not only does the relative predominance of each oxidative pathway vary with the specific cell type within each organ, but the physiological role of individual metabolites may also differ somewhat within each different cell type.

Interactions of this AA metabolic pathway with the other two known pathways are also known to occur (FITZPATRICK and MURPHY, 1989; MCGIFF, 1991; HIRT and JACOBSON, 1991). In some tissues, the products of the cyclooxygenase (prostaglandins) or lipoxygenase (leukotrienes) pathways may be further activated or inactivated by P450-mediated oxidations. In others, P450 oxidized AA products may serve as substrates for the subsequent production of physiologically active secondary metabolites (HIRT et al., 1989). A case in point is the secondary endothelial cyclooxygenation of either 20-HETE or 5,6-EETs generated in the renal tubular cells (where they apparently control ion transport), to yield vasoactive prostanoids which influence renal blood flow (SCHWARTZMAN et al., 1989; MCGIFF, 1991). Additional examples include the vasodilatation of rat caudal artery (CARROLL et al., 1988), the vasoconstriction of rat aorta (ESCALENTE et al., 1989) or that of cerebral arterioles (ELLIS et al., 1990) following the secondary endothelial cyclooxygenation of either 5,6-EET or 20-HETE. Thus, P450-dependent AA metabolism may not only directly influence local physiological processes by generating the relevant ultimate modulators in situ, but by supplying the critical intermediates for alternative *non-P450 dependent* pathways of AA metabolism, it can also indirectly affect other distally located complementary functions. Thus, P450-dependent AA metabolites may effectively regulate the concerted functions of an organ by synchronizing the tasks of its individual cellular structures.

More importantly, alterations in P450-dependent AA metabolite levels and function in many tissues correlate with the corresponding changes in P450 content, consistent with P450 participation in these processes. To date, the biological effects of several P450 dependent AA metabolites have been elucidated, and undoubtedly more will be uncovered in the near future. Because such assessments have often relied on pharmacological doses of the individual metabolites, their physiological significance may be open to question. A common criterion employed in assessment of physiological significance is the biosynthesis and/or constitutive presence of the AA metabolite implicated in a particular physiological function at the site of action (FITZPATRICK and MURPHY, 1989). By this criterion, the better characterized biologically relevant metabolites include:

(i) 5,6 EET, an established potent vasodilator, whose involvement in intestinal hyperemia (PROCTOR et al., 1987), counteraction of burn-induced ischemia after thermal skin injury (PROCOR et al., 1988), modulation of cerebral microcirculation (ELLIS et al., 1990), and closure of the ductus arteriosus (COCEANI et al., 1984) has been documented. It also is proposed to be the second messenger in the angiotensin II induced natriuresis in the proximal tubular epithelium (ROMERO et al., 1990).

(ii) 12-HETEs appear to mediate key physiological processes in several tissues. Specifically, the more physiologically active and/or potent (R)-isomer is the principal P450-dependent AA metabolite formed in the corneal epithelium, wherein its potent and relatively selective inhibition of Na^+,K^+-ATPase may modulate fundamental ocular functions such as deturgescence of the cornea (ocular transparency) or aqueous humor secretion by the ciliary body, thereby playing a critical role in maintenance of ocular hypertension (SCHWARTZMAN et al., 1987, 1990). 12-HETEs are also the major P450-dependent AA metabolites in the rat kidney (JIM et al., 1982), including renal function by inhibiting renin release and increasing Na^+ excretion (QUILLEY and MCGIFF, 1990). The R-isomer is considerably more potent and longer acting and its potent inhibiton of renal Na^+,K^+-ATPase is believed to result in increased urinary volume and electrolyte excretion in rats (TAKAHASHI et al., 1988). Furthermore, 12(R)-HETE is also a potent renal vasoconstrictor in dogs (MA et al., 1990). Additionally, synthesis of 12-HETEs is also increased 5–10 fold by canine coronary artery stenosis along with increases in other AA metabolites (EETs and their vicinal DHTs, and 15-HETEs) (ROSOLOWSKY et al., 1990).

12(R)-hydroxy-5,8,14-eicosatrienoic acid [12(R)-HETrE], a secondary metabolite of 12(R)HETE appears to have considerably different biological properties, which may implicate it in pathophysiological ocular processes. The marked increases in the aqueous humor protein concentration and the 20-fold superior potency over PGE_2 in breaking down the ocular blood-aqueous barrier (MURPHY et al., 1988; MASFERRER et al., 1989; SCHWARTZMAN et al., 1990a,b), reveals its potent pro-inflammatory properties. Furthermore, it may also behave as an angiogenic factor in the induction of corneal vascularization and leukocytic infiltration, hallmarks of chronic corneal inflammation. Indeed, progression of the inflammatory response in a rabbit model of acute and chronic corneal inflammation was accompanied by dramatic increases in corneal 12(R)-HETrE levels (Schwartzman et al., 1990a,b).

(iii) ω- and ω-1 products: P450-dependent AA metabolism in the kidney yields ω- and ω-1 oxidized 19- and 20-HETEs as the major products (Morrison and Pascoe, 1981; Oliw et al., 1981; Capdevila et al., 1981a). 19- and 20-HETEs reportedly account for 36 and 90% of the total P450-oxidized AA products formed by the kidney cortex and medulla, respectively (Takahashi et al., 1990). The ω-oxidized 20-HETE and its secondary metabolite 20-COOH-AA are found to be the principal P450-dependent metabolites of AA in the renal medullary thick ascending limb of the loop of Henle (mTALH) (Escalante et al., 1991). Both 20-HETE and 20-COOH-AA impair ion transport by inhibiting the mTALH Na^+,K^+-ATPase as well as its furosemide-sensitive Na^+,K^+2Cl^- cotransporter (Escalante et al., 1991). Since 20-HETE is the precursor for 20-COOH-AA, the common effects of both AA-metabolites may be due to 20-COOH-AA formation. Moreover, as discussed above, 20-HETE, after subsequent endothelial cyclooxygenation, is converted to potent vasoconstricting metabolites (Escalente et al., 1989), identified as the labile prostaglandin endoperoxides (20-hydroxy-$PGF_{2\alpha}$ and 20-hydroxy-PGH_2; Schwartzman et al., 1989). On the other hand, the ω-1 oxidation product 19(S)-HETE (but not its R-enantiomer) is a stimulator of partially purified renal cortical Na^+,K^+-ATPase activity (Escalante et al., 1988), a property that qualifies 19(S)-HETE as a potential endogenous modulator of renal salt and water excretion.

The combined Na^+,K^+-ATPase stimulatory properties of 19-HETE and the potent vasoconstrictor properties of 20-HETE-derived metabolites have considerable pathophysiological implications (vide infra). Excessive generation/accumulation of these AA metabolites in the kidney would increase renal salt retention and contribute to renal hypertension. Indeed, the production of ω- and ω-1 oxidized AA metabolites, along with that of 11,12 EETs, is increased in the spontaneously hyertensive rat (SHR), a purported model of essential hypertension in humans (Sacerdoti et al., 1988). In these rats, renal P450 content and the formatin of ω- and ω-1 oxidized AA metabolites increase in parallel with enhanced Na^+ retention and the development of hypertension (Sacerdoti et al., 1988). Furthermore, consistent with the involvement of renal P450 in the generation of these hypertensive AA metabolites, renal P450 depletion with either $CoCl_2$ or $SnCl_2$ significantly reduces the renal formation of ω- and ω-1 oxidized AA metabolites (but not that of EETs) while increasing renal Na^+ excretion and normalizing the elevated blood pressure in these animals (Sacerdoti et al., 1989). Because impairment of renal excretion is believed to pivotally contribute to the development of systemic hypertension, reduction of P450-dependent AA metabolism with consequent stimulation of Na^+ excretion and the normalization of blood pressure is of therapeutic import. Furthermore, in this context it is noteworthy that heme arginate, a widely accepted agent (except in the US where it is currently an "orphan drug") in the clinical treatment of the acute hepatic porphyrias, and a considerably safer compound than either metal chloride, is found to also lower blood pressure in the SHR rats via a similar mechanism (LEVERE et al., 1990).

Additional examples of other alterations in P450-dependent AA metabolism exist, although their mechanistic significance, if any, in the pathogenesis of the particular condition is speculative. For instance, pregnancy-induced hypertension in women is associated with dramatic urinary elevations of the vicinal diols of 14,15 EET and 11,12 EET (CATELLA et al., 1990). Increased P450-dependent AA metabolism is also reported in hypertension induced by suprarenal aortic coarctation in rabbits (CARROLL et al., 1988b). As discussed above, 11,12 EET is also markedly elevated in the SHR model (SACERDOTI et al., 1988) and may also contribute to SHR development. In addition, both EETs and ω- and ω-1 metabolites are also increased in diabetes induced by streptozotocin or alloxan (BADR et al., 1988), and in rats with cyclosporine-induced renal vasoconstriction and nephrotoxicity (SACERDOTI et al., 1990). It remains to be established whether the elevations of P450-dependent AA metabolites represent the cause or consequence of the underlying pathologies.

Collectively, the findings from several research teams strengthen the evidence that P450-dependent AA metabolites play pivotal modulatory roles in normal physiological functions and by extension, in pathophysiological processes. Although, impairment of P450-dependent AA metabolism can clearly lead to disruption of many vital normal physiological systems in several organs and tissues, cumulative evidence also indicts elevated P450-dependent AA metabolism in pathogenesis. Specific examples of pathogenic contributions of P450-dependent AA metabolism include (i) the 12(R)-HETrE-induced corneal inflammation, increased membrane permeability, chemotaxis and angiogenesis (SCHWARTZMAN et al., 1989, 1990) and (ii) the hypertensive effects of the ω- and ω-1 oxidized AA metabolites and possibly of 11,12 EETs in essential hypertension (SACERDOTI et al., 1989), as well as of 14,15 EETs in pregnancy-induced hypertension (CATELLA et al., 1990). In some of these instances, the beneficial effects of either lowering P450 or inhibiting its catalytic function have been documented in experimental animal models. Although not all the agents used for that purpose can be therapeutically exploited, nevertheless the concept that inhibiton of P450-dependent AA metabolism can be clinically beneficial has considerable merit.

In addition to SKF 525-A, metyrapone, ketoconazole and clotrimazole, several inhibitors of P450-dependent AA metabolism, have been reported. These include specifically designed 5,6-NC (5-methoxymethyleneoxy-6-isocyanoeicosatrienoic acid and its regioisomer), 8,9-NC, (8-methoxy-methyleneoxy-9-isocyanoeicosatrienoic acid and its regio-isomer), 11,12-NC (11-methoxy-methyleneoxy-12-

isocyanoeicosatrienoic acid and its regioisomer), 14,15-NC, 14-methoxy-methyleneoxy-15-isocyanoeicosatrienoic acid and its regioisomer), and 14,15-EET aziridine. These compounds bind rat liver microsomal P450s yielding Type II difference spectra, and potently inhibit several P450-dependent AA oxidative pathways, albeit in a non-selective fashion (CAPDEVILA et al., 1988). Since many of these P450-dependent AA metabolites are physiologically relevant, their unwarranted inhibition could compromise vital physiologic pathways, in addition to those targeted for inhibition. This problem has been largely circumvented with the development of suicide inhibitors with greater selectivity for specific P450s through judicious incorporation of a latent reactive moiety such as an acetylenic function (ORTIZ DE MONTELLANO and REICH, 1984). Several of these inhibitors have been synthesized and found to be highly promising as relatively selective inactivators of P450-dependent rat liver fatty acid ω- and ω-1 hydroxylases (ORTIZ DE MONTELLANO and REICH, 1984), human PMN leukotriene B_4 ω-hydroxylases (SHAK et al., 1985), or rabbit lung prostaglandin ω- and ω-1 oxidases (MUERHOFF et al., 1989; MASTERS et al., 1989). Specifically, the 15-hexadecynoic acid (15-HDYA), the 18-carbon 17-octadecynoic acid (17-ODYA) and the 12-hydroxy-16-heptadecynoic acid (12-HHDYA) are excellent inhibitors of pregnant rabbit lung microsomal prostaglandin ω-hydroxylase. 12-HHDYA is also partially active in reducing the development of active tension in dog renal arteries in response to an elevation in transmural pressure (KAUSER et al., 1990). These preliminary findings suggest that inhibitors such as these or closely related analogs could in principle be strategically designed and developed as specific inactivators of individual P450-dependent AA metabolic pathways. Thus, in addition to their therapeutic promise, such inhibitors could serve as valuable probes for dissection of the physiological contributions of 19- and 20-HETEs versus those of the EETs to the development of SHR.

4.5.3. Prospects for P450 inhibitors in the treatment of hyperaldosteronism and hypertension

Aldosterone, a mineralocorticoid, critically regulates plasma electrolyte balance by promoting renal Na^+ retention and K^+ excretion by the distal tubule. It also increases Na^+ reabsorption by other cell membranes. Hypersecretion of this mineralocorticoid either due to adrenal tumors/ aldosterone-producing adenomas (i.e., primary hyperaldosteronism), or in response to diuretic-induced hypovolemia (i.e., secondary hyperaldosteronism), results in characteristic hypertension and edematous conditions. It also results in hypokalemia, which if severe, increases the incidence of cardiac arrhythmias. Although spironolactone has been the anti-mineralocorticoid of choice, its side-effects, specially hyperkalemia, have prompted the search for more specific agents to reduce and/or counteract aldosteronemia.

Aldosterone is biosynthesized from cholesterol in the zona glomerulosa of the adrenal cortex through a series of oxidations catalyzed by distinct steroidogenic P450s [side chain cleavage (XIA1); 21-hydroxylase (P450XXI); and 11β-hydroxylase (P450XIB)/18-hydroxylase)], with progesterone, 11-deoxycorticosterone (DOC) and corticosterone as essential intermediates (WATERMAN et al., 1986; and references therein). In attempts to reduce its synthesis, a rationally designed active-site directed 18-acetylenic deoxycorticosterone *[21-hydroxy-*13(-2-) propynyl)-18-nor-preg-4-ene-3,20-dione, MDL 19,347] was evaluated and found to be a promising suicide inactivator of the adrenal corticosterone 18-methyl oxidase (hydroxylase) (Ki ≈ 38 nM; $t_{1/2}$, 4.6 min) from rats and rhesus monkeys (JOHNSTON et al., 1990a, 1991). The compound is also effective in reducing plasma aldosterone levels in vivo (JOHNSTON et al., 1990a, 1991). Furthermore, the failure of MDL 19,347 to inhibit the 11β-hydroxylation of corticosterone and DOC in parallel not only indicates that this compound is a relatively selective inhibitor of aldosterone biosynthesis, but such functional dissociation of the C18-methyl oxidase and 11β-hydroxylase activities provides key evidence for their catalyses by two separate P450s (JOHNSTON et al., 1990a, 1991). Other rationally designed 18-substituted steroids (18 vinyl- and 18-ethynyl progesterone analogs) are also potent mechanism-based inactivators for the C-18 hydroxylase (VIGER et al., 1988a,b; 1989). These rationally developed inhibitors represent potential therapeutic agents in the management of conditions associated with hyperaldosteronism such as hypertension, hypokalemia and edema.

The enhanced urinary excretion of 19-nor-deoxycorticosterone (19-norDOC) and its secondary oxidized metabolites in experimental, genetic and human hypertension has implicated this potent mineralocoriticoid in the pathogenesis of hypertension (DALE et al., 1982; MELBY et al., 1985, 1987; GRIFFING et al., 1988, 1989, 1991). Demethylation of DOC to 19-norDOC via 19-hydroxylation enhances its mineralocorticoid and hypertensive effects. Because aromatizaton of androgens also entails 19-hydroxylation, aromatase inhibitors have been examined as prospective inactivators of 19-norDOC formation (MELBY et al., 1987; GRIFFING et al., 1988, 1989). Indeed, the clinically established aromatase inhibitor 4-OHA, selectively inactivated mitochondrial 19-norDOC formation in rat adrenal and human aldosterone-producing adenoma (GRIFFING et al., 1988). Subcutaneous administration of 4-OHA to SHR not only significantly lowered the increased 19-norDOC but also normalized the underlying hypertension in these rats (GRIFFING et al., 1988). Essentially similar effects were observed with another well tested aromatase inhibitor, 10-propargyl-androst-4-ene, 3,17-dione in SHR (MELBY et al., 1987). For similar reasons, 19-acetylenic androstenedione and 19-difluoroandrostene-dione when probed as active site-directed inactivators of adrenal mitochondrial 19-hydroxylase were found to be excellent inactivators (IC_{50} values, 1 and 13 μM, respectively), whereas 19-acetylenic-DOC was even more potent (IC_{50} values, 0.08 μM). Several C-4-hydroxylated steroids have been synthesized and tested for their potential to inhibit hamster adrenal 19-

hydroxylase, but only 4-OH-DOC is a reasonably good, albeit competitive, inhibitor of the enzyme (JOHNSTON et al., 1990b).

4.5.4. P450 inhibitors in other current uses or for prospective therapy

Space constraints do not permit detailed discussion of several other P450 inhibitors currently in therapeutic use or used for purposes other than those stated above. For instance, AG, in addition to its clinical use as an aromatase inhibitor, is also used to reduce steroid secretion in patients with adrenocortical malignancy. Similarly, the antifungal ketoconazole, because of its general inhibition of several steroidogenic P450s, is used for the treatment of CUSHING's disease, hypercortisolemia, male breast cancer, ectopic ACTH production and LHRH unresponsive testotoxicosis (VANDEN BOSSCHE et al., 1987; and references therein; GOLDFIEN, 1990). Metyrapone reduces cortisol and corticosterone synthesis through inhibition of P450 dependent-11β-hydroxylation and, although not widely used in treatment of CUSHING's disease, is useful in lowering glucocorticoid production in patients with adrenal tumors, ectopic ACTH syndromes and hyperplasia (GOLDFIEN, 1990). Mitotane (o,p'-DD; dichlorodiphenyl-dichloroethane, Lysodren), is principally used for the treatment of neoplasms of the adrenal cortex and the hyperadrenocorticism (CUSHING's syndrome) resulting from adrenal tumor or hyperplasia, as well as for the palliation of inoperable adrenocortical carcinoma (CALABRESI and CHABNER, 1991).

P450scc (XIA1), which converts cholesterol to pregnenolone (Fig. 3), and P450XXI, a key enzyme in the biosynthesis of either corticosterone or cortisol (WATERMAN et al., 1986), have also been targeted for inactivation as an approach to lowering steroidogenesis. A number of synthetic AG-congeners (azoxy- and hydrazo-derivatives) have been found to be relatively selective inhibitors of P450XIA1 since they failed to inhibit either aromatase or the 17,20 lyase (JARMAN et al., 1988). Similarly, the aminosteroids *(22, amino-23,24-bisnor-5-cholen-3β-ol or 23, amino-24-nor-5-cholen-3β-ol)* are potent competitive inhibitors of the adrenal enzyme from several species (Fig. 6; KELLIS et al., 1984). In addition, several excellent mechanism-based inhibitors of P450XIA have been rationally designed and evaluated for their ability to inactivate the enzyme. These include: (i) *20-(1,5-hexadiynyl)-5-pregnen-3β,20α-diol* (NAGAHISA et al., 1983); (ii) *20-(2-trimethyl-silyl)ethyl-pregn-5-en-3β,20α-diol* (NAGAHISA et al., 1984; KRUEGER et al., 1985); (iii) *20-(4-tetrahydropyranyl-1-butynyloxy)pregn-5-en-3β,20α-diol and 20-(3-tetrahydropyranyl-1-butynyloxy)pregn-5-en-3β,20α-diol* (OLAKANMI and SEYBERT, 1990); and (iv) *(20S)-22-Nor-22-thiacholesterol* (MIAO et al., 1990).

Specifically designed inhibitors of P450XXI include 21,21 dihaloprogesterones such as 21, 21-difluoroprogesterone which suicidally inactivates bovine adrenal C-21 hydroxylase activity ($K_{inact} \approx 0.22\,min^{-1}$) but not the corresponding rabbit enzyme, possibly through prosthetic heme alkylation (STEVENS et al., 1991). On the other hand, 17β-ethynylprogesterone also suicidally inactivates bovine adrenal C-21 hydroxylase, apparently through apocytochrome modification (STEVENS et al., 1991).

5. Conclusions

It should be evident from the foregoing discussion that the therapeutic exploitation of P450 inhibitors is clinically meritorious. It is particularly valuable in those instances wherein either the pathophysiological role of P450-dependent processes is well defined such as in the etiology of breast and prostate cancer, or the biochemical rationale for the intervention with P450 inhibitors is irrefutable, such as in the chemotherapy of fungal infections. Moreover, the increasing recognition that peripheral estrogens contribute to the pathophysiology of BPH, as well as the compelling evidence for the beneficial effects of aromatase inhibitors in experimental BPH, are currently championing the clinical exploitation of these inhibitors in BPH chemotherapy.

On the other hand, the potential etiological contributions of P450-dependent pathways in many other pathophysiological processes (hypertension, platelet aggregation, ocular inflammation) are only just beginning to be questioned and understood at an empirical level. Better definition of the inherent pathophysiological implications of P450-mediated reactions in these conditions should rationalize the strategic exploitation of P450 inhibitors in their clinical therapy.

Acknowledgement

The authors wish to thank Dr. H. VANDEN BOSSCHE, Janssen Research Foundation, Beerse, Belgium and Dr. J. O'NEAL JOHNSTON, Merrill-Dow Research Institute, Cincinnati, OH, for generously providing helpful published and preliminary information on P450 inhibitors. They also thank Professor Nader G. ABRAHAM, New York Medical College, Valhalla, N.Y., for his critical review of the section on P450-dependent arachidonic acid metabolism and Ms. Christine ZERWEZK for typing portions of this manuscript and literature retrieval. The authors also acknowledge the support of NIH grants DK-26506 (MAC) and GM-25515 (POM).

6. References

ABUL-HAJJ, Y. J., (1986), J. Med. Chem. **29**, 582–584.

ABUL-HAJJ, Y. J., (1989), J. Steroid Biochem. **34**, 439–442.

ABUL-HAJJ, Y. J., (1990), J. Steroid Biochem. **35**, 139–143.

ADAMS, J. L., M. J. BOSSARD, R. F. MAYER, T. F. GALLAGHER, T. A. TOMASZEK Jr., and B. W. METCALF, (1990), 199th ACS Meeting, MEDI Abst. No. 10.

Akhtar, M., M. R. Calder, D. L. Corina, and J. N. Wright, (1982), Biochem. J. **201**, 569–580.
Amery, W. K., R. De Coster, and I. Caers, (1986), Drug Dev. Res. **8**, 299–307.
Angelastro, M. R., M. E. Laughlin, G. L. Schatzman, P. Bey, and T. R. Blohm, (1989), Biochem. Biophys. Res. Commun. **162**, 1571–1577.
Aoyama, Y., Y. Yoshida, and R. Sato, (1984), J. Biol. Chem. **259**, 1661–1666.
Aoyama, Y., Y. Yoshida, and R. Sato, (1987a), J. Biol. Chem. **262**, 1239–1243.
Aoyama, Y., Y. Yoshida, Y. Sonoda, and Y. Sato, (1987b), Biochimica et Biophysica Acta **922**, 270–277.
Arndt, C. A. S., T. J. Walsh, C. L. McCully, F. M. Balis, P. A. Pizzo, and D. G. Poplack, (1988), J. Inf. Dis. **157**, 178–180.
Augusto, O., H. S. Beilan, and P. R. Ortiz de Montellano, (1982), J. Biol. Chem. **257**, 11288–11295.
Ayub, M. and M. J. Levell, (1989), J. Steroid Biochem. **32**, 515–524.
Badr, K., K. Takahashi and J. R. Falck, J. Capdevila, and H. R. Jacobson, (1988), Clin. Res. **36**, 513A.
Banting, L., P. J. Nicholls, M. A. Shaw, and H. J. Smith, (1989), in: Progress in Medicinal Chemistry, (G. P. Ellis and G. B West, eds.), Elsevier Science Publishers, 253–297.
Barone, R. M., I. M. Shamonki, P. K. Siiteri, and H. L. Judd, (1979), J. Clin. Endocrinol. Metab. **49**, 672–676.
Barrett-Bee, K., J. Lees, P. Pinder, J. Campbell, and L. Newboult, (1988), Ann. N.Y. Acad. Sci. **544**, 231–244.
Bartsch, W., H. Klein, H. Stürenburg, and K. Voigt, (1987), J. Steroid Biochem. **27**, 557–564.
Bartsch, W., H. Klein, U. Schiemann, H. W. Bauer, and K. D. Voigt, (1990), Ann. N. Y. Acad. Sci. **595**, 53–66.
Bednarski, P. J. and S. D. Nelson, (1989), J. Med. Chem. **32**, 203–213.
Bélanger, C., R. Veilleux, and F. Labrie, (1990), Ann. N. Y. Acad. Sci. **595**, 399–402.
Bennett, G. A., P. A. Swift, G. B. Mullen, S. D. Allen, J. T. Mitchell, C. R. Kinsolving, and V. St. Georgiev, (1988a), Ann. N.Y. Acad. Sci. **544**, 369–372.
Bennett, G. A., G. B. Mullen, J. T. Mitchell, S. D. Allen, C. R. Kinsolving, and V. St. Georgiev, (1988b), Ann. N. Y. Acad. Sci. **544**, 384–386.
Bennett, G. A., G. B. Mullen, T. R. DeCory, S. D. Allen, J. T. Mitchell, and V. St. Georgiev, (1988c), Ann. N. Y. Acad. Sci. **544**, 396–399.
Bennett, G. A., G. B. Mullen, S. D. Allen, J. T. Mitchell, and V. St. Georgiev, (1988d), Ann. N. Y. Acad. Sci. **544**, 400–402.
Bennett, G. A., P. A. Swift, G. B. Mullen, J. T. Mitchell, S. D. Allen, C. R. Kinsolving, and V. St. Georgiev, (1988), Ann. N.Y. Acad. Sci. **544**, 375–377.
Bhatnagar, A. S., K. Schieweck, A. Hausler, L. J. Browne, and R. E. Steele, (1989), Proc. Roy. Soc. Edin. **95B**, 293–303.
Bhatnagar, A. S., A. Häusler, K. Schieweck, L. J. Browne, R. Bowman, and R. E. Steele, (1990), J. Steroid Biochem. **37**, 363–367.
Birkoff, J. D., (1983), in: Benign Prostatic Hypertrophy (Hinman Jr., F., ed.), Springer, New York, 5–9.
Blankestein, M. A. and G. H. Bakker, (1985), in: Therapeutic principles in Metastatic Prostatic Cancer, (F. H. Schroeder and B. Richard, eds.), AR Liss, New York, 161–185.
Block, M., D. Trump, D. P. Rose, K. B. Cummings, and T. F. Hogan, (1984), Cancer Treat. Rep. **68**, 719–722.
Bodey, G. P., (1982), in: Cancer Medicine, (J. F. Holland and E. Frei, eds.), Lea & Febiger, Philadelphia, 1339–1372.
Bodey, G. P., (1984), Am. J. Med. **77(4D)**, 13–19.

BORNHEIM, L. M. and M. A. CORREIA, (1990), Mol. Pharmacol. **38**, 319–326.
BOYLE, F. T., D. D. GILMAN, M. B. GRAVESTOCK, and J. M. WARDLEWORTH, (1988), Ann. N. Y. Acad. Sci. **544**, 86–100.
BRANDT, M. K., D. PUETT, R. GAROLA, K. FENDL, D. F. COVEY, and S. J. SIMNISKI, (1988), Prog. Clin. Biol. Res. **262**, 65–84.
BRASSEUR, R., C. VANDENBOSCH, H. VANDEN BOSSCHE, and J. M. RUYSSCHAERT, (1983), Biochem. Pharmacol. **32**, 2175–2180.
BRITTON, H., Z. SHEBAB, E. LIGHTNER, M. NEW, and D. CHOW, (1988), J. Pediatrics **112**, 488–492.
BRODIE, A. M. H., D. A. MARSH, and H. J. BRODIE, (1979), J. Steroid Biochem. **10**, 423–429.
BRODIE, A. M. H., W. M. GARRETT, J. R. HENDRICKSON, C.-H. TSAI-MORRIS, P. A. MARCOTTE, and C. H. ROBINSON, (1981), Steroids **38**, 393–402.
BRODIE, A. M. H., H. J. BRODIE, W. M. GARRETT, J. R. HENDRICKSON, D. A. MARSH, and C. H. TSAI-MORRIS, (1982), Biochem. Pharmacol. **31**, 2017–2023.
BRODIE, A. M. H., W. M. GARRETT, and J. R. HENDRICKSON, (1983), J. Steroid. Biochem. **19**, 53–58.
BRODIE, A. M. H., L. Y. WING, P. GOSS, M. DOWSETT, and R. C. COOMBES, (1986), J. Steroid Biochem. **25**, 859–865.
BRODIE, A. M. H. and R. J. SANTEN, (1986), CRC Crit. Rev. Oncol. **5**, 361–396.
BRODIE, A. M. H., R. C. COOMBES, and M. DOWSETT, (1987), J. Steroid Biochem. **27**, 899–903.
BRODIE, A. M. H. and L. Y. WING, (1987), Steroids **50**, 89–103.
BRODIE, A. M. H., M. DOWSETT, and R. C. COOMBES, (1988), Cancer Treat. Res. **39**, 51–65.
BRODIE, A. M. H., C. SON, D. A. KING, K. M. MEYER, and S. E. INKSTER, (1989), Cancer Res. **49**, 6551–6555.
BRODIE, A. M. H., P. K. BANKS, S. E. INKSTER, M. DOWSETT, and R. C. COOMBES, (1990), J. Steroid Biochem. Molec. Biol. **37**, 327–333.
BRUEGGEMEIER, R. W., P. K. LI, C. E. SNIDER, M. V. DARBY, and N. E. KATLIC, (1987), Steroids **50**, 163–178.
BRUEGGEMEIER, R. W. and P. K. LI, (1988), Cancer Res. **48**, 6808–6810.
BRUEGGEMEIER, R. W., P. K. LI, H. H. CHEN, P. P. MOH, and N. E. KATLIC, (1990), J. Steroid Biochem. Molec. Biol. **37**, 379–385.
BRUEGGEMEIER, R. W. and N. E. KATLIC, (1990), Cancer Res. **50**, 3652–3656.
BRUYNSEELS, J., R. DE COSTER, P. VAN ROOY, W. WOUTERS, A. RAEYMAEKERS, E. FREYNE, G. SANZ, G. VANDEN BOSSCHE and P. A. J. JANSSEN, (1990), Prostate **16**, 345–357.
BURKHART, J. P., N. P. PEET, C. L. WRIGHT, and J. O'NEAL JOHNSTON, (1991), J. Med. Chem. **34**, 1348–1750.
BURSTEIN, S., B. S., MIDDLEDITCH, and M. GUT, (1975), J. Biol. Chem. **250**, 9028–9037.
BYON, C.-Y. and M. GUT, (1980), Biochem. Biophys. Res. Commun. **94**, 549–552.
CAJACOB, C. A., W. K. CHAN, E. SHEPHARD, and P. R. ORTIZ DE MONTELLANO, (1988), J. Biol. Chem. **263**, 18640–18649.
CALABRESI, P. and B. A. CHABNER, (1990), in: The Pharmacological Basis of Therapeutics, (A. G. GILMAN, T. W. RALL, A. S. NIES, P. TAYLOR, eds.), Pergamon Press, New York, 1202–1208.
CAPDEVILA, J., L. PARKHILL, N. CHACOS, R. OKITA, B. S. S. MASTERS, and R. W. ESTABROOK, (1981), Biochem. Biophys. Res. Commun. **101**, 1357–1363.
CAPDEVILA, J., N. CHACOS, J. WERRINGLOER, R. A. PROUGH, and R. W. ESTABROOK, (1981), Proc. Natl. Acad. Sci. USA **78**, 5362–5366.
CAPDEVILA, J., L. GIL, M. ORELLANA, L. J. MARNETT, J. I. MASON, P. YADAGIRI, and J. R. FALCK, (1988), Arch. Biochem. Biophys. **261**, 257–263.
CAPDEVILA, J. H., A. KARARA, D. J. WAXMAN, M. V. MARTIN, J. R. FALCK, and F. P. GUENGUERICH, (1990), J. Biol. Chem. **265**, 10865–10871.

Carroll, M. A., M. L. Schwartzmann, J. Capdevila, J. R. Falck, and J. C. McGiff, (1987), Eur. J. Pharm. **138**, 281–283.
Carroll, M. A., M. Schwartzman, M. Baba, M. J. S. Miller, and J. C. McGiff, (1988), Am. J. Physiol. **255**, 151–157.
Catella, F., J. A. Lawson, D. J. Fitzgerald, and G. A. Fitzgerald, (1990), Proc. Natl. Acad. Sci. USA **87**, 5893–5897.
Cauwenbergh, G. and H. Vanden Bossche, (1987), in: Clinical perspectives: terconazole, an advance in vulvovaginal candidiasis therapy, (J. D. Sobel, ed.), BMI/McGraw-Hill, Laguna Niguel, California, 19–26.
Cavalli, F. (1986), in: Endocrine therapy of breast cancer, Springer-Verlag, New York.
Chandler, F. W., (1985), in: Current Topics in Medical Mycology, (M. R. McGinnis, ed.), Springer-Verlag, New York, 1–23.
Chevalier, S., G. Turcotte, G. McKercher, P. Boulanger, and A. Chapdelaine, (1990), Ann. N. Y. Acad. Sci. **595**, 173–183.
Childers, W. E. and C. H. Robinson, (1987a), J. Chem. Soc. Chem. Commun. 320–321.
Childers, W. E., M.-J. Shih, P. S. Furth, and C. H. Robinson, (1987b), Steroids **50**, 121–134.
Chytil, F., (1984), Pharmacol. Rev. **36**, 93S.
Coceani, F., N. C. Hamilton, J. Labuc, and P. M. Olley, (1984), Am. J. Physiol. **246**, H640–643.
Cohen, M. P. and P. P. Foa, (1969), Proc. Soc. Exp. Biol. Med. **127**, 1086–1090.
Cole, P. A. and C. H. Robinson, (1988), J. Am. Chem. Soc. **110**, 1284–1285.
Cole, P. A., J. M. Bean, and C. H. Robinson, (1990), Proc. Natl. Acad. Sci. U.S.A. **87**, 2999–3003.
Cole, P. A. and C. H. Robinson, (1990), J. Med. Chem. **33**, 2933–2942.
Collman, J. P., J. I. Brauman, B. Meunier, S. A. Raybuck, and T. Kodadek, (1984), Proc. Natl. Acad. Sci. U.S.A. **81**, 3245–3248.
Collman, J. P., J. I. Brauman, P. D. Hampton, H. Tanaka, D. S. Bohle, and R. T. Hembre, (1990), J. Am. Chem. Soc. **112**, 7980–7984.
Coombes, R. C., P. E. Goss, M. Dowsett, G. Hutchinson, D. Cunningham, M. Jarman, and A. M. H. Brodie, (1987), Steroids **50**, 245–252.
Cooper, A. B., J. J. Wright, A. K. Ganguly, J. Desai, D. Loebenberg, R. Parmegiani, and D. S. Feingold, (1988), Ann. N. Y. Acad. Sci. **544**, 109–112.
Cooper, A. B., J. J. Wright, A. K. Ganguly, J. Desai, D. Loebenberg, R. Parmegiani, D. S. Feingold, and I. J. Sud, (1989), J. Chem. Soc., Chem. Commun. 898–900.
Covey, D. F., W. F. Hood, and V. D. Parikh, (1981), J. Biol. Chem. **256**, 1076–1079.
Covey, D. F. and W. F. Hood, (1981), Endocrinology **108**, 1597–1599.
Covey, D. F. and W. F. Hood, (1982). Mol. Pharmacol. **21**, 173–180.
Covey, D. F. and W. F. Hood, (1982), Cancer Res. **42 (suppl.)**, 3327–3333.
Covey, D. F., W. F. Hood, D. D. Beusen, and H. L. Carrell, (1984), Biochemistry **23**, 5398–5406.
Covey, D. F., (1988), in: Sterol Biosynthesis Inhibitors, (D. Berg and M. Plempel, eds.), Ellis Horwood Ltd., Chichester, England, 534–571.
Craven, P. C., J. R. Graybill, J. H. Jorgensen, W. E. Dismukes, and B. E. Levine, (1983), Ann. Intern. Med. **98**, 160–167.
Dale, S. L., M. Holbrook, P. Komanicky, and J. C. Melby, (1982), Endocrinology **110**, 1989–1993.
Daly, M. J., G. W. Jones, P. J. Nicholls, H. J. Smith, M. G. Rowlands, and M. A. Bunnet, (1986), J. Med. Chem. **229**, 520–523.
Dawson M. I. and W. H. Okamura, eds., (1990), Chemistry and Biology of Synthetic Retinoids, CRC Press, Inc., Boca Raton, FL, and references therein.

De Coster, R., D. Beerens, J. Dom, and G. Willemsens, (1984), Acta Endocrinol. **107**, 275–000.
De Coster, R., I. Caers, M. C. Coene, W. Amery, D. Beerens, and C. Haelterman, (1986), Clin. Endocrinol. **24**, 657–664.
De Coster, R., C. Mahler, L. Denis, M.-C. Coene, I. Caers, W. Amery, C. Haelterman, and D. Beerens, (1987), Acta. Endocrinol. **115**, 265–271.
De Coster, R., R. Van Ginckel, W. Wouters, N. Goeminne, W. Vanherck, and M. Byloos, (1990a), J. Enzyme Inhibition **4**, 159–167.
De Coster, R., W. Wouters, C. R. Bowden, H. Vanden Bossche, J. Bruynseels, R. W. Tuman, R. Van Ginckel, E. Snoeck, A. Van Peer, and P. A. J. Janssen, (1990b), J. Steroid Biochem. Molec. Biol. **37**, 335–341.
De Felice, R., D. G. Johnson, and J. N. Galgiani, (1981), Antimicrob. Agents Chemother. **19**, 1073–1074.
De Luca, H. F., (1985), in: Retinoids: new trends in Research and Therapy, (J. H. Saurat, ed.), Basel, Karger, 12–19.
Decker, C. J., M. S. Rashed, T. A. Baillie, D. Maltby, and M. A. Correia, (1989), Biochemistry **28**, 5128–5136.
DeCory, T. R., G. B. Mullen, J. T. Mitchell, S. D. Allen, C. R. Kinsolving, and V. St. Georgiev, (1988), Ann. N. Y. Acad. Sci. **544**, 363–365.
DeCory, T. R., P. A. Swift, G. B. Mullen, J. T. Mitchell, S. D. Allen, C. R. Kinsolving, and V. St. Georgiev, (1988), Ann. N. Y. Acad. Sci. **544**, 366–368.
De Matteis, F., C. Hollands, A. H. Gibbs, N. de Sa, and M. Rizzardini, (1982), FEBS Lett. **145**, 87–92.
Demers, L. M., J. C. Melby, T. E. Wilson, A. Lipton, H. A. Harvey, and R. J. Santen, (1990), J. Clin. Endocrinol. Metab. **70**, 1162–1166.
Denis, L. ed. (1988a), in: The Medical Management of Prostate Cancer, Springer-Verlag. New York. 1.
Denis L., (1988b), in: The Medical Management of Prostate Cancer, (L. Denis ed.), Springer-Verlag, New York, 59–64.
Deresinski, S. C., R. B. Lilly, H. G. Levine, J. N. Galgiani, and D. A. Stevens, (1977), Arch. Intern. Med. **137**, 1180–1185.
Dexter, R. N., L. M. Fishman, and G. W. Liddle, (1967), J. Clin. Endocrinol. Metab. **27**: 473–480.
Dickson, R. B., E. W. Thompson, and M. E. Lippman, (1990), J. Steroid Biochem. Molec. Biol. **37**, 305–316.
DiSalle, E., G. Briatico, D. Giudici, G. Ornati, and T. Zaccheo, (1989), J. Steroid Biochem. **34**, 431–434.
DiSalle, E., D. Giudici, G. Briatico, and G. Ornati, (1990a), Ann. N. Y. Acad. Sci. **595**, 357–367.
DiSalle, E., D. Giudici, G. Ornati, R. Briatico, R. D'Alessio, V. Villa, and P. Lombardi, (1990b), J. Steroid Biochem. Molec. Biol. **37**, 369–374.
Distlerath, L. M. and F. P. Guengerich, (1987), in: Mammalian Cytochromes P-450, Vol. I, (F. P. Guengerich, ed.), CRC Press, Boca Raton, Florida, 133–198.
Dolle, R. E., H. S. Allaudeen, and L. I. Kruse, (1990), J. Med. Chem. **33**, 877–880.
Dowling, J. E. and G. Wald, (1960), Proc. Natl. Acad. Sci. USA **46**, 587–608.
Dowsett, M., P. E. Goss, T. J. Powles, G. Hutchinson, A. M. H. Brodie, S. L. Jeffcoate, and R. C. Coombes, (1987), Cancer Res. **47**, 1957–1961.
Drago, J. R., R. J. Santen, A. Lipton, T. J. Worgul, H. A. Harvey, A. Boucher, A. Manni, and T. J. Rohner, (1984), Cancer **53**, 1447–1450.
Dreizen, S., (1984), Am. J. Med. **77(4D)**, 28-33.
EBCTCG Early Breast Cancer Trialists' Collaborative Group, (1990), Treatment of early breast cancer, Oxford University Press, New York.

EICHENBERGER, T., J. TRACHTENBERG, T. P., and A. KEATING, (1989), J. Urol. **141**, 190–191.
ELLIS, E. F., S. C. AMRUTHESH, R. J. POLICE, and L. M. YANCEY, (1990), Adv. Prostaglandin, Thromboxane and Leukotriene Res. **21**, 201–204.
ESCALANTE, B., J. R. FALCK, P. YADAGIRI, L. SUN and M. LANIADO-SCHWARTZMAN, (1988), Biochem. Biophys. Res. Commun. **152**, 1269–1274.
ESCALANTE, B., W. C. SESSA, J. R. FALCK, P. YADAGIRI, and M. L. SCHWARTZMAN, (1989), J. Pharmacol. Exp. Ther. **248**, 229–232.
ESCALANTE, B., D. ERLIJ, J. R. FALCK, and J. C. MCGIFF, (1991), Science **251**, 799–802.
FALCK, J. R., S. LUMIN, I. BLAIR, E. DISHMAN, M. V. MARTIN, D. J. WAXMAN, F. P. GUENGERICH, and J. H. CAPDEVILA, (1990), J. Biol. Chem. **265**, 10244–10249.
FARR, P. M., L. B. DRAUSE, and J. M. MARKS, (1985), Lancet **2**, 921.
FAVENNEC, L. and M.-J. CALS, (1988), J. Clin. Chem. Clin. Biochem. **26**, 479–489.
FISCHER, R. T., J. M. TRZASKOS, R. L. MAGOLDA, S. S. KO, C. S. BROSZ, and B. LARSEN, (1991), J. Biol. Chem. **266**, 6124–6132.
FISHMAN, J. and E. F. HAHN, (1987), Steroids **50**, 339–345.
FITZPATRICK, F. A. and R. C. MURPHY, (1989), Pharmacol. Rev. **40**, 229–241.
FONTANA, J. A., (1987), Exp. Cell. Biol. **55**, 136–144.
FOSTER, A. B., M. JARMAN, and C.-S. LEUNG, (1983), J. Med. Chem. **26**, 50–54.
FOSTER, A. B. (1985), Adv. Drug Res. **14**, 1–40.
FOSTER, A. B., M. JARMAN, C.-S. LEUNG, M. G. ROWLANDS, G. N. TAYLOR, R. G. PLEVEY, and P. SAMPSON, (1985), J. Med. Chem. **28**, 200–204.
FOSTER, A. B., M. JARMAN, and C.-S. KWAN, (1986a), Br. Pat. 8517999, Chem. Abstr. **105**, 54601.
FOSTER, A. B., M. JARMAN, J. MANN, and I. B. PARR, (1986b), J. Steroid Biochem. **24**, 607–617.
FRAKER, L. D., S. A. HALTER, and J. T. FORBES, (1984), Cancer Res. **44**, 5757–5763.
FROLICK, C. A., (1981), Ann. N. Y. Acad. Sci. **359**, 37–44.
FRYE, L. and C. H. ROBINSON, (1988), J. Chem. Soc., Chem. Comm., 129–131.
FRYE, L. L. and C. H. ROBINSON (1990), J. Org. Chem. **55**, 1579–1584.
FURR, B. J. A. and V. C. JORDAN, (1984), Pharm. Ther. **25**, 127–205.
FURTH, P. S., J. ROSENBERGER, P. A. MARCOTTE, and C. H. ROBINSON, (1990), J. Enzym. Inhib. **4**, 131–135.
GALLIANI, G., M. NALI, B. RINDONE, S. TOLLARI, M. ROCCHETTI, and M. SALMONA, (1986), Xenobiotica **16**, 511–517.
GANGULY, J. (1989), in: Biochemistry of vitamin A, CRC Press Inc, Boca Raton, Florida, 115–144; and 195–212; and references therein.
GARRISON, J. M., D. OSTOVIC, and T. C. BRUICE, (1989), J. Am. Chem. Soc. **111**, 4960–4966.
GEELEN, J. A. A., G. H. DECKERS, J. T. H. VAN DER WARDT, H. J. J. LOOZEN, L. J. W. TAX, and H. J. KLOOSTERBOER, (1991), J. Steroid Biochem. Molec. Biol. **38**, 181–188.
GOLDENBERG, I. S., N. WATERS, R. S. RAVADIN, F. J. ANSFIELD, and A. SEGALOFF, (1973), JAMA **223**, 1267–1268.
GOLDFIEN, A., (1989), in: Basic and Clinical Pharmacology, (B. G. KATZUNG, ed.), 490–492.
GOLDHIRSCH, A. (ed.), (1990), in: Endocrine therapy of breast cancer. Springer-Verlag, Berlin.
GOSS, P. E., M. JARMAN, J. R. WILKINSON, and R. C. COOMBES, (1986), J. Steroid Biochem. **24**, 619–622.
GOWER, D. B., (1974), J. Steroid Biochem. **5**, 501–510.
GRAVES, P. E. and H. A. SALHANICK, (1979), Endocrinology **105**, 52–57.
GRAYBILL, J. R., (1990), in: Chemotherapy of fungal diseases, (J. F. RYLEY, ed.), Springer-Verlag, New York, 455–482.

Grayhack, J. T. and R. W. Sadlowski, (1975), in: Benign Prostatic Hyperplasia (J. T. Grayhack, J. D. Wilson, and M. J. Scherbenske, eds.), U.S. Government Printing Office, Washington, D.C., 125–134.
Griffing, G. T., M. Holbrook, J. C. Melby, and A. H. Brodie, (1988), Clin. Physiol. Biochem. **6**, 171–178.
Griffing, G. T., M. Holbrook, J. C. Melby, J. Alberta, and N. Orme-Johnson. (1989), Am. J. Med. Sci. **298**, 83–88.
Griffing, G. T., J. C. Melby, M. Holbrook, and J. O. Johnston, (1991), Hypertension, **17**, 771–775.
Groves, J. T., G. A. McClusky, R. E. White, and M. J. Coon, (1978), Biochem. Biophys. Res. Commun. **81**, 154–160.
Groves, J. T. and D. V. Subramanian, (1984), J. Am. Chem. Soc. **106**, 2177–2181.
Groves, J. T. and Y. Watanabe, (1986), J. Am. Chem. Soc. **108**, 507–508.
Habenicht, U. F., K. Schwarz, H. U. Schweikert, F. Neumann, and M. F. El Etreby, (1986), Prostate **8**, 181–194.
Habenicht, U., K. Schwarz, F. Neumann, and M. F. El Etreby, (1987), Prostate **11**, 313–326.
Habenicht, U. F. and M. F. El Etreby, (1989), Prostate **14**, 309–322.
Halpert, J. and R. A. Neal, (1980), Molec. Pharmacol. **17**, 427–431.
Halpert, J. and R. A. Neal, (1981), Drug Metab. Rev. **12**, 239–259.
Halpert, J., (1981), Biochem. Pharmacol. **30**, 875–881.
Halpert, J. R., N. E. Miller, and L. D. Gorsky, (1985), J. Biol. Chem. **260**, 8397–8403.
Halpert, J. R.. C. Balfour, N. E. Miller, and L. S. Kaminsky, (1986), Molec. Pharmacol. **30**, 19–24.
Halpert, J., J.-Y. Jaw, and C. Balfour, (1989), Molec. Pharmacol. **34**, 139–147.
Hammons, G. J., W. L. Alworth, N. E. Hopkins, F. P. Guengerich, and F. F. Kadlubar, (1989), Chem. Res. Toxicol. **2**, 367–374.
Hanzlik, R. P. and R. H. Tullman, (1982), J. Am. Chem. Soc. **104**, 2048–2050.
Hardwick, J. P., B.-J. Song, E. Huberman, and F. J. Gonzalez, (1987), J. Biol. Chem. **262**, 801–810.
Harris, A. L., M. Dowsett, S. L. Jeffcoate, J. A. Mckinna, M. Morgan, and I. E. Smith, (1982), J. Clin. Endocrinol. Metab. **55**, 718–722.
Hartmann, R. W. and C. Batzl, (1986), J. Med. Chem. **29**, 1362–1369.
Harvey, H. A., A. Lipton, and R. J. Santen, (1984), in: Aminoglutethimide as an Aromatase Inhibitor in the Treatment of Cancer, (G. A. Nagel and R. J. Santen, eds.), Hans Huber Publishers, Vienna, 74–89.
Havlin, K., V. C. Jordan, K. Cummings, E. Messing, and D. L. Trump, (1987), Proc. Am. Soc. Clin. Oncol. **6**, 106.
Havlin, K. A. and D. L. Trump, (1988), Cancer Treat. Res. **39**, 83–96.
Hector, F. R. and E. Yee, (1990), Antimicrob. Agents Chemother. **34**, 448–454.
Henderson, D., G. Norbisrath, and U. Kerb, (1986a), Endocrinology **115**, 776–785.
Henderson, D., G. Norbisrath, and U. Kerb, (1986b), J. Steroid Biochem. **24**, 303–306.
Henderson, D., U. Habenicht, Y. Nishino, U. Kerb, and M. F. El Etreby, (1986c), J. Steroid Biochem. **25**, 867–876.
Henderson, D., G. Norbisrath, and U. Kerb, (1986d), J. Steroid Biochem. **25**, 867–876.
Henderson, D., U. F. Habenicht, Y. Nishino, and M. F. El Etreby, (1987), Steroids **50**, 219–233.
Henderson, D., (1987), J. Steroid Biochem. **27**, 905–914.
Heyns, W., A. Drochmans, F. Van der Schueren, and G. Verhoeven, (1985), J. Urol. **110**, 276–283.
Hirsch, K. S., E. R. Adams, D. G. Hoffman, J. K. Markham, and N. Owen, (1986), Tox. Appl. Pharmacol. **86**, 391–399.

Hirsch, K. S., C. D. Jones, T. D. Lindstrom, N. B. Stamm, G. P. Sutton, H. M. Taylor, and D. E. Weaver, (1987), Steroids **50**, 201–217.
Hirt, D. L., J. Capdevila, J. R. Falck, M. D. Breyer, and H. R. Jacobson, (1989), J. Clin. Invest. **84**, 1805–1812.
Hirt, D. L. and H. R. Jacobson, (1991), Semin. Nephrol. **11**, 148–155.
Hjelmeland, L. M., L. Aronow, and J. R. Trudell, (1977), Biochem. Biophys. Res. Commun. **76**, 541–549.
Höffken, K., W. Jonat, K. Possinger, M. Kölbel, T. Kunz, H. Wagner, R. Becher, R. Callies, P. Friederich, W. Willmanns, H. Maass, and C. G. Schmidt, (1990), J. Clin. Oncol. **8**, 875–880.
Hudson, R. W. and D. Wherrett, (1990), J. Steroid Biochem. **35**, 231–236.
Humphrey, M., J. S. Jevons, and M. H. Tarbit, (1985), Antimicrob. Agents Chemother. **28**, 648–653.
James, V. H. T., M. J. Reed, E. F. Adams, M. W. Ghilchick, L. C. Lai, N. G. Coldham, C. J. Newton, A. Purohit, A. M. Owen, A. Singh, and S. Islam, (1989), Proc. Roy. Soc. Edin. **95B**, 185–193.
Janssen, P. A. J., H. Vanden Bossche, J. P. Van Wauwe, G. Cauwenbergh, and H. J. Degreef, (1989), Int. J. Dermatol. **28**, 493–496.
Jarman, M., S. E. Barrie, C. S. Leung, and M. G. Rowlands, (1988), Anti-Cancer Drug Des. **3**, 185–190.
Jarman, M., S. E. Barrie, J. J. Deadman, J. Houghton, R. McCague, and M. G. Rowlands, (1990), J. Med. Chem. **33**, 2452–2455.
Jefcoate, C. R. (1978), Methods Enzymol. **52**, 258–279.
Jefcoate, C. R., (1986), in; Cytochrome P450: Structure, Mechanism, and Biochemistry, (P. R. Ortiz de Montellano, ed.), Plenum Press, New York, 387–428.
Jim, K., A. Hassid, F. Sun, and M. J. Dunn, (1982), J. Biol. Chem. **257**, 10294–10299.
Johnston, J. O. and B. W. Metcalf, (1984), in: Novel Approaches to Cancer Therapy, (S. S. Prasad, ed.), Academic Press, New York, 307–328.
Johnston, J. O., C. L. Wright, and B. W. Metcalf, (1984), Endocrinology **115**, 776–785.
Johnston, J. O., (1987), Steroids **50**, 105–120.
Johnston, J. O., C. L. Wright, and G. W. Holbert, (1990a), Endocrinology **126 (suppl. abst. 7)**.
Johnston, J. O., P. M. Weintraub, and J. C. Melby, (1990b), Steroids **55**, 378–382.
Johnston, J. O., C. L. Wright, R. A. Bohnke, and P. R. Kastner, (1991), Endocrinology **128 (Suppl Abst. 24)**.
Jones, C. D., M. A. Winter, K. S. Hirsch, N. Stamm, H. M. Taylor, H. E. Holden, J. D. Davenport, E. V. Krumkalns, and R. G. Suhr, (1990), J. Med. Chem. **33**, 416–429.
Jordan, V. C., (1986), in: Endocrine Therapy of Breast Cancer; Concepts and Strategies, (F. Cavalli eds.), Springer-Verlag, New York, 41–49.
Jutley, J. K., S. Reaney, J. Kelleher, and P. Whelan, (1990), Prostate **16**, 299–304.
Kalb, V. F., J. C. Loper, C. R. Dey, C. W. Woods, and T. R. Sutter, (1986), Gene **45**, 237–245.
Kalb, V. F., C. W. Woods, T. G. Turi, C. R. Dey, T. R. Sutter, and L. C. Loper, (1987), DNA **6**, 529–537.
Karara, A., E. Dishman, I. Blair, J. R. Falck, and J. H. Capdevila, (1989), J. Biol. Chem. **264**, 19822–19827.
Karara, A., E. Dishman, H. Jacobson, J. R. Falck, and J. H. Capdevila, (1990), FEBS Lett. **268**, 227–230.
Karr, J. P., L. S. Chai, U. Kim, G. P. Murphy, J. A. Resko, A. A. Sandberg, and S. Schneider, (1984), Urology **23**, 276–289.
Karr, J. P., Y. Kaburagi, C. F. Mann, and A. A. Sandberg, (1987), Steroids **50**, 449–457.

Kauser, K., J. E. Clark, B. S. S. Masters, D. R. Harder, and R. J. Roman, (1990), FASEB J.**4**, A1112.
Kellis, J. T. and L. E. Vickery, (1984), Endocrinology **114**, 2128–2137.
Kellis, J. T., J. J. Sheets, and L. E. Vickery, (1984), J. Steroid Biochem. **20**, 671–676.
Kellis, J. T., W. E. Childers, C. H. Robinson, and L. E. Vickery, (1987), J. Biol. Chem. **262**, 4421–4426.
Kerridge, D. and H. Vanden Bossche, (1990), in: Chemotherapy of Fungal Diseases, (J. F. Ryley, ed.), Springer-Verlag, New York, 31–76.
Kirkpatrick, C. H., (1984), Am. J. Med. **77(4D)**, 1–12.
Ko, S. S., H. Chen, M. F. Favata, R. T. Fischer, J. L. Gaylor, P. R. Johnson, R. L. Magolda, S. H. Stam, and J. M. Trzaskos, (1987), 194th ACS Meeting, MEDI Abst. No. 77.
Komives, E. A. and P. R. Ortiz de Montellano, (1987), J. Biol. Chem. **262**, 9793–9802.
Kovacs, J. A., A. A. Kovacs, M. Polis, W. C. Wright, V. J. Gill, C. U. Tuazon, E. P. Gehmann, H. C. Lane, R. Longfield, G. Overturf, A. M. Macher, A. S. Fauci, J. E. Parillo, J. E. Bennett, and H. Masur, (1985), Ann. Intern. Med. **103**, 533–538.
Krieg, M. and Tunn, S., (1990), in: Testosterone Action Deficiency Substitution, (E. Nieschlag, and H. M. Behre, eds.), Springer-Verlag, New York, 219–244.
Krueger, R. J., A. Nagahisa, M. Gut, S. R. Wilson, and W. H. Orme-Johnson, (1985), J. Biol. Chem. **260**, 852–859.
Kunze, K. L., B. L. K. Mangold, C. Wheeler, H. S. Beilan, and P. R. Ortiz de Montellano, (1982), J. Biol. Chem. **258**, 4202–4207.
Lamberts, S. W. J., H. A. Bruining, H. Marzouk, J. Zuiderwijk, P. Uitterlinden, J. J. Blijd, W. H. L. Hackeng, and F. H. De Jong, (1989), J. Clin. Endocrinol. Metab. **69**, 896–901.
Laniado-Schwartzman, M., K. L. Davis, J. C. McGiff, R. D. Levere, and N. G. Abraham, (1988), J. Biol. Chem. **263**, 2536–2542.
Laughton, C. A., R. McKenna, S. Neidle, M. Jarman, R. McCague, and M. G. Rowlands, (1990a), J. Med. Chem. **33**, 2673–2679.
Laughton, C. A. and S. Neidle, (1990b), J. Med. Chem. **33**, 3055–3060.
Lee, J. S., N. E. Jacobsen, and P. R. Ortiz de Montellano, (1988), Biochemistry **27**, 7703–7710.
Lee, J., W. Ritter, J. Leccione, P. Kelly, P. Pizzo, and T. J. Walsh, (1989), 29th Intersci. Cong. Antimicrob. Agents Chemother., Abst. 727.
Leo, M. A., S. Iida, and C. S. Lieber, (1984), Arch. Biochem. Biophys. **234**, 305–311.
Leo, M. A. and C. S. Lieber, (1985), J. Biol. Chem. **260**, 5228–5231.
Leo, M. A., J. M. Lasker, J. L. Raucy, C.-I. Kim, M. Black, and C. S. Lieber, (1989), Arch. Biochem. Biophys. **269**, 305–312.
Leung, C. S., M. G. Rowlands, M. Jarman, A. B. Foster, L. S. Griggs, and D. E. V. Wilman, (1987), J. Med. Chem. **30**, 1550–1554.
Levere, R. D., P. Martasek, B. Escalante, M. L. Schwartzman, and N. G. Abraham, (1990), J. Clin. Invest. **86**, 213–219.
Li, P. K. and R. W. Brueggemeier, (1990), J. Med. Chem. **33**, 101–105.
Liebler, D. C. and F. P. Guengerich, (1983), Biochemistry **22**, 5482–5489.
Lin, C., H. Kim, E. Radwanski, M. Affrime, and S. Symchowicz, (1989), 29th Intersci. Conf. Antimicrobial Agents Chemother.
Lindstrom, T. D. and G. W. Whitaker, (1987), Fundam. Appl. Toxicol. **8**, 595–604.
Lipton, J. and R. J. Santen, (1974), Cancer **33**, 503–505.
Lipton, A., H. A. Harvey, L. M. Demers, J. R. Hanagan, M. T. Mulagha, G. M. Kochak, S. Fitzsimmons, S. I. Sanders, and R. J. Santen, (1990), Cancer **65**, 1279–1285.
Longcope, C., (1987), Steroids **50**, 253–267.
Longcope, C., A. Femino, and J. O. Johnston, (1988), Endocrinology **122**, 2007–2011.

LOOSE, D. S., P. B. KAN, M. A. HIRST, R. A. MARCUS, and D. FELDMAN, (1983), J. Clin. Invest. **71**, 1495–1499.
LOTAN, R., M. G. NEUMANN, and D. LOTAN, (1981), Ann. N. Y. Acad. Sci. **359**, 150–159.
LOTAN, R., (1986), Cancer Bull. **38**, **3**, 113–116.
LØNNING, P. E., D. C. JOHANNESSEN, T. THORSEN, and D. EKSE, (1989), J. Steroid Biochem. **33**, 541–545.
LØNNING, P. E., M. DOWSETT, and T. J. POWLES, (1989), Br. J. Cancer **60**, 5–8.
MA, Y.-H., D. R. HARDER, J. E. CLARK, and R. J. ROMAN, (1990), FASEB J. **4**, 3361.
MACDONALD, T. L., K. ZIRVI, L. T. BURKA, P. PEYMAN, and F. P. GUENGERICH, (1982), J. Am. Chem. Soc. **104**, 2050–2052.
MACDONALD, T. L., W. G. GUTHEIM, R. B. MARTIN, and F. P. GUENGERICH, (1989), Biochemistry **28**, 2071–2077.
MACINDOE, J. H., G. R. WOODS, L. A. ETRE, and D. F. COVEY, (1982), Cancer Res. **42**, 3378–3381.
MAGOLDA, R. L., H. CHEN, M. F. FAVATA, R. T. FISCHER, J. L. GAYLOR, P. R. JOHNSON S. S. KO, S. H. STAM, and J. M. TRZASKOS, (1987), 194th ACS Meeting, MEDI Abst. No. 78.
MAKSYMIUK, A. W., S. THONGPRASET, R. HOPFER, M. LUNA, V. FAINSTAIN, and G. P. BODEY, (1984), Am. J. Med. **77(4D)**, 20–27.
MANNI, A. and O. H. PEARSON, (1980), Cancer Treat. Rep. **64**, 111–116.
MANNI, A. and R. J. SANTEN, (1988), Cancer Treat. Res. **39**, 67–81.
MARCOTTE, P. A. and C. H. ROBINSON, (1981), Fed. Proc. **40**, 1867.
MARSH, D. A., H. J. BRODIE, W. GARRETT, C. TSAI-MORRIS, and A. M. H. BRODIE, (1985), J. Med. Chem. **28**, 788–795.
MARTINI, L., F. CELOTTI, M. J. LECHUGA, R. C. MELCANGI, M. MOTTA, P. NEGRI-CESI, A. POLETTI, and S. ZOPPI, (1990), Ann. N. Y. Acad. Sci. **595**, 184–198.
MARYNIAK, D. M., G. B. MULLEN, J. T. MITCHELL, S. D. ALLEN, C. R. KINSOLVING, and V. ST. GEORGIEV, (1988), Ann. N. Y. Acad. Sci. **544**, 373–374.
MARYNIAK, D. M., G. B. MULLEN, S. D. ALLEN, J. T. MITCHELL, C. R. KINSOLVING, and V. ST. GEORGIEV, (1988), Ann. N. Y. Acad. Sci. **544**, 378–380.
MASFERRER, J. L., R. C. MURPHY, P. J. PAGANO, M. W. DUNN and M. L. SCHWARTZMAN, (1989), Invest. Opthalmol. Vis. Sci. **30**, 454–460.
MASTERS, B. S. S., R. T. OKITA, A. S. MUERHOFF, M. T. LEITHAUSER, A. GEE, S. WINQUIST, D. L. ROERIG, J. E. CLARK, R. C. MURPHY, and P. ORTIZ DE MONTELLANO, (1989), Adv. Prostaglandin Thromboxane Leuk. Res. **19**, 335–338.
MCCAGUE, R., M. G. ROWLAND, S. E. BARIE, and J. HOUGHTON, (1990), J. Med. Chem. **33**, 3050–3055.
MCCARTHY, D. J., C. LINDAMOOD III, and D. L. HILL, (1987), Cancer Res. **47**, 5014–5020.
MCGIFF, J. C., (1991), Annu. Rev. Pharmacol. Toxicol. **31**, 339–369.
MCGUIRE, W. L., (1980), in: Hormones and Cancer, Vol **15**, (S. IACOBELL, ed.), Raven Press, New York, 337–343.
MCNEAL, J. E., (1981), Urology **17 (suppl.)**, 11-16.
MCNEAL, J. E., (1983), in: Benign Prostatic Hypertrophy (F. HINMAN Jr., ed.), Springer, New York, 153–166.
MCNEIL, J. M., M. J. REED, P. A. BERANEK, R. C. BONNEY, M. W. GHILCHICK, D. T. ROBINSON, and V. H. T. JAMES, (1986), Int. J. Cancer **38**, 193–196.
MELBY, J. C., S. L. DALE, M. HOLBROOK, T. E. WILSON, G. T. GRIFFING, and B. H. ARISON, (1985), Endocr. Res. **10**, 591–607.
MELBY, J. C., M. HOLBROOK, G. T. GRIFFING, and J. O. JOHNSTON, (1987), Hypertension **10**, 484–487.
MENARD, R. H. and H. F. MARTIN, (1974), Life Sci. **15**, 1639–1648.
MENARD, R. H., T. M. GUENTHNER, H. KON, and J. R. GILLETTE, (1979a), J. Biol. Chem. **254**, 1726–1733.

Menard, R. H., T. M. Guenthner, A. M. Taburet, H. Kon, L. R. Pohl, J. R. Gillette, H. V. Gelboin, and W. F. Trager, (1979b), Molec. Pharmacol. **16**, 997–1010.
Metcalf, B. W., C. L. Wright, J. P. Burkhart, and J. O'Neal Johnston, (1981), J. Am. Chem. Soc. **103**, 3221–3222.
Miao, E., C. Zuo, A. Nagahisa, B. J. Taylor, S. Joardar, C. Byon, S. R. Wilson, and W. H. Orme-Johnson, (1990), Biochemistry **29**, 2199.
Miller, S. L., J. N. Wright, D. L. Corina, and M. Akhtar, (1991), J. Chem. Soc., Chem. Commun., 157–159.
Miller, W. R., (1990), J. Steroid Biochem. Molec. Biol. **37**, 467–480.
Miller, W. R. and J. S. O'Neill, (1990), J. Steroid Biochem. Molec. Biol. **37**, 317–325.
Miyairi, S. and J. Fishman, (1986), J. Biol. Chem. **261**, 6772–6777.
Moon, R. C. and L. M. Itri, (1984), in: The Retinoids, Vol. **2**, (M. B. Sporn, A. B. Roberts, and D. S. Goodman, eds.), Academic Press, Orlando, FL, 327–372.
Morrison A. R. and N. Pascoe, (1981), Proc. Natl. Acad. Sci. USA **78**, 7375–7378.
Muerhoff, A. S., D. E. Williams, N. O. Reich, C. A. CaJacob, P. Ortiz de Montellano, and B. S. S. Masters, (1989), J. Biol. Chem. **264**, 749–756.
Mullen, G. B., D. M. Maryniak, P. A. Swift, S. D. Allen, J. T. Mitchell, C. R. Kinsolving, and V. St. Georgiev, (1988a), Ann. N. Y. Acad. Sci. **544**, 359–362.
Mullen, G. B., P. A. Swift, G. A. Bennett, T. R. DeCory, D. M. Maryniak, P. G. Dormer, and V. St. Georgiev, (1988b), Ann. N. Y. Acad. Sci. **544**, 32–45.
Murphy, R. C., J. R. Falck, S. Sumin, P. Yadagiri, J. A. Zirrolli, M. Balazy, J. L. Masferrer, N. G. Abraham, and M. L. Schwartzman, (1988), J. Biol. Chem. **263**, 17197–17202.
Murray, M., (1987), Drug Metab. Rev. **18**, 55–81.
Murray, R. I. and S. G. Sligar, (1985), J. Am. Chem. Soc. **107**, 2186–2187.
Murray, R. and P. Pitt, (1985), Eur. J. Cancer Clin. Oncol. **21**, 153–158.
Nagahisa, A., R. W. Spencer and W. H. Orme-Johnson, (1983), J. Biol. Chem. **258**, 6721–6723.
Nagahisa, A. and W. H. Orme-Johnson, (1984), J. Am. Chem. Soc. **106**, 1166–1167.
Nakajin, S., M. Shinoda, M. Haniu, J. E. Shively, and P. F. Hall, (1984), J. Biol. Chem. **259**, 3971–3976.
Napoli, J. L. and A. M. McCormick, (1981), Biochim. Biophys. Acta **660**, 165–175.
Napoli, J. L., (1990), in: Chemistry and Biology of Synthetic Retinoids, (M. I. Dawson and W. H. Okamura, eds.), CRC Press, Boca Raton, 229–249.
Needleman, P., J. Turk, B. Jakshick, A. Morrison, and J. Lefkowith, (1986), Annu. Rev. Biochem. **55**, 69–102.
Nelson, S. D. and P. J. Harvison, (1987), in: Mammalian Cytochromes P-450, Vol. II (F. P. Guengerich, ed.), CRC Press, Boca Raton, Florida, 19–79.
Nicholls P. J., M. J. Daly, and H. J. Smith, (1986), Breast Cancer Res. Treat. (Suppl.) **7**, 55–68.
Nishino, Y., M. R. Schneider, H. Michna, and M. F. El Etreby, (1989), J. Steroid Biochem. **34**, 435–437.
O'Neill, J. S., R. A. Elton, and W. R. Miller, (1988), Br. Med. J. **296**, 741–743.
O'Neal Johnston, J., C. L. Wright, G. W. Holbert, and H. D. Benson, (1990), J. Enzyme Inhib. **4**, 137–142.
Oesterling, J. E., P. E. Juniewicz, J. R. Walters, J. D. Strandberg, R. E. Steele, L. L. Ewing, and D. S. Coffey, (1988), J. Urol. **139**, 832–839.
Ohba, M., R. Sato, Y. Yoshida, T. Nishino and H. Katsuki, (1978), Biochem. Biophys. Res. Commun. **85**, 21–27.
Okamoto, T., K. Sasaki, and S. Oka, (1988), J. Am. Chem. Soc. **110**, 1187–1196.
Okuda, T., T. Tania, K. Ichise, and T. Nakajima, (1988), 29th Intersci. Conf. Antimicrob. Agents Chemother. Abst. 174.

Olakanmi, O. and D. W. Seybert, (1990), J. Steroid Biochem. **36**, 273–280.
Oldham, H. G., (1989), in: Sulfur-Containing Drugs and Related Organic Compounds, (L. A. Damani, ed.), Vol. **2**, John Wiley, New York, 9–45.
Oliw, E. H., F. P. Guengerich, and J. A Oates, (1981), J. Biol. Chem. **257**, 3771–3781.
Ong, D. E. and F. Chytil, (1983), Vitam. Horm. **40**, 105–144.
Ortiz de Montellano, P. R. and J. M Mathews, (1981), Biochem. J. **195**, 761–764.
Ortiz de Montellano, P. R., K. L. Kunze, H. S. Beilan, and C. Wheeler, (1982), Biochemistry **21**, 1331–1339.
Ortiz de Montellano, P. R., O. Augusto, F. Viola, and K. L. Kunze, (1983), J. Biol. Chem. **258**, 8623–8629.
Ortiz de Montellano, P. R., J. M. Mathews, and K. C. Langry, (1984), Tetrahedron **40**, 511–519.
Ortiz de Montellano, P. and N. O. Reich, (1984), J. Biol. Chem. **259**, 4136–4141.
Ortiz de Montellano, P. R., (1985), in: Bioactivation of Foreign Compounds, (M. W. Anders, ed.), Academic Press, New York, 121–155.
Ortiz de Montellano, P. R., (1986), in: Cytochrome P450: Structure, Mechanism, and Biochemistry, (P. R. Ortiz de Montellano, ed.), Plenum Press, New York, 217–271.
Ortiz de Montellano, P. R., and N. O. Reich, (1986), in: Cytochrome P450: Structure, Mechanism, and Biochemistry, (P. R. Ortiz de Montellano, ed.), Plenum Press, New York, 273–314.
Ortiz de Montellano, P. R., and R. A. Stearns, (1987), J. Am. Chem. Soc. **109**, 3415–3420.
Ortiz de Montellano, P. R., (1988), in: Progress in Drug Metabolism, (G. G. Gibson, ed.), Taylor & Francis, New York, 99–148.
Osawa, Y., Y. Osawa, and M. J. Coon, (1987), Endocrinology **121**, 1010–1016.
Palmer, G. C., J. M. Ordy, R. D. Simmons, J. C. Strand, G. B. Mullen, C. R. Kinsolving, V. St. Georgiev, and S. D. Allen, (1988), Ann. N. Y. Acad. Sci. **544**, 386–391.
Parnell, J. P. and B. S Sherman, (1962), in: Fundamentals of Keratinization, A symposium, (E. O. Butcher and R. F. Sognnaes, eds.), American Association for Advancement of Science, Washington, DC, 114–131.
Pasanen, M., T. Taskinen, M. Iscan, E. A. Sotaniemi, M. Kairaluoma, and O. Pelkonen, (1988), Biochem. Pharmacol. **37**, 3861–3866.
Peck, G. L. (1984), in: The Retinoids, Vol. **2**, (M. B. Sporn, A. B. Roberts, and D. S. Goodman, eds.), Academic press, Orlando, FL. 391–412.
Perel, E., S. P. Davis, D. F. Covey, and D. W. Killinger, (1981), Steroids **38**, 397–405.
Peterson, J. A. and R. A. Prough, (1986), in: Cytochrome P450: Structure, Mechanism, and Biochemistry, (P. R. Ortiz de Montellano, ed.), Plenum Press, New York, 89–119.
Plempel, M. and K. H. Büchel, (1988), Revista Ibérica de Micológia **5 (Suppl. 1)**, Abst. O–7.
Polak, A., (1990), in: Chemotherapy of fungal diseases, (J. F. Ryley, ed.), Springer-Verlag, New York, 153–182.
Ponder, B. A. J., R. J. Shearer, R. D. Pocock, J. Miller, D. Easton, C. E. D. Chilvers, M. Dowsett, and S. L. Jeffcoate, (1984), Br. J. Cancer **50**, 757–763.
Pont, A., P. L. Williams, S. Azhar, R. E. Reitz, C. Bochra, E. R. Smith, and D. A. Stevens, (1982), Arch. Intern. Med. **142**, 2137–0000.
Pont, A., J. R. Graybill, P. C. Craven, J. N. Galgiani, W. E. Dismukes, R. E. Reitz, and D. A. Stevens, (1984), Arch. Intern. Med. **144**, 2150–2153.
Pont, A., (1987), J. Urol. **137**, 902–904.
Proctor, K. G., J. R. Falck, and J. Capdevila, (1987), Circ. Res. **60**, 50–59.
Proctor, K. G., S. Shatkin Jr, P. M. Kaminski, J. R. Falck, and J. H. Capdevila, (1988), Circulation **77**, 1185–1196.

QUILLEY, C. P. and J. C. MCGIFF, (1990), J. Pharmacol. Exp. Ther. **254**, 774–780.
RALLISON, M. L., F. H. TYLER, L. F. KUMAGI, and J. KUMAGI, (1964), Pediatrics **65**, 1095–1099.
RANE, D. F., J. A. DESAI, D. LOEBENBERG, R. PARMEGIANI, and R. E. PIKE, (1988), Ann. N. Y. Acad. Sci. **544**, 105–108.
REICH, M. F., P. J. FABIO, V. J. LEE, N. A. KUCK, and R. T. TESTA, (1988), Ann. N. Y. Acad. Sci. **544**, 101–104.
RICHARDSON, K., K. COOPER, M. S. MARRIOTT, M. H. TARBIT, P. R. TROKE, and P. J. WHITTLE, (1988), Ann. N. Y. Acad. Sci. **544**, 4–11.
RITTER, W., G. HOLMWOOD, H.-J. AHR, K. DETZER, U. KRAATZ, M. PLEMPEL, D. SCHERLING, and H.-M. SIEFERT, (1988), Ann. N. Y. Acad. Sci. **544**, 74–85.
ROBINSON, M. G. R., R. J. SHEARER, and J. D. FERGUSSON, (1974), Br. J. Urol. **46**, 555–559.
ROBERTS, A. B., (1981), Ann. N. Y. Acad. Sci. **359**, 45–52.
ROMERO, M. F., Z. T. MADHUN, U. HOPFER, and J. G. DOUGLAS, (1990), Adv. Prostaglandin, Thromboxane Leukotriene Res. **21**, 205–208.
ROSOLOWSKY, M., J. R. FALCK, J. T. WILLERSON, and W. B. CAMPBELL, (1990), Circ. Res. **66**, 608–621.
ROSTOM, A. Y., A. FOLKES, C. LORD, R. G. NOTLEY, F. A. W. SCHWEITZER, and W. F. WHITE, (1982), Br. J. Urol. **54**, 552–555.
ROWLANDS, M. G., J. MANN, P. PIETRZAK, and R. C. COOMBES, (1986), Eur. J. Cancer Clin. Oncol. **22**, 1–40, 739.
ROWLANDS, M. G., M. A. BUNNET, A. B. FOSTER, M. JARMAN, J. STANEK, and E. SCHWEISER, (1988), J. Med. Chem. **31**, 971–976.
RYLEY, J. F., S. MCGREGOR, and R. G. WILSON, (1988), Ann. N. Y. Acad. Sci. **544**, 310–328.
RYLEY, J. F., R. S. GORDEE, and D. LOEBENBERG, (1990), in: Chemotherapy of fungal diseases, (J. F. RYLEY, ed.), Springer-Verlag, New York, 525.
SACERDOTI, D., N. G. ABRAHAM, J. C. MCGIFF, and M. L. SCHWARTZMAN, (1988), Biochem. Pharmacol. **37**, 521–527.
SACERDOTI, D., B. ESCALANTE, N. G. ABRAHAM, J. C. MCGIFF, R. D. LEVERE, and M. L. SCHWARTZMAN, (1989), Science **243**, 388–390.
SAKSENA, A. K., V. M. GINJAVALLABHAN, A. B. COOPER, and D. LOEBENBERG, (1989), Ann. Rep. Med. Chem. **24**, 111–120.
SALHANICK, H. A., (1982), Cancer Res. **42 (Suppl.)**, 3315s.
SAMOJLIK, E., M. A. KIRSCHNER, and A. J. LIPPMAN, (1985), Proc. Am. Soc. Clin. Oncol. **4**, 97.
SANFORD, E. J., J. R. DRAGO, T. J. ROHNER, R. SANTEN, and A. LIPTON, (1976), J. Urol. **115**, 170–174.
SANTEN, R. J., A. LIPTON, and J. KENDALL, (1974), JAMA **230**, 1661–1665.
SANTEN, R. J., S. SANTNER, B. DAVIS, J. VELDHUIS, E. SAMOJLIK, and E. RUBY, (1978), J. Clin. Endocrinol. Metab. **55**, 1257–1265.
SANTEN, R. J., N. COHN, R. MISBIN, E. SAMOJLIK, and E. FOLTZ, (1979), J. Clin. Endocrinol. Metab. **49**, 631–634.
SANTEN, R. J., T. J. WORGUL, E. SAMOJLIK, A. INTERRANTE, A. E. BOUGHER, A. LIPTON, H. A. HARVEY, D. S. WHITE, E. SMART, C. COX, and S. A. WELLS, (1981), N. Engl. J. Med. **305**, 545–551.
SANTEN, R. J. and A. M. H. BRODIE, (1982), in: Clinics in Oncology, (B. J. A. FURR, ed.), Vol **1**, Philadelphia, Saunders, 77–130.
SANTEN, R. J., E. BADDER, S. LERMAN, H. HARVEY, A. LIPTON, A. E. BOUCHER, A. MANNI, H. ROSEN, and S. A. WELLS, (1982), Breast Cancer Res. Treat. **2**, 375–383.
SANTEN, R. J., H. VANDEN BOSSCHE, J. SYMOENS, J. BRUGMANS, and R. DE COSTER, (1983), J. Clin. Endocrin. Metab. **57**, 732–736.

SANTEN, R. J., (1986), Breast Cancer Res. Treat. **7 (Suppl.)**, 23–36.
SANTEN, R. J., L. M. DEMERS, H. ADLERCREUTZ, H. HARVEY, S. SANTNER, S. SANDERS, and A. LIPTON, (1989), J. Clin. Endocrinol. Metab. **68**, 99–106.
SCHAUDE, M., G. PETRANY, J. G. MEINGASSNER, H. ACKERBAUER, and H. MIETH, (1988), Revista Ibérica de Micológia **5 (Suppl. 1)**, Abst. P-151.
SCHENKMAN, J. B., S. G. SLIGAR, and D. L. CINTI, (1981), Pharmacol. Ther. **12**, 43–71.
SCHIEWECK, K., A. S. BHATNAGAR, and A. MATTER, (1988), Cancer Res. **48**, 834–838.
SCHLEICHER, R. L., R. C. MOON, M. K. PATEL, and C. W. BEATTIE, (1988), Cancer Res. **48**, 1465–1469.
SCHRÖDER, F. H., (1988), in: The Medical Management of Prostate Cancer, (L. DENIS, ed.), Springer-Verlag, New York, 3–8.
SCHWARTZMAN, M. L., N. G. ABRAHAM, M. A. CAROLL, R. D. LEVERE, and J. C. MCGIFF, (1986), Biochem. J. **238**, 283–290.
SCHWARTZMAN, M. L., M. BALAZY, J. MASFERRER, N. G. ABRAHAM, J. C. MCGIFF, and R. C. MURPHY, (1987), Proc. Natl. Acad. Sci. USA **84**, 8125–8129.
SCHWARTZMAN, M. L., J. R. FALCK, P. YADAGIRI, and B. ESCALANTE, (1989), J. Biol. Chem. **264**, 11658–11662.
SCHWARTZMAN, M. L., (1990a), Adv. Prostaglandin, Thromboxane Leukotriene Res. **20**, 241–249.
SCHWARTZMAN, M. L., K. L. DAVIS, M. NISHIMURA, N. G. ABRAHAM, and R. C. MURPHY, (1990b), Adv. Prostaglandin, Thromboxane Leukotriene Res. **21**, 185–191.
SCHWEIKERT, H. U. and U. W. TUNN, (1987), Steroids **50**, 191–200.
SEAGO, A., M. H. BAKER, J. HOUGHTON, M. JARMAN, C.-S. LEUNG, and M. G. ROWLANDS, (1988), Biochem. Pharmacol. **37**, 2167–2172.
SEGALOFF, A., J. B. WEETH, K. K. MEYER, E. L. RONGONE, and M. E. G. CUNNINGHAM, (1962), Cancer **15**, 633–635.
SHAK, S., N. O. REICH, I. M. GOLDSTEIN, and P. ORTIZ DE MONTELLANO, (1985), J. Biol. Chem. **260**, 13023–13028.
SHAW, M. A., P. J. NICHOLLS, and H. J. SMITH, (1988), J. Steroid Biochem. **31**, 137–146.
SHEARER, R. J., J. H. DAVIES, M. DOWSETT, P. R. MALONE, A. HEDLEY, D. CUNNINGHAM, and R. C. COOMBES, (1990), Br. J. Cancer **62**, 275–276.
SHEETS, J. J. and L. E. VICKERY, (1982), Proc. Natl. Acad. Sci. U.S.A. **79**, 5773–5777.
SHEETS, J. J. and L. E. VICKERY, (1983), J. Biol. Chem. **258**, 11446–11452.
SHERMAN, M. I. and G. A. TRUITT, (1988), Eur. J. Cancer Clin. Oncol. **24**, 359–362.
SHERWIN, P. F., P. C. MCMULLEN, and D. F. COVEY, (1989), J. Med. Chem., **32**, 651–658.
SHESSEL, F. S., N. L. BLOCK, B. STOVER, A. CLAFLIN, T. I. MALININ, and V. A. POLITANO, (1980), Invest. Urol. **17**, 529–533.
SHIH, M.-J., M. H. CARRELL, H. L. CARRELL, C. L. WRIGHT, J. O. JOHNSTON, and C. H. ROBINSON, (1987), J. Chem. Soc., Chem. Commun., 213–214.
SHIH, T. W. and D. L. HILL, (1991), Drug Metab. Dispos. **19**, 332–335.
SILVERBERG, E. and J. LUBERA, (1986), Cancer **36**, 9–25.
SISLER, H. D. and N. N. RAGSDALE, (1983), On: Mode of action of antifungal agents (A. P. J. TRINCI and J. F. RYLEY, eds.) British Mycological Society Symposium **9**, 256–282, Cambridge University Press.
SMITH, I.E., A. L. HARRIS, M. MORGAN, J. C. GAZET, and J. A. MCKINNA, (1982), Cancer Res. **42**, 3430S–3433S.
SOBEL, J. D., (1988), Clinical perspectives: terconazole, an advance in vulvovaginal candidiasis therapy. Proceedings from a symposium, Laguna Niguel, California, 1987, BMI/McGraw-Hill, New York, NY.
SOMOJLIK, E., R. J. SANTEN, and S. A. WELLS, (1977), J. Clin. Endocrinol. Metab. **45**, 480–487.
ST. GEORGIEV, V., (1988), Ann. N. Y. Acad. Sci. **544**, 1–3.

STEELE, R. E., L. B. MELLOW, W. K. SAWYER, J. M. WASVARY, and L. J. BROWNE, (1987), Steroids **50**, 147–161.
STEIN, R. C., M. DOWSETT, J. DAVENPORT, A. HEDLEY, H. T. FORD, J. GAZET, and R. C. COOMBES, (1990), Cancer Res. **50**, 1381–1384.
STEVENS, D. A., (1977), Am. Rev. Resp. Dis. **116**, 801–806.
STEVENS, D. A., (1983a), Drugs **26**, 334–336.
STEVENS, D. A., (1983b), Drugs **26**, 347–354.
STEVENS, J. C., J.-Y. JAW, C.-T. PENG, and J. HALPERT, (1991), Biochemistry **30**, 3649–3658.
STEVENSON, D. E., J. N. WRIGHT, and M. AKHTAR, (1988), J. Chem. Soc., Perkin Trans. **I**, 2043–2052.
STONE, N. N., W. R. FAIR, and J. FISHMAN, (1986), Prostate **9**, 311–318.
STONE, N. N., V. P. LAUDANE, W. R. FAIR, and J. FISHMAN, (1987), Urol. Res. **15**, 165–167.
SUGAR, A. M., S. ALSIP, J. N. GALGIANI, J. R. GRAYBILL, W. E. DISMUKES, G. A. CLOUD, P. C. CRAVEN, and D. A. STEVENS, (1987), Antimicrob. Agents Chemother. **31**, 1874–1878.
SWANSON, B. N., C. A. FROLIK, D. W. ZAHAREVITZ, P. P. ROLLER, and M. D. SPORN, (1981), Biochem. Pharmacol. **30**, 107–113.
TAKAHASHI, K. J. CAPDEVILA, J. R. FALCK, H. R. JAKOBSON, and K. F. BADR, (1988), FASEB J. **2**, 5620.
TAKAHASHI, K., J. CAPDEVILA, A. KARARA, J. R. FALCK, H. R. JACOBSON, and K. F. BADR, (1990), Am. J. Physiol. **258**, F781–789.
TAN, L. P., E. G. HRYCAY and K. MATSUMOTO (1983), J. Steroid Biochem. **19**, 1329–1338.
TAPAZOGLOU, E., M. G. SUBRAMANIAN, M. AL-SAFFAR, C. KRESGE, and D. A. DECKER, (1986), Am. J. Clin. Oncol. **9**, 369–375.
TAYLOR, H. M., C. D. JONES, J. D. DAVENPORT, K. S. HIRSCH, T. J. KRESS, and D. E. WEAVER, (1987), J. Med. Chem. **30**, 1359–1365.
TENNISWOOD, M., (1986), Prostate **9**, 375–385.
TESTA, B. and P. JENNER, (1981), Drug Metab. Rev. **12**, 1–117.
THIENPONT, D., J. VAN CUTSEM, F. VAN GERVEN, J. HEERES, and P. A. JANSSEN, (1979), Experientia **35**, 606–607.
TRACHTENBERG, J., N. HALPERN, and A. PONT, (1983), J. Urol. **130**, 152–153.
TRACHTENBERG, J., (1984), J. Urol. **132**, 61–63.
TRAYLOR, T. G. and F. XU, (1988), J. Am. Chem. Soc. **110**, 1953–1958.
TRZASKOS, J. M., R. T. FISCHER, and M. F. FAVATA, (1986), J. Biol. Chem. **261**, 16937–16942.
TRZASKOS, J., S. KAWATA, and J. L. GAYLOR, (1986), J. Biol. Chem. **261**, 14651–14657.
TUCK, S. F. and C. H. ROBINSON, (1988), FASEB **2**, A563.
TUCKER, R. M., L. H. HANSON, E. BRUMMER, and D. A. STEVENS, (1989), Antimicrob. Agents Chemother. **33**, 573–575.
TULLMAN, R. H. and R. P. HANZLIK, (1984), Drug Metab. Dispos. **15**, 1163–1182.
TUNN, U. W., T. SENGE, F. NEUMANN, and H. U. SCHWEIKERT, (1981), in: Antihormone: Bedeutung in der urologie, (J. E. ALTWEIN, G. BARTSCH, and G. H. JACOBI, eds.), W. Zuckschwerdt Verlag, Munich, 41–50.
TURKES, A. O. and K. GRIFFITHS, (1989), Prog. Med. Chem. **26**, 299–321.
UNO, J., M. L. SHIGEMATSU, and T. ARAI, (1982), Antimicrob. Agents Chemother. **21**, 912–918.
VAHLQUIST, A., (1980), Experientia **36**, 317–318.
VAN CAUTEREN, H., J. HEYKANTS, R. DE COSTER, and G. CAUWENBERGH, (1987), Rev. Inf. Dis. **9**, S43–S46.
VAN CUTSEM, J., (1983), Am. J. Med. **74 (Suppl. 1B)**, 9–15.
VAN CUTSEM, J., F. VAN GERVEN, and P. A. J. JANSSEN, (1987), Rev. Inf. Dis. **9**, S15–S32.
VAN CUTSEM, J., F. VAN GERVEN, and P. A. J. JANSSEN, (1988), Revista Ibérica de Micológia **5 (Suppl. 1)**, Abst. 0–2.

Van Cutsem, J., F. Van Gerven, and P. A. J. Janssen, (1989), Antimicrob. Agents Chemother. **33**, 2063–2068.
Van Ginckel, R., R. De Coster, W. Wouters, W. Vanherck, R. van der Veer, N. Goenminne, E. Jagers, H. Van Cauteren, L. Wouters, W. Distelmans, and P. A. J. Janssen, (1990), Prostate **16**, 313–323.
Van Landeghem, A. A. J., J. Poortman, M. Nabuurs, and J. H. H. Thyssen, (1985), Cancer Res. **45**, 2900–2904.
Van Tyle, J. H., (1984), Pharmacotherapy **4**, 343–373.
Van Wauwe, J. P., M.-C. Coene, J. Goosens, G. Van Nijen, W. Cools, and W. Lauwers, (1988), J. Pharmacol. Exp. Ther. **245**, 718–722.
Van Wauwe, J. P. and P. A. J. Janssen, (1989), J. Med. Chem. **32**, 2231–2239.
Van Wauwe, J. P., M.-C. Coene, J. Goossens, W. Cools, and J. Monbaliu, (1990), J. Pharmacol. Exp. Ther. **252**, 365–369.
Vanden Bossche, H., J. M. Ruysschaert, F. Defrise-Quertain, G. Willemsens, F. Cornelissen, P. Maricahl, W. Cools, and J. Van Cutsem, (1982), Biochem. Pharmacol. **31**, 2509–2517.
Vanden Bossche, H., D. Bellens, W. Cools, J. Gorrens, P. Marichal, H. Verhoeven, G. Willemsens, R. De Coster, D. Beerens, C. Haeltermann, M. C. Coene, W. Lauwers, and L. Le Jeune, (1986), Drug Dev. Res. **8**, 287–298.
Vanden Bossche, H., R. De Coster, and W. K. Amery, (1987), in: Pharmacology and clinical uses of inhibitors of hormone secretion and action., (F. J. A. Furr and A. E. Wakeling, eds.), Bailliere Tindall, London, 288–307.
Vanden Bossche, H., (1988a), in: Sterol Biosynthesis Inhibitors, (D. Berg, and M. Plempel, eds.), Ellis Horwood Limited, Weinheim, 79–119.
Vanden Bossche, H., P. Marichal, J. Gorrens, H. Geerts, and P. A. J. Janssen, (1988b), Ann. N. Y. Acad. Sci. **544**, 191–207.
Vanden Bossche, H., G. Willemsens, and P. A. J. Janssen, (1988c), Skin Pharmacol. **1**, 176–185.
Vanden Bossche, H., P. Marichal, J. Gorrens, M.-C. Coene, G. Willemsens, D. Bellens, I. Roels, H. Moereels, and P. A. J. Janssen, (1989), Mycoses **32 (suppl. 1)**, 35–52.
Vanden Bossche, H., P. Marichal, G. Willemsens, D. Bellens, J. Gorrens, I. Roels, M.-C. Coene, L. Le Jeune, and P. A. J. Janssen, (1990a), Mycoses **33**, 335–352.
Vanden Bossche, H., P. Marichal, J. Gorrens, D. Bellens, M.-C. Coene, W. Lauwers, L. Le Jeune, H. Moereels, and P. A. J. Janssen, (1990b), in: Mycoses in AIDS Patients, (H. V. Bossche, D. W. R. Mackenzie, G. Cauwenbergh, J. V. Cutsem, E. Drouhet, and B. Dupont, eds.), Plenum Press, New York, 223–243.
Vanden Bossche, H., G. Willemsens, D. Bellens, I. Roels, and P. A. J. Janssen, (1990c), Biochem. Soc. Trans. **18**, 10–13.
Vanden Bossche, H., G. Willemsens, I. Roels, D. Bellens, H. Moereels, M. Coene, L. Le Jeuene, W. Lauwers, and P. A. J. Janssen, (1990d), Biochem. Pharmacol. **40**, 1707–1718.
Viger, A., S. Coustal, S. Perard, B. Chappe, and A. Marquet, (1988), J. Steroid Biochem. **30**, 469–472.
Viger, A., S. Coustal, S. Perard, and A. Marquet, (1988), Tetrahedron **44**, 1127–1134.
Viger, A., S. Coustal, S. Perard, A. Piffeteau, and A. Marquet, (1989), J. Steroid Biochem. **33**, 119–124.
Volk, H., R. H. Deupree, I. S. Goldenberg, R. C. Wilde, R. A. Carabasi, and G. C. Esher, (1974), Cancer **33**, 9–15.
Watanabe, Y. and Y. Ishimura, (1989), J. Am. Chem. Soc. **111**, 8047–8049.
Waterman, M. R., M. E. John, and E. R. Simpson, (1986), in: Cytochrome P450: Structure, Mechanism, and Biochemistry, (P. R. Ortiz de Montellano, ed.), Plenum Press, New York, 345–386.
Williams, J. B. and J. L. Napoli, (1987), Biochem. Pharmacol. **36**, 1386–1388.

Wolfbach, S. B. and P. R. Howe, (1925), J. Exp. Med. **47**, 753–777.
Worgul, T. J., R. J. Santen, E. Samojlik, J. D. Feldhuis, A. Lipton, H. A. Harvey, J. R. Drago, and T. J. Rohner, (1983), J. Urol. **129**, 51–55.
Wouters, W., R. De Coster, M. Krekels, J. Van Dun, D. Beerens, C. Haelterman, A. Raeymaekers, E. Freyne, J. Van Gelder, M. Venet, and P. A. J. Janssen, (1989a), J. Steroid Biochem. **32**, 781–788.
Wouters, W., R. De Coster, R. W. Tuman, C. R. Bowden, J. Bruynseels, H. Vanderpas, P. Van Rooy, W. K. Amery, and P. A. J. Janssen, (1989b), J. Steroid Biochem. **34**, 427–430.
Wouters, W., R. De Coster, D. Beerens, R. Doolaege, J. A. Gruwez, K. Van Camp, H. Van Der Pas, and B. Van Herendael, (1990a), J. Steroid Biochem. **36**, 57–65.
Wouters, W., R. De Coster, J. Van Dun, and M. D. W. G. Krekels, (1990b), J. Steroid Biochem. Molec. Biol. **37**, 1049–1054.
Wright, J. N., M. R. Calder, and M. Akhtar, (1985), J. Chem. Soc., Chem. Commun., 1733–1735.
Wright, J. N., P. T. van Leersum, S. G. Chamberlin, and M. Akhtar, (1989), J. Chem. Soc., Perkin Trans. **1**, 1647–1655.
Wright, C. L., G. H. Holbert, E. Safi, C. H. Robinson, and J. O. Johnston, (1991), Endocrinology **128 (Suppl. Abs. 670)**.
Yang, C. S. and A. Y. H. Lu, (1987), in: Mammalian Cytochromes P-450, Vol. **II**, (F. P. Guengerich, ed.), CRC Press, Boca Raton, Florida, 1–17.
Yoshida, Y. and Y. Aoyama, (1984), J. Biol. Chem. **259**, 1655–1660.
Yoshida, Y. and Y. Aoyama, (1987), Biochem. Pharmacol. **36**, 229–235.
Yoshida, Y., (1988), in: Current topics in medical mycology, (M. R. McGinnis, ed.), Springer-Verlag, New York, 388–418.
Zimniski, S. J., M. E. Brandt, M. H. Melner, D. F. Covey, and D. Puett, (1985), Cancer Res. **45**, 4883–4889.
Zimniski, S. J., M. E. Brandt, D. F. Covey, and D. Puett, (1987), Steroids **50**, 135–146.
Zirngibl, L., J. Fischer, U. Jahn, and K. Thiele, (1988), Ann. N. Y. Acad. Sci. **544**, 63–73.
Zuber, M. X., E. R. Simpson, and M. R. Waterman, (1986), Science **234**, 1258–1261.

Chapter 4

Ergosterol Synthesis: The Major Target in Antimycotic Therapy

I. SCHUSTER

1. Fungal infections—a serious medical problem

Fungi are ubiquitous in the biosphere and all living organisms are exposed to them; infections affect both plant and animal kingdoms. In agriculture the prevention of fungal infections is of outstanding importance; in recent times soils have been excessively exploited and in areas of high disease pressure, fungicides have to cope with continuously evolving resistant strains.

With respect to human mycoses physicians encounter a wide spectrum of diseases, ranging from superficial mycoses such as dermatomycoses—including infections of skin appendages—and genital mycoses in a large fraction of the population, up to severe life-threatening systemic mycoses. Predisposing factors are the increasing life-span, changing life-styles and a decreased host resistance due to aggressive cytotoxic chemotherapy in neoplastic diseases, increasing numbers of organ transplantations followed by immunosuppressive therapy and the propagation of acquired immunodeficiency syndrome.

A reasonable therapy of fungal infections was not started earlier than some 30 years ago. In order to treat the whole range of diseases there is still a strong demand for effective novel drugs which lack adverse side effects and penetrate to the sites of infections (HAY, 1990). With respect to superficial mycoses particularly persistent and frequently recurrent infections such as onychomycosis have to be considered. The number of affected persons with systemic mycoses is increasing and effective and safe treatment is still a problem (WARNOCK, 1989; PERFECT, 1990). Systemic mycoses caused by highly virulent endemic fungi such as strains of Histoplasma, Coccidioides, Blastomyces and Paracoccioides represent a smaller fraction of treated cases. However, in the rising population of immunocompromised patients opportunistic pathogens such as *Candida spp.*, *Aspergillus sp.* and *Cryptococcus sp.* gain importance which are ubiquitous and may cause fatal infections. Neutropenia is one of the main predisposing factors and disseminated fungal infections have been observed in up to 50% of deceased patients with hematologic neoplasms (HÖFFKEN, 1989).

Antifungal chemotherapy is directed against eucariotic cells which are biochemically similar to the host cells. An ideal drug should therefore show a highly selective action on fungal pathways only leading to a rapid onset of high fungicidal activity. With respect to these criteria the present medication is still far from being satisfying and requires the discovery and development of novel more active and safer antifungals.

To date a spectrum of compounds is used which attack the fungal cells by several different mechanisms, such as inhibition of nucleic acid synthesis (e.g. with flucytosine), microtubular stability (e.g. with griseofulvin), impairment of cell wall integrity (with polyene antibiotics) and inhibition of ergosterol synthesis.

Abbreviations used: MIC: minimum drug concentration which inhibits fungal growth by 50%; Ki: inhibition constant; Ks: substrate binding constant.

From these mechanisms inhibition of ergosterol biosynthesis has become the predominant rationale in the prevention and treatment of all kinds of fungal infections ranging from plant diseases to human and animal mycoses (BERG et al., 1986; KATO, 1986). Attempts to interfere effectively and selectively have been undertaken at different stages of its biosynthetic pathway. Two enzymes—14α-lanosterol demethylase and squalene epoxidase—turned out to be especially suitable targets for human therapy and several inhibitors of these enymes were successfully marketed as novel antifungal drugs. The research in this field has greatly improved our understanding of the underlying molecular processes and enabled us also to cope with some adverse side effects associated with the use of the inhibitors. Moreover, it led to major contributions in apparently distant areas such as drug metabolism, cell differentiation, endocrinology and cancer. Besides discussing the main classes of ergosterol synthesis inhibitors it is the aim of this review to also show the mutual stimulation which resulted from applying molecular concepts to pharmaceutical needs.

2. Ergosterol synthesis as a target for antifungals

Ergosterol is the dominant sterol in most fungi and an indispensable component of their membrane structures (KATO, 1986; RODRIGUEZ et al., 1985). It differs from cholesterol, the predominant mammalian sterol, by the presence of a 24-methyl group and $\Delta 7$ and $\Delta 22$-double bonds (Fig. 1). Early studies in *S. cerevisiae* showed the essential requirement of ergosterol for membrane biogenesis (MORPURGO et al., 1964; WALLACE et al., 1968) and the failure of cholesterol to substitute for the main ergosterol functions (PINTO et al., 1983).

Ergosterol is thought to exert more than a single function within the membranes; as a quasi-planar molecule it associates with the fatty acid chains of the phospholipids in an analogous way to cholesterol in mammalian membranes and stabilizes the membrane phase (ROTTEM et al., 1973). However, besides exhibiting bulk function and being part of critical domains—functions which might be satisfied also by other sterols (PINTO et al., 1983)—nanomolar levels of ergosterol

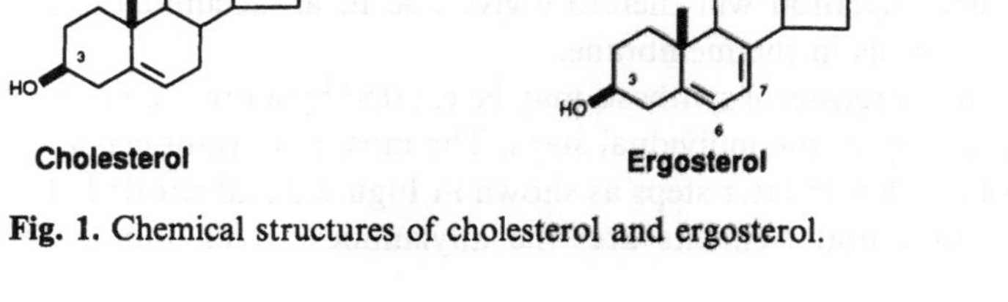

Fig. 1. Chemical structures of cholesterol and ergosterol.

Table 1. Sterol biosynthesis inhibitors in medicine and agriculture. Launched compounds with their primary biochemical targets and mode of application. All compounds for systemic use in medicine are listed, whereas a selection of topically applied therapeutics and agricultural compounds is given

Class	**Mode of action (inhibition of)**	**Application in medicine**		**Application in agriculture**
		topical	**systemic**	
azoles	C-14-demethylation	clotrimazole bifonazole miconazole econazole ketoconazole	miconazole ketoconazole itraconazole fluconazole	triadimefon propiconazole prochloraz
allylamines	squalene epoxidase	naftifine terbinafine	terbinafine	
morpholines	Δ14-reduction, Δ8–Δ7 isomerization	amorolfine		dodemorph tridemorph fenpropimorph

were shown to induce fungal growth pointing to a role designated as growth sparking (Rodriguez et al., 1985). These levels were clearly too low to modulate physical membrane properties.

The concept of reducing fungal ergosterol by interfering with its biosynthesis appeared as a useful approach to cope with fungal infections in therapy and also in plant protection (Berg et al., 1986; Kato, 1986). Table 1 shows a selection of compounds which have been introduced successfully to the market with their primary targets of ergosterol biosynthesis and the mode of application.

The biosynthesis of ergosterol is a multistep process which starts with acetyl-CoA and runs via mevalonic acid and squalene as key intermediates to the tetracyclic lanosterol. A simplified scheme showing also the far-reaching analogy to cholesterol synthesis is given in Figure 2. In the conversion of lanosterol to ergosterol several processes of demethylation, isomerisation, saturation/desaturation and alkylation take place and the sequence of events can differ. With respect to the order of transmethylation at C-24, different routes are observed in yeasts and filamentous fungi: methylation at C-24 takes place at the lanosterol level in a wide range of fungi, however, at a later level in *S. cerevisiae* (Kato, 1986). Inhibition of 14α-demethylation will therefore give rise to an accumulation of different 14α-methyl-sterols in the membrane.

Effective inhibition of ergosterol synthesis may be exerted by a diverse range of chemical structures at any of the individual steps. The most important classes of inhibitors listed from earlier to later steps as shown in Figure 2 and entitled after their characteristic structural elements are: the allylamines which inhibit the

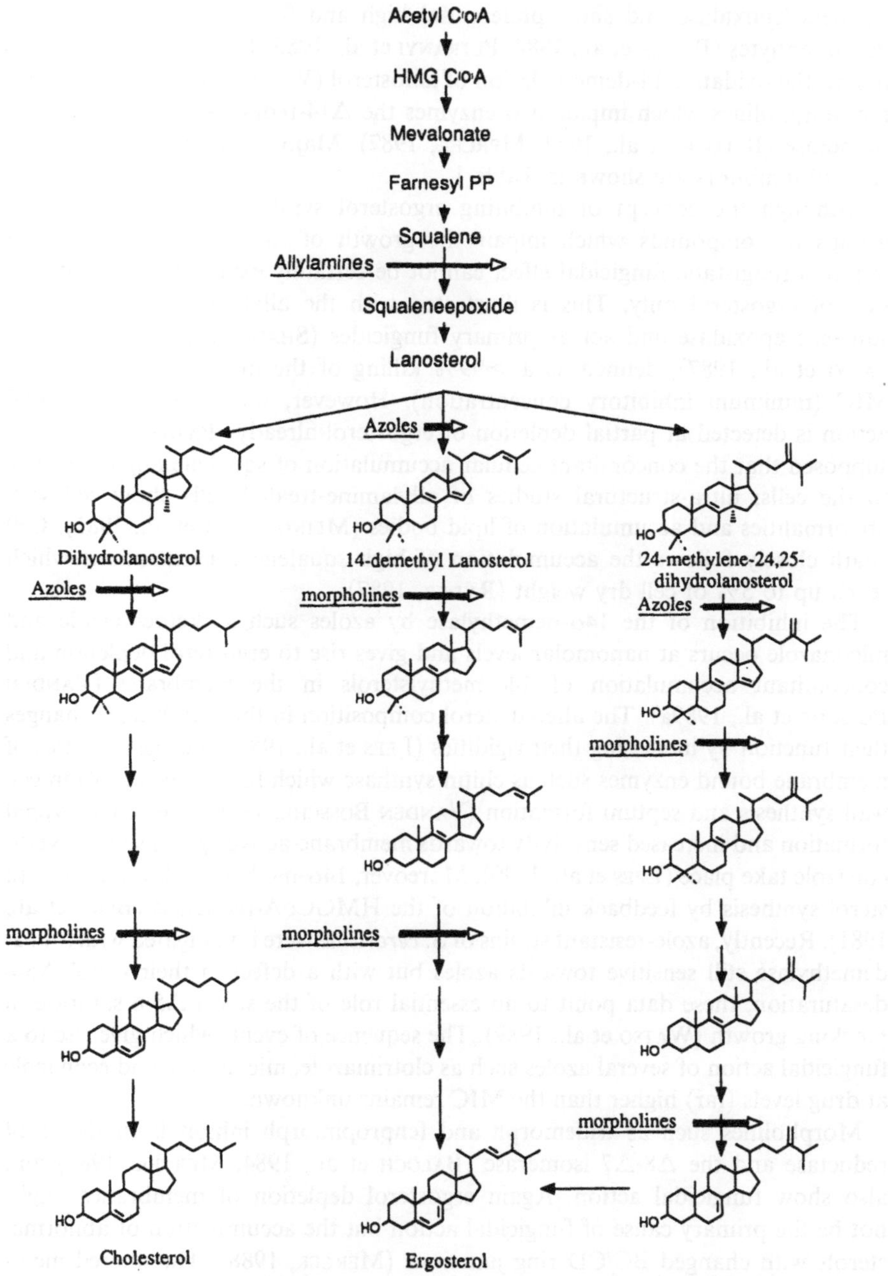

Fig. 2. Biosynthesis of cholesterol and ergosterol. A highly simplified scheme showing the steps attacked by the allylamines, the azoles and the morpholines (See also VANDEN BOSSCHE, 1988; MERCER, 1988).

squalene epoxidase and show preferential high and fungicidal activity against dermatophytes (RYDER et al., 1984; PETRANYI et al., 1985, 1987), the azoles which inhibit the oxidative 14-demethylation of lanosterol (VANDEN BOSSCHE, 1985) and the morpholines which impair two enzymes the Δ14-reductase and the Δ8-Δ7-isomerase (BALOCH et al., 1984; MERCER, 1987). Major representatives of each class of inhibitors are shown in Table 1.

Although the concept of inhibiting ergosterol synthesis has led to several groups of compounds which impair the growth of fungi very efficiently the eventual fungistatic/fungicidal effect cannot be certainly explained by the depletion of ergosterol only. This is illustrated with the allylamines which inhibit squalene epoxidase and act as primary fungicides (SHADOMY et al., 1985; PETRANYI et al., 1987); defined as a >99% killing of the initial inoculum at the MIC (minimum inhibitory concentration). However, the onset of fungicidal action is detected at partial depletion of ergosterol already (RYDER, 1987). It is supposed that the concomitant cellular accumulation of squalene might be toxic to the cells; ultra-structural studies of allylamine-treated cells show cell wall abnormalities and accumulation of lipid bodies (MEINGASSNER et al., 1981). Cell death closely follows the accumulation of high squalene concentrations which reach up to 5% of cell dry weight (RYDER, 1987).

The inhibition of the 14α-demethylase by azoles such as ketoconazole and miconazole occurs at nanomolar levels and gives rise to ergosterol depletion and concomitant accumulation of 14α-methylsterols in the membrane (VANDEN BOSSCHE et al., 1989a). The altered sterol composition in the membranes changes their function by increasing their rigidities (LEES et al., 1988) and the activities of membrane bound enzymes such as chitin synthase which has a crucial role in cell wall synthesis and septum formation (VANDEN BOSSCHE, 1988). Defective hyphal formation and increased sensitivity towards membrane-active agents such as ketoconazole take place (LEES et al., 1990). Moreover, 14α-methylsterols interfere with sterol synthesis by feedback inhibition of the HMGCoA-reductase (BERG et al., 1981). Recently, azole-resistant strains of *S. cerevisiae* were investigated with a 14α-demethylase still sensitive towards azoles but with a defect in their sterol Δ5-6 desaturation; these data point to an essential role of the sterol Δ5-6 saturase in blocking growth (WATSO et al., 1989). The sequence of events which give rise to a fungicidal action of several azoles such as clotrimazole, miconazole and econazole at drug levels (far) higher than the MIC remains unknown.

Morpholines such as tridemorph and fenpropimorph inhibit both the Δ-14 reductase and the Δ8-Δ7 isomerase (BALOCH et al., 1984; MERCER, 1987) and also show fungicidal action. Again ergosterol depletion of membranes might not be the primary cause of fungicidal action but the accumulation of abnormal sterols with changed BC/CD ring junctions (MERCER, 1988). The altered membrane properties become evident from a hyperfluidity and changed permeability (MERCER, 1988) and the observed irregularities in the distribution of chitin in the cell wall might be the consequence of the hyperfluidity (KERKENAAR et al., 1984).

From these examples it is evident that the killing effect on fungi may not be caused primarily by ergosterol depletion of membranes but that it is the particular impaired target enzyme which dictates the changes in membrane composition and the resulting toxic consequences. Considering the intermediary products in ergosterol synthesis (Fig. 2) which accumulate on the impairment of different steps they all are highly lipophilic and will be preferentially distributed into membranes. From their different chemical structures and properties a diversity of membrane associated changes might be suspected.

The biosynthesis of ergosterol has many steps in common with cholesterol synthesis (Fig. 2). Inhibitors of a particular fungal enzyme may affect also its mammalian counterpart and changes in cholesterol homeostasis as well as accumulation of potentially dangerous intermediates may occur. Antifungals especially designed for systemic use, should therefore possess a high degree of selectivity towards the fungal enzymes and therapeutic tissue levels should be far lower than the Ki's for the mammalian enzymes. This was proven for the antifungal allylamines which inhibit in vitro hepatic squalene epoxidase—depending on the source—by 100 to 2000 fold less than the fungal enzyme (RYDER, 1988) and presumptive tissue levels during therapy will not approach the required heights (SCHUSTER et al., 1988).

With respect to the azoles itraconazole, ketoconazole and miconazole a selectivity of 29–88 fold was shown for the fungal C-14 demethylase with Ki-values for the respective mammalian enzyme in the micromolar range (VANDEN BOSSCHE, 1988). In fact, in subcellular systems and in rat hepatocyte cultures the accumulation of several oxygenated sterols, especially of the C-32 aldehyde intermediate was observed at moderate levels of ketoconazole (IGLESIAS and GIBBONS, 1989). In vivo, local tissue levels of azoles after therapeutic doses might exceed the micromolar range (RITTER, 1988) and therefore a potential for a substantial inhibition of the mammalian 14α-demethylase exists.

Morpholines are of particular importance for agricultural needs (Table 1) and only a single compound, amorolfine, has been developed and recently marketed for topical therapy of superficial mycoses in man (POLAK and DIXON, 1987; POLAK and ZAUG, 1990). The drug is lacking systemic activity (POLAK and ZAUG, 1990) and, therefore, a possible impairment of mammalian sterol synthesis is of minor importance. A detailed description of this class of ergosterol synthesis inhibitors will not be given in this review.

3. Squalene epoxidase as an antifungal target enzyme

3.1. General properties of squalene epoxidase

The 30-carbon open-chain hydrocarbon squalene is one of the most lipophilic compounds produced in cells which does not occur in a free form but exclusively

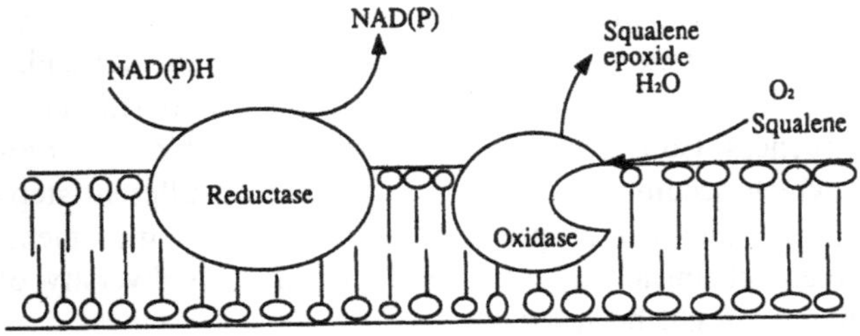

Fig. 3. Components of the squalene epoxidase complex. The enzyme complex from mammalian and fungal sources is located in the microsomal fraction and consists of a NAD(P)H-cytochrome reductase and a terminal oxidase which is not a cytochrome P-450; oxygen, NAD(P)H and FAD are required for enzyme activity (Ono et al., 1977, 1982; Ryder and Dupont, 1984).

partitioned into hydrophobic spaces. Its conversion to the tetracyclic lanosterol (Fig. 2) takes place in two steps and is catalyzed by two discrete enzymes, the squalene epoxidase and the subsequent cyclase (Van Tamelen et al., 1966).

Squalene epoxidase has been detected in most eucaryotic cells. It appears to be an important site in the control of cholesterol homeostasis since its activity is regulated in vivo by the abundance of precursors and is decreased on cholesterol feeding (Eilenberg and Shechter, 1987). The purification and extensive investigation of the enzyme from liver has been reported (Ono and Bloch, 1975; Ono et al., 1977, 1982). As an antifungal target the enzyme has been also characterized from different Candida strains (Ryder and Dupont, 1984, 1985) and from *Saccharomyces cerevisiae* (M'Baya and Karst, 1987).

Squalene epoxidases from the mammalian and the fungal sources have many properties in common. They are located in the respective microsomal fractions and consist of a NAD(P)H-cytochrome reductase and a terminal oxidase (Fig. 3; Ono et al., 1977, 1982). For exerting catalytic activity molecular oxygen, FAD and NAD(P)H are required. However, the terminal oxidase is not a heme protein and is not impaired by inhibitors of cytochrome P-450 (Ono et al., 1982; Ryder and Dupont, 1984).

3.2. Allylamines as inhibitors of squalene epoxidase

The antifungal activity of a novel structure—naftifine—was discovered in a screening program accidentally (Georgopoulos et al., 1981). Suspecting some resemblance between the chemical structure of naftifine (Fig. 4) and sterols biochemical studies on a possible interference with ergosterol synthesis were started and showed inhibition at a single step, the squalene epoxidation (Paltauf et al., 1982; Ryder and Troke, 1982). Naftifine itself was found to be highly effective in the topical treatment of dermatomycoses and was launched 1985 (Exoderil[R]).

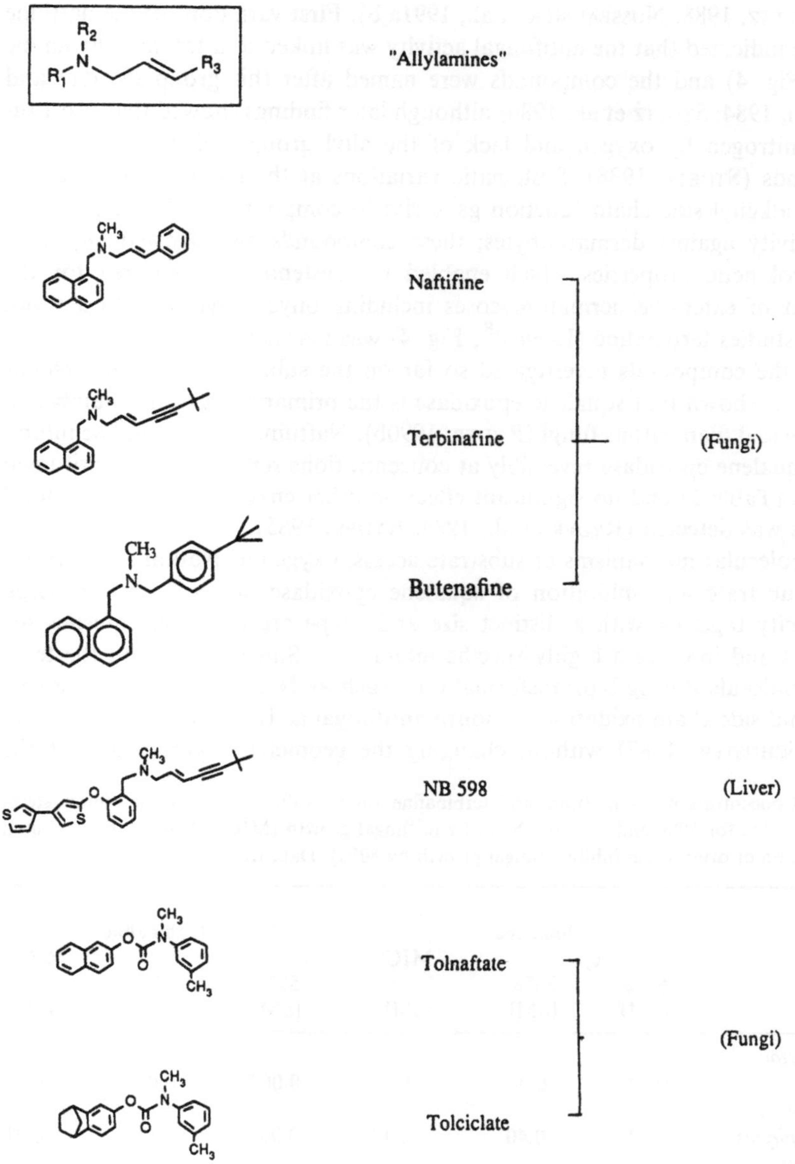

Fig. 4. Chemical structures of squalene epoxidase inhibitors.

The novel mechanism and a high primarily fungicidal activity against a range of human pathogens in vitro and in vivo gave rise to an extensive program with the synthesis of more than a thousand new compounds and the establishment of structure-activity relationships (PETRANYI et al., 1984; STUETZ et al., 1986; STUETZ,

1987, STUETZ, 1988; NUSSBAUMER et al., 1991a,b). First variations of the naftifine molecule indicated that the antifungal activity was linked to a tertiary allylamine group (Fig. 4) and the compounds were named after this group (STUETZ and PETRANYI, 1984; STUETZ et al., 1986) although later findings showed that substitution of nitrogen by oxygen and lack of the allyl group still produced active compounds (STUETZ, 1988). Systematic variations at the naphthalene ring and at the n-alkenyl side chain function gave rise to compounds with exceptionally high activity against dermatophytes; these compounds also showed improved pharmacokinetic properties which enabled the systemic use required for the treatment of extensive dermatomycoses including onychomycoses. As a result of these studies terbinafine (Lamisil[R]; Fig. 4) was marketed recently.

From the compounds investigated so far on the subcellular and the cellular level it was shown that squalene epoxidase is the primary target in a number of yeast-like and filamentous fungi (RYDER, 1990b). Naftifine and terbinafine inhibit fungal squalene epoxidase reversibly at concentrations reflecting their antifungal activities (Table 2) and no significant effect on other enzymes involved in sterol synthesis was detected (RYDER et al., 1984; RYDER, 1985).

The molecular mechanisms of substrate access, oxygen activation and transfer to the substrate and inhibition of squalene epoxidase are still unclear. High lipophilicity together with a distinct size and shape are clear prerequisites for inhibitors and indicate a highly specific interaction. Single attacks on the terbinafine molecule during biotransformation—such as N-demethylation, ring oxidation and side chain oxidation—abolish antifungal activity (NUSSBAUMER et al., 1991b; SCHUSTER, 1987) without changing the geometrical parameters of the

Table 2. Concentrations of naftifine and terbinafine (mean values) required to inhibit sterol biosynthesis (c_i; for 50% and 95% inhibition) and fungal growth (MIC). (MIC is the minimum concentration of drug which inhibits fungal growth by 50%). Data from RYDER, 1990b.

Fungi	**Naftifine** c_i 50% [μM]	95% [μM]	MIC [μM]	**Terbinafine** c_i 50% [μM]	95% [μM}	MIC [μM]
Trichophyton rubrum	0.02	0.36	0.17	0.002	0.07	0.01
Trichophyton mentagrophytes	0.02	0.40	0.17	0.007	0.13	0.01
Aspergillus fumigatus				0.230	3.96	2.64
Candida parapsilosis	0.76	13.20	5.28	0.020	1.00	1.32
Candida albicans	1.16	24.75	165.00	0.026	0.66	10.23
Candida glabrata	1.12	37.29	330.00	0.132	2.97	330.00

molecule too much. Despite their high lipophilicity allylamines are still up to 10 orders of magnitude less hydrophobic than squalene and also lack a structural resemblance to squalene; from their non-competitive inhibition kinetics a binding to the substrate site of the fungal enzyme was thought to be unlikely (RYDER, 1990b). However, the involvement of an independent regulatory site appears questionable and would imply a different type of regulation in the mammalian enzyme which is inhibited by naftifine and terbinafine in a competitive manner (the enzymes from different mammalian species are far less sensitive to allylamine inhibition (Ki's >(>) micromolar range) than the fungal enzymes (RYDER, 1990b)). The selectivity towards the fungal enzymes is a clear asset for this type of inhibitors: presumptive organ levels during therapy will be far too low to interfere with cholesterol synthesis. In fact, in the course of extended multi-center clinical trials with terbinafine particular emphasis was also given to cholesterol levels but no significant changes are known.

The lack of interference with cholesterol homeostasis by terbinafine and naftifine does not indicate a general failure of squalene epoxidase inhibitors to work in mammals. Recent data on inhibition of the mammalian enzyme by structures resembling to some extent to the allylamines (e.g. NB-598; Fig. 4) showed competitive kinetics at nanomolar levels (HORIE et al., 1990). As a consequence of the strong inhibition oral doses of 10 mg/kg · d evoked a strong hypocholesteremic effect in dogs.

Thiocarbamate derivatives such as tolnaftate and tolciclate had been used as topical antifungals for a long time. Besides being less hydrophobic the structures show some similarity in size and shape to the allylamines (Fig. 4). After establishing the mode of action for allylamines the thiocarbamates were checked in the squalene assay and found to be inhibitory (MORITA and NOZAWA, 1985; RYDER et al., 1986).

3.3. Antifungal spectrum of allylamines and in vivo activity

In vitro allylamines are highly active against a wide range of medically important fungi (Tables 2,3) (GEORGOPOULOS et al., 1981; PETRANYI, 1985; SHADOMY et al., 1985; PETRANYI et al., 1987; CLAYTON, 1987; SCHUSTER et al., 1988; RYDER and MIETH, 1991). Dermatophytes are the most susceptible species with MIC-values of terbinafine in the nanomolar range. The activities against aspergilli, dematiaceous and dimorphic fungi are also very high with MIC-values in the subμmolar range for terbinafine. The activity against yeasts is subject to wider variations and depends on the strain and growth form. The mycelial growth form of *C. albicans* which is considered to be the pathogenic form is inhibited by terbinafine in the subμmolar range (SCHAUDE et al., 1987).

Allylamines are fungicidal against all dermatophytes, some aspergilli,

Table 3. In vitro activities (MIC) of terbinafine and naftifine against a wide range of fungi (data from many different groups, summarized in SCHUSTER et al., 1988). MIC is the minimum concentration of drug which inhibits fungal growth by 50%

Fungi	MIC [μM] Terbinafine	Nafitifine
dermatophytes *Trichophyton* spp., *Epidermophyton* sp., *Microsporum* spp.	0.0033–0.033	0.033–3.30
filamentous fungi *Aspergillus* spp., *Mucor* spp. et al.	0.017–>500.00	0.800–660.00
dimorphic fungi *Blastomyces* sp., *Histoplasma* sp., *Sporothrix* sp.	<0.165–6.60	<0.165–26.40
pathogenic yeasts *Candida* spp., *Cryptococcus* sp., *Pityrosporum* spp.	0.330–>300.00	0.660–660.00

dimorphic fungi and yeasts; fungistatic effects were observed against *C. albicans* (SHADOMY et al., 1985; PETRANYI et al., 1987).

Fungicidal action of allylamines has been proven also in vivo (MIETH, 1990) and together with a favourable distribution and a lack of adverse effects these properties make allylamines first choice drugs especially, in the treatment of dermatomycoses including nail infections. Compounds with basically fungistatic properties (azole antimycotics) improve clinical symptoms but after discontinuing treatment still viable fungi may be left which cause the relapse of the infection.

Allylamines are highly lipophilic and penetrate easily to the sites of infection (SCHUSTER et al., 1988; SCHUSTER, 1990). After topical application naftifine and terbinafine show high drug levels in the stratum corneum and invade quickly and deeply the hair follicles where residual fungi frequently occur and from where recurrence of infection may take place; both drugs exert a fungicidal action (MIETH and PETRANYI, 1989). Orally applied allylamines become rapidly and entirely absorbed and distribute preferentially to adipose tissue and skin; for terbinafine high, long-lasting and bioavailable drug levels are achieved at the target sites of infection (BATTIG et al., 1987; SCHUSTER et al., 1988). A higher rate of biotransformation (GRIMUS and SCHUSTER, 1984) together with generally lower antifungal potency (Table 3) prevents the establishment of sufficiently high antifungal levels after oral administration of naftifine.

From the experience with the spectrum of marketed antifungals, especially with the azole-derivatives, particular emphasis has to be directed to a potential interference with physiological processes in the host. As already mentioned antifungal allylamines do not interfere with cholesterol synthesis (RYDER, 1990b) and there is no evidence for a different target enzyme in the host. Toxicological studies of terbinafine in animals relevant to man, including acute and chronic protocols, showed a lack of substantial negative side effects (VILLARS and JONES, 1990; JONES and VILLARS, 1990).

Compared to azoles a major advantage of allylamines is their lacking potential to inhibit cytochromes P-450 (Table 4). As discussed in the next chapter azole antifungals act by inhibiting 14α-lanosterol demethylase, a member of the superfamily of cytochrome P-450 (e.g. VANDEN BOSSCHE, 1988). Considering the host of different families with different subtypes it is obvious that specificity for a particular cytochrome P-450 will be a limited one. Impairment of other related enzymes involved e.g. in drug metabolism, steroidal hormone and prostaglandin synthesis becomes plausible. Almost every step in steroidal hormone synthesis is impaired by azoles such as ketoconazole, clotrimazole and miconazole; allylamines fail to interact with the respective enzymes (SCHUSTER, 1985, 1987b; Table 4). No inhibitory binding (type II-interaction) of allylamines to other cytochromes P-450, also involved in prostacylin synthesis and in drug metabolism could be observed (SCHUSTER, 1985, 1987b; BACK et al., 1989); neither the binding of endogenous substrates nor of carbon monoxide to the reduced cytochromes P-450 were impeded by allylamines.

As a consequence of their lipophilic character allylamines bind as substrates in type-I manner to a fraction of hepatic cytochromes P-450 with moderate affinity (Table 4) and become degraded by this enzyme (SCHUSTER, 1985, 1987; SCHUSTER et al., 1988). The potential to compete with other substrates of the same cytochrome P-450 is considered to be low: presumptive hepatic terbinafine levels after therapeutic doses will not be adequate to saturate a major fraction of the degrading enzyme (Ks > 10μM; Table 4).

3.4. Clinical efficacy and tolerance of allylamines

Naftifine and terbinafine are both very effective in the topical treatment of dermatophyte infections such as tinea corporis, tinea cruris and pedis, cutaneous candidiasis and pityriasis (RYDER and MIETH, 1991). Summarizing data from some 50 studies on roughly 2,900 patients treated with 1% cream of either drug at standard duration mycological cure rates between 79 and 100% were achieved in the different indications; adverse events (such as irritation, erythema, burning or dryness) occurred at the same low frequency as in the placebo groups (JONES, 1990a,b; JONES and VILLARS, 1990). As a result of the fungicidal action of the drugs and their extensive distribution to skin tissue short-term treatment of 1 or 2 weeks (half of that currently used) was sufficient to yield eventually similar high

Table 4. The interaction of allylamines and azoles with cytochromes P-450 involved in steroidal hormone synthesis and in liver

Cyt P-450 in	**Nafti**	**Terbi**	**Keto**	**Clo**	**Itra**	**Fluco**	**SDZ**
adrenals							
scc, 11β, 18							
type	nb	nb	II	II	II	nt	II
n [% P-450]			>65	>65	>30		nt
CO-inhib.	0	0	+++	+++	+		0
K'_d [μM]			<0.30	<0.50	1.00		18.5; 90
17α, 21							
type	nb	nb	II	II	II	nt	II
n [% P-450]			>50	>85	>50		nt
CO-inhib.	0	0	+	+++	+		0
K'_d [μM]			1.00	0.25	1.10		142.80
testis							
17α-20lyase							
type	nb	nb	II	II	nt	nt	II
n [% P-450]			>50	>50			nt
CO-inhib.	0	0	+++	++			0
K'_d [μM]			<0.04	<0.08			21.70
placenta							
aromatase							
type	nb	nb	II	II	nb	II	II
n [% P-450]			>80	>80		>50	>50
CO-inhib.	0	0	+	+		+	0
K'_d [μM]			2.00	<0.40		0.95	1.49
keratinocytes							
VitD-1-OHase							
K'_d	nt	nt	<0.20	<0.20	0.15	>10.0	>10.0
VitD-24-OHase							
K'_d	nt	nt	0.40	<0.20	1.00	>10.0	>10.0
Liver (man)							
total P-450's							
type	I	I	II	II	II	II	II
n [% P-450]	<5	<5	>60	>60	>10	>35	>35
CO-inhib.	0	0	++	++	(+)	+	+
K'_d [μM]	20.0	>10.0	0.33sp	0.14sp	6.67	<0.3sp	0.52sp

The abbreviated names stand for the allylamines naftifine and terbinafine and for the azoles ketoconazole, clotrimazole, itraconazole, fluconazole and SDZ 89-48J (cp. Fig. 8; p. 166).

Drug interactions with cytochromes P-450 in adrenal mitochondria and microsomes from bull (scc: cholesterol side chain cleavage, 11β-hydroxylase, 18-hydroxylase, 17α-hydroxylase, 21-hydroxylase), in testis microsomes from bull (17α-20-lyase), in human placenta (aromatase) and in human liver microsomes (several gene families of cytochrome P-450 with different subtypes) were followed by spectrophotometric titrations at pH 7.4, 25°C as described (Schuster, 1985):

Type I-binding indicates the formation of an enzyme-substrate complex (allylamine binding to hepatic P-450), Type II-binding the direct ligation to the heme iron (general type of azole binding to P-450). The fraction of involved cytochrome P-450 – n [%P-450]—was calculated for Type-I substrates from the size of the maximal spectral change. For Type II inhibitors a minimal n was assessed from the fraction of antifungal efficacy; after stopping treatment healing continued without relapse (JONES, 1990).

Dermatophyte infections are often associated with an inflammatory response and clinical improvement is enhanced when combining antifungal drugs with weak corticosteroids. Naftifine (but not terbinafine) shows intrinsic anti-inflammatory properties in addition to its antifungal action (SOWDEN and ALLEN, 1990; JONES, 1990). Short-term treatment together with anti-inflammatory action without potential problems of corticoid use are clearly substantial advantages over other marketed drugs.

Dermatophyte infections which cause widespread lesions of the skin and/or affect the nails and the scalp require systemic therapy. Oral terbinafine meets the requirements for the treatment of these medical indications exceptionally well and shows therapeutic advantages over the existing medication, especially, in the treatment of chronic forms of tinea pedis and onychomycoses; this follows from the outcome of 32 clinical studies including more than 1,000 patients in which the high efficacy and tolerance were proven (VILLARS and JONES, 1989, 1990). The favourable distribution properties—yielding high drug levels in areas of damaged thickened skin and in nails (FAERGEMANN et al., 1990) and the rapid onset of a principal fungicidal action give rise to clinical cures even after short-term treatment (6–12 weeks in onychomycoses) and to an unusually low relapse rate (VILLARS and JONES, 1990).

4. 14α-Lanosterol demethylase as an antifungal target enzyme

4.1. 14α-Lanosterol demethylase—a member of the cytochrome P-450 superfamily

Lanosterol 14α-demethylase is a constitutive enzyme common to fungal ergosterol synthesis and mammalian cholesterol synthesis. A functional 14α-demethylase is essential for the aerobic growth of fungi. Mutants from *S. cerevisiae* which lack this activity and therefore ergosterol are obligate anaerobes (KALB et al. 1987).

---

cytochrome P-450 retarded and/or blocked in carbon monoxide binding (CO-inhib. ranging from very strong inhibition = + + + to lacking inhibition = 0). K_d': apparent dissociation constant calculated after Scatchard from the titration curves. In case of hepatic microsomes a spectrum (sp) of Kd's was observed and the strongest interaction is listed only.

Vitamin D-hydroxylases were determined in human foreskin keratinocytes according to BIKLE et al. (1986).

nb = no binding observed.

nt = not tested.

Data from SCHUSTER, 1985, 1987; SCHUSTER and MEINGASSNER, 1988 and SCHUSTER and BIKLE, unpublished (vitamin D3-hydroxylases).

Early data on the inhibition of cholesterol synthesis by carbon monoxide showed accumulation of lanosterol indicating the 14α-demethylation as the vulnerable step and the reversal of this effect by light of 450 nm wavelength (GIBBONS and MITROPOULOS, 1973). Moreover, the potential involvement of a cytochrome P-450 enzyme in the oxidative removal of the 14α-methyl group from lanosterol was substantiated by the strict requirement of NADPH and oxygen and the inhibition by antibodies against the P-450 reductase (OHBA et al., 1978).

The mammalian as well as the fungal enzymes are present in the respective microsomal fractions at low levels only; they have been purified from both sources to homogeneity (YOSHIDA and AOYAMA, 1984; TRZASKOS et al., 1986; HITCHCOCK et al., 1989).

14α-Demethylase was isolated from Sprague-Dawley rat liver and identified as a single cytochrome P-450 isozyme with a molecular weight of 51 kD and a Soret peak at 448 nm for the reduced CO binding complex (TRZASKOS et al., 1986). The lanosterol 14α-demethylase has been also purified from a strain of *Candida albicans* (HITCHCOCK et al., 1989) and from *Saccharomyces cerevisiae* and its amino acid sequence has been deduced from its DNA-sequence (YOSHIDA and AOYAMA, 1984; LAI and KIRSCH, 1989).

For exerting its catalytic function the mammalian enzyme shows an absolute requirement for molecular oxygen and NADPH-cytochrome c-reductase. Other electron transport components such as cytochrome b5 and NADH-cytochrome reductase were not needed (TRZASKOS et al., 1986). In fungal cells the function of the NADPH-cytochrome P-450 reductase might be substituted by a different system (cytochrome b-5?) since mutants defective in this enzyme are still capable of aerobic growth (SUTTER and LOPER, 1989).

14α-Demethylase shows a high degree of substrate specificity, especially, with respect to the position of double bonds in the steroid nucleus; it is less sensitive to the size and shape of the side chain (SEKIGAWA et al., 1988; SONODA et al., 1989). 14α-Demethylase does not catalyze other attacks on the sterol nucleus; the microsomal 7α-hydroxylation of cholesterol could be clearly assigned to a different constitutive isozyme of cytochrome P-450 (TRZASKOS et al., 1986). The purified enzyme also did not exhibit significant catalytic activity towards standard drugs such as aminopyrine, p-nitroanisole and aniline (TRZASKOS et al., 1986).

4.2. 14α-Demethylation—a complex multistep process

In a reconstituted system containing the pure enzyme from the mammalian or fungal source, NADPH-reductase, dilaurylphosphatidylcholine, NADPH and ^{3}H-dihydrolanosterol as substrate it could be shown that the single cytochrome P-450 was responsible for the entire stereospecific process of 14α-demethylation

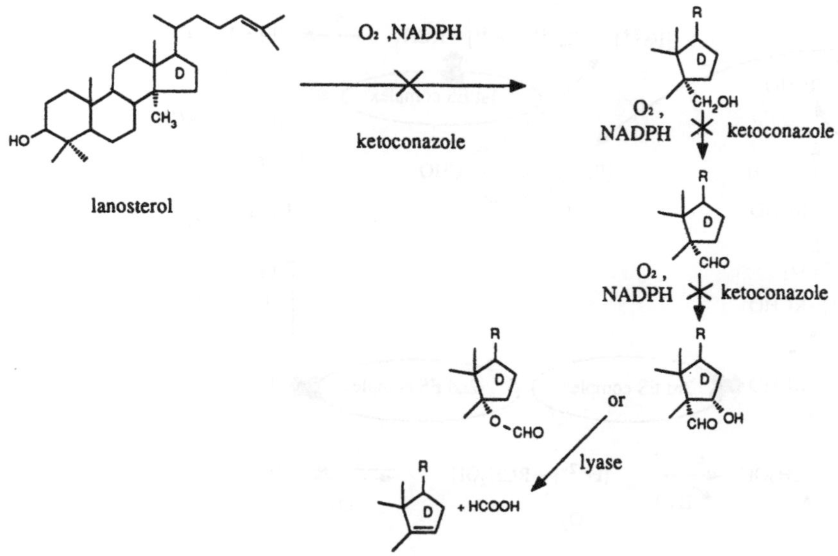

Fig. 5. Pathway for the 14α-demethylation of lanosterol. The three sequential oxidation steps require oxygen and NADPH and are inhibited by ketoconazole. The final lytic step requires the enzyme but is not impaired by the azole indicating a lytic site distant from the heme site. Data from TRZASKOS et al. (1986) and FISCHER et al. (1991).

consisting of three consecutive oxidations and a final lytic step with a Δ8,14 diene sterol formation (Fig. 5) (TRZASKOS et al., 1986; AOYAMA et al., 1987). In the first step the C-32 methyl becomes hydroxylated to the corresponding alcohol which can be detected by HPLC-analysis. The aldehyde produced in the subsequent step is detectable as well. The third oxidative intermediate prior to elimination of formic acid has not yet been detected with the mammalian enzyme (SHAFIEE et al., 1986) but has been identified recently for the fungal enzyme; FISCHER et al. (1991) succeeded in isolating a 14α-formyloxy intermediate (Fig. 5). Whether the same product or a 15α-hydroxyl sterol (Fig. 5) should be postulated for the liver enzyme is unclear. Carbon monoxide inhibits the entire oxidative process (TRZASKOS et al. 1986) and all three oxidative steps are also sensitive to ketoconazole inhibition; the final lytic step requires the enzyme but is independent of NADPH and oxygen and is not impaired by ketoconazole, thus, suggesting that a different catalytic site of the enzyme is involved in this process (FISCHER et al. 1991).

The overall features of a P-450 reaction cycle (reviewed e.g. by GUENGERICH (1991)) are shown in a simplified form in Figure 6, including also the characteristics of 14α-demethylation (Fig. 5). In short, the lipophilic substrate (RCH_3) binds to the ferric form of the enzyme—as shown in Figure 7—changing the spin

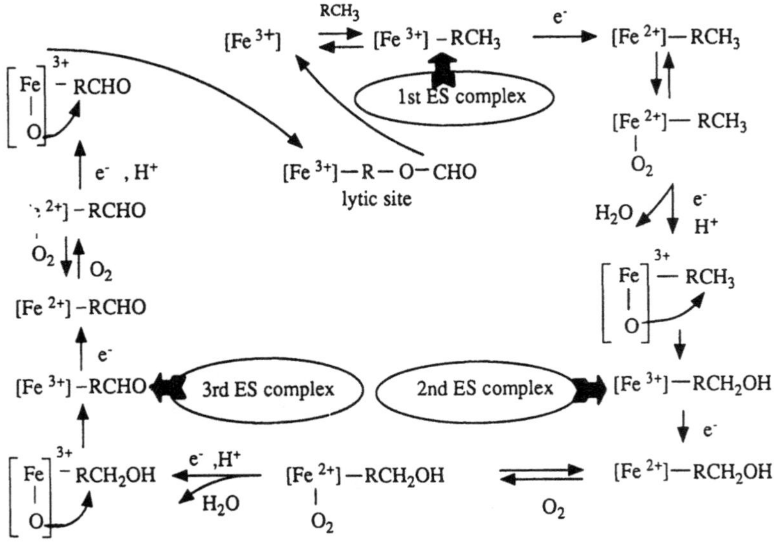

Fig. 6. The catalytic cycles of 14α-demethylase. Proposal based on the data from TRZASKOS et al. (1986) and FISCHER et al. (1991). It is supposed that the product of each individual oxidative P-450 cycle stays in the active site as a substrate for the next cycle.

equilibrium to the high spin state which generally becomes evident from the Type I-spectral change. The resulting complex receives an electron from the NADPH-dependent reductase and forms a ferrous enzyme-substrate complex. Molecular oxygen is associated to the ferrous heme iron to give a dioxygen-enzyme-substrate complex which is then reduced by transfer of a second electron from the reductase or from cytochrome b-5. Dioxygen cleavage in the reduced complex gives rise to the insertion of one oxygen into the substrate and reduction of the other to water. In case of lanosterol the less hydrophobic product of the first cycle, R-CH_2OH, may not leave but reside in the active site as an enzyme-substrate complex and enter the second cycle—again consuming two electrons and one oxygen. R-CHO, the product of the second cycle is still substrate for the enzyme and a third cycle will lead to the final oxidation product which is transferred to a different lytic site within the enzyme to form the final Δ8,14 diene sterol which is released from the enzyme (FISCHER et al., 1991).

Multifunctional enzymes which catalyze sequential processes at distant catalytic sites, such as tryptophan synthase or fatty acid synthase, have been well described (KIRSCHNER and BISSWANGER, 1974; WAKIL, 1989). The peculiar situation that a series of subsequent steps is catalysed at the same active site is not unique to 14α-demethylase but is shared by other steroid hydroxylases, e.g. by P450XIA (which converts cholesterol to pregnenolone; HEYL et al., 1986), by P450XIB (conversion of 11-deoxycorticosterone to aldosterone; IKUSHIRO et al.,

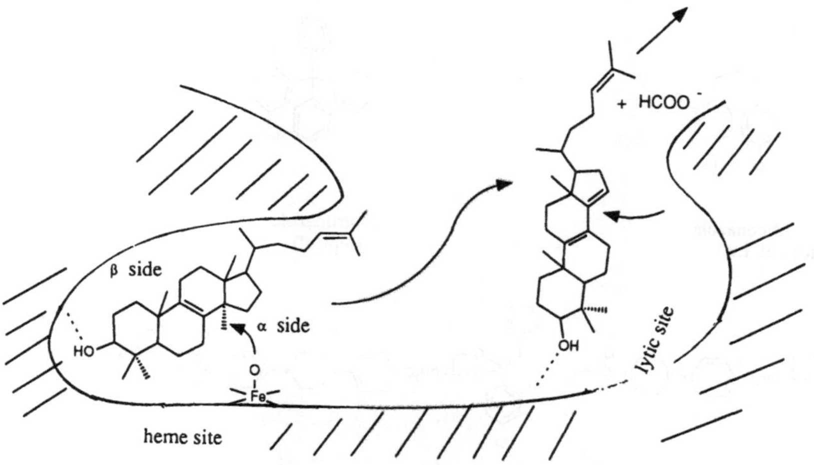

Fig. 7. Schematic proposal of the organisation of the oxidative and the lytic site of 14α-lanosterol demethylase. Scheme based on Figures 5 and 6.

1989), P450XVII (conversion of progesterone to androstenedione; NAKAJIN and HALL, 1981) and P450XIX (androstenedione to estrone). A high degree of similarity is observed in the 3-step oxidation mechanism catalyzed by the aromatase (STEVENSON et al., 1988) and 14α-lanosterol demethylase.

The intermediates of lanosterol demethylation may occupy almost identical positions to the heme and enter directly the next oxidative cycle without prior dissociation. Channeling could lead the final oxidation product to the lytic site which is distant to the heme and the ketoconazole binding region (FISCHER et al., 1991; Fig. 7). The regulatory advantage which a protein may gain from a restrainment of intermediates has been described for P450XVII recently (KUEHN-VELTEN et al., 1991).

The spatial organisation of the hydroxylating and the lytic site is still a matter of speculation. As with other membrane-associated cytochromes P-450, no relevant X-ray structural data are available and a mapping of the active sites can be attempted by fitting of substrate analogues and inhibitors. Summarizing such data from YOSHIDA et al., (1989) and FISCHER et al., (1991) the sterol is proposed to be accommodated in the heme site by an essential binding of its 3-OH group, possibly, to an oxyamino acid residue of the protein and by an interaction of the residual molecule via its β-surface with a different hydrophobic part of the active site (Fig. 7). This binding of the substrate—presenting its α-side in an almost invariant position to the heme iron throughout the 3-step process—is critically influenced by the conformation of the lanostene ring and the double bond within. The final oxidation product is then transferred to a distant lytic site.

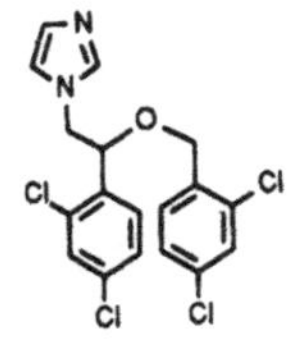

Miconazole
(topical, i.v.)

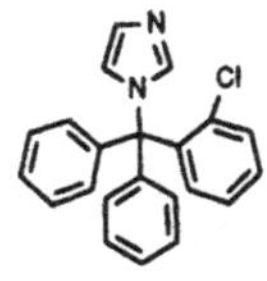

Clotrimazole
(topical)

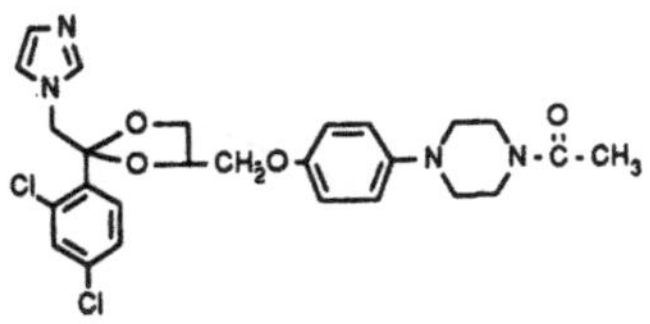

Ketoconazole
(oral, topical)

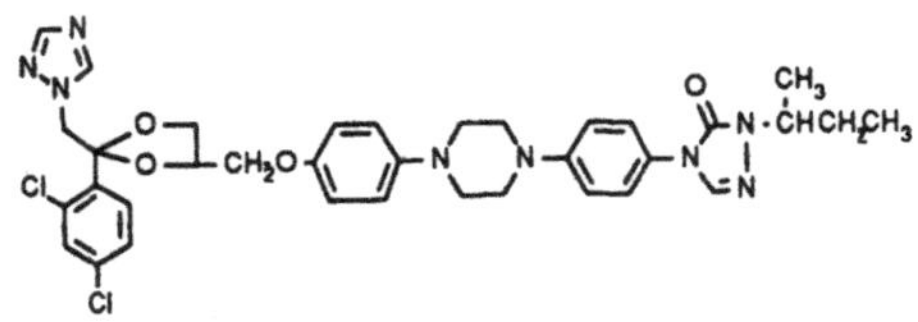

Itraconazole
(oral)

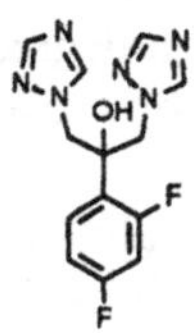

Fluconazole
(oral)

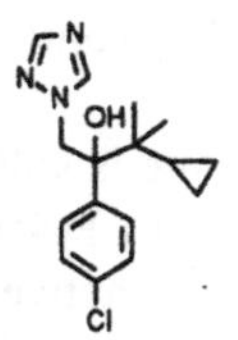

SDZ 89-485
(oral)
in development

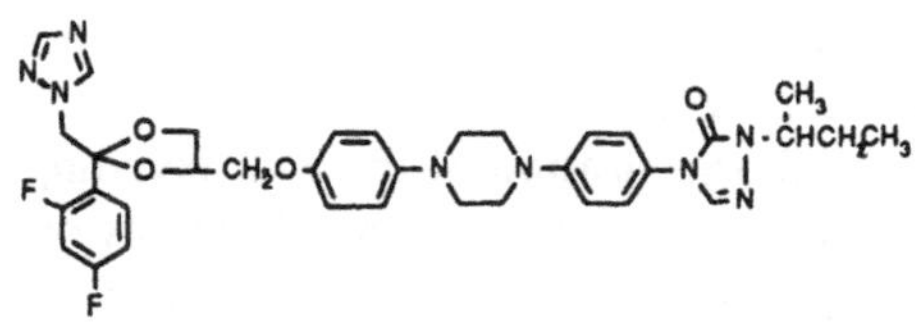

Saperconazole

in development

Fig. 8. Chemical structures of important azole therapeutics and mode of application.

4.3. Azole derivatives as inhibitors of 14α-lanosterol demethylase

Considering the complex sequential scheme of lanosterol demethylation inhibitors may interfere with individual steps of a single oxidative cycle (ORTIZ DE MONTELLANO and REICH, 1986), of different cycles and also with the lytic

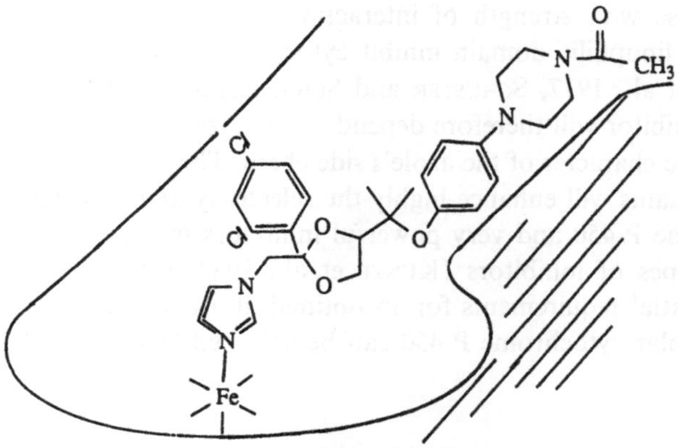

Fig. 9. Binding of an azole antifungal to the heme site of 14α-demethylase. The azole (ketoconazole) binds directly with its accessible nitrogen to the sixth position of the heme iron and with the residual molecule to the substrate site or a different hydrophobic site in the heme domain. See text.

reaction. The inhibitors applied in antimycotic therapy belong to the structural class of azoles and there exists evidence about how these compounds interact with cytochromes P-450 (Testa and Jenner, 1981; Ortiz de Montellano and Reich, 1986). The term azoles comprehends compounds with nitrogen bearing 5-membered heterocycles and most compounds for medical and agricultural use have been developed from the imidazole and triazole derivatives. The chemical structures of important therapeutics on the market and in development are given in Figure 8.

Potent azole containing derivatives interfere with the first steps prior to the oxygen transfer in a reversible way. They inhibit the association of molecular oxygen to the heme iron and simultaneously substrate approach and binding (Fig. 9). When diffusing into the active domain of the enzyme azoles meet a pentacoordinated heme iron or a weak ligand at the sixth position of the heme iron. Coordination of the nucleophilic nitrogen(s) of the azole to the sixth position of the iron gives rise to a so-called type II-difference spectrum with a Soret maximum around 430 nm indicative of the low spin form of the enzyme (Jefcoate, 1978; Schenkman et al., 1981). The change in the spin state is paralleled by a change in the redox potential making the reduction by the NADPH-reductase more difficult (Guengerich, 1983). Moreover, the formation of the crucial adduct of molecular oxygen with the ferrous heme may be severely impeded as shown by the retarded association of the marker carbon monoxide followed by the emergence of the 450 nm absorption band (Hitchcock et al., 1990).

Besides ligation to the heme iron residual parts of an azole molecule may bind simultaneously also to lipophilic regions of the protein (Ortiz de Montellano,

1986) and thereby increase with strength of interaction. Azoles that lack an additional binding to the lipophilic domain inhibit cytochrome P-450 function only weakly (Rogerson et al., 1977; Schuster and Schneeberger, 1989). The accommodation of the inhibitor will therefore depend on the size, shape, stereochemistry and hydrophobic character of the azole's side chain. Thorough screening for appropriate side chains will enhance highly the selectivity and specificity for a particular cytochrome P-450 and very powerful inhibitors may result. As exemplified with other types of inhibitors (Kunze et al., 1983; Murray and Reidy, 1990) from the spatial requirements for an optimal fit a mapping of the active domain of a particular cytochrome P-450 can be achieved (Hecht et al., 1989).

4.3.1. Assay of azole interactions with fungal 14α-lanosterol demethylase

The inhibition of cytochromes P-450 by azoles can be generally assayed at levels of different complexity ranging from in vivo models to cellular, subcellular and (partially) purified, reconstituted systems. Several different methods are usually applied.

a) The binding of azoles can be directly followed by the emergence of the type II binding spectrum. Spectrophotometric titrations of the enzymes with increasing azole concentrations allow the determination of the binding constant (Schenkman et al., 1981; Sheets and Mason, 1984; Schuster, 1985). The binding constant is a true measure for the competition which the substrate encounters on entering the active domain of the enzyme preoccupied by the azole; moreover, it is a highly valuable tool in the probing of the active site. However, relatively high enzyme concentrations are required and very strong binding ($Kd < 0.1 \mu M$) as observed with the fungal 14α-demethylase can be assessed on a qualitative level only (see below).

b) Inhibition of carbon monoxide binding to the ferrous enzyme-azole complex is a rapid, simple and therefore widely applied method (e.g. Vanden Bossche, 1985, 1988). However, it might not be indicative for the true inhibitory potential of a particular azole. A high-affinity azole binding may lock the enzyme in the ferric low spin state which scarcely becomes reduced then by the reductase and is prevented from entering the catalytic cycle. The failure to relate azole potency against the 14α-demethylase to its interference with CO binding has recently been reported by Hitchcock et al. (1990).

c) The inhibition of the catalytic activity is a very common method; for 14α-demethylase it is monitored either by the incorporation of 14C-mevalonate into 14α-demethylated sterols or by the conversion of lanosterol into the 14α-demethylated products (Pye and Marriott, 1982; Hitchcock et al., 1990).

Recent investigations on the inhibitory ability of the most important azole antifungals—including ketoconazole, clotrimazole, miconazole, fluconazole, itraconazole, SDZ 89-485 (for chemical structures see Figure 8) and ICI 153066—have been undertaken by several groups using whole-cells, cell-free preparations and (partially) purified and reconstituted systems derived from *Candida* and *S. cerevisiae* (Yoshida and Aoyama, 1987; Hitchcock et al., 1990; Ryder, 1990a). Summarizing their results it is obvious that all these compounds bind in type II manner to the demethylase with very high affinity giving rise to a 1 : 1 stoichiometry at 0.1μM enzyme concentration (Ki therefore estimated < 0.01μM). They inhibit the demethylase roughly with the same high potency if tested by the same laboratory (Table 5). (Different data obtained for a single drug by different laboratories may be due to the differing testing conditions with respect to incubation time and enzyme concentration). This appears very astonishing since the used compounds have highly different chemical structures, sizes and polarities (Fig. 8) which rule out a similar excellent accommodation within the active site. However, the extrapolated results may be an upper limit of the true inhibitory concentrations which are obscured by the (almost) stoichiometric interaction. Therefore, a sound ranking of the compounds cannot be made on the basis of these measurements on the target enzyme. Other properties with respect to partitioning, metabolic stability, antifungal spectrum, type and onset of antifungal action and potential side effects will help to distinguish between apparently equipotent compounds.

4.3.2. Antifungal spectrum of azole derivatives

In general, azoles show a broad antifungal spectrum which includes almost all human fungal pathogens such as dermatophytes (*trichophyton spp.*, *microsporum spp.*, *epidermophyta sp.*), yeasts (*candida spp.*, *cryptococcus sp.*, *torulopsis sp.*),

Table 5. Inhibition of 14α-lanosterol demethylase by azoles in cell free systems of Candida albicans and in rat liver microsomes. Inhibitory potencies (IC_{50} = drug concentration which causes 50% inhibition) determined from the reduced incorporation of 14C-mevalonate into 14α-demethylated sterols. Mean values; data from a) Hitchcock et al., 1990; b) Ryder, 1990a; c) Vanden Bossche, 1988.

Drug	IC_{50} [μM]		
	C. albicans		**rat liver**
	a)	b)	
ketoconazole	0.06	0.0087	2.00[c)]
itraconazole		0.0086	3.20[c)], 7.00[b)]
miconazole	0.11		6.00[c)]
clotrimazole	0.061		
fluconazole	0.074	0.0203	
SDZ 89-485		0.0059	2.02[b)]
ICI 153066	0.078		

moulds (*aspergillus sp.*) and dimorphic fungi (*sporothrix sp.*, *histoplasma sp.*, *blastomyces sp.*).

In vitro MIC-values (MIC: minimum concentration which inhibits growth by 50%) of most major azole representatives against these fungi are reported in the literature to be in the range of nM to > μM. (MARRIOTT and RICHARDSON, 1987; GRANT and CLISSOLD, 1989; SCHAUDE et al., 1990). A ranking of different azoles according to their in vitro activities against particular species is, however, difficult. Data from individual laboratories have been obtained either with too few drugs or on a too small range of fungal strains and interlaboratory data cannot be compared: No standardisation has been introduced yet into antifungal testing procedures, in particular, with respect to the selection of tested strains, to their physiological status (active or dormant cells) and to the testing conditions (PLEMPEL et al., 1988). The problems discussed with the IC-50 values measured at the molecular level of 14α-demethylase inhibition (see 4.3.1.) are obviously encountered at least to some extent also with MIC-testing.

4.3.3. Selectivity for fungal 14α-lanosterol demethylase

Compared to its mammalian counterpart highly selective inhibition of the fungal enzyme has been reported for the main antimycotic azoles (SHAW et al., 1987; VANDEN BOSSCHE, 1988; RYDER, 1990a) (Table 5).

However, specificity for a particular cytochrome P-450 is an incomplete one. The rationale to inhibit 14α-demethylase by azole binding is a general principle applying also to other cytochromes P-450. In the host organism the inhibitory action of azoles will not remain restricted to the fungal cytochrome P-450 but they will also interfere with cytochromes P-450 from the host and potentially lead to (toxic) side effects. Therefore, a very high degree of selectivity for the fungal enzyme is mandatory in the development of novel azole antifungals. As discussed below particular selectivity should be achieved with respect to cytochromes P-450 involved in cholesterol synthesis and degradation, in steroidal hormone synthesis and degradation, in the synthesis and degradation of products from the arachidonic acid cascade and in drug metabolism.

On the other hand, investigations on the interference of azoles with host enzymes have led and will lead to major contributions in the field of cytochrome P-450. In particular, the applicability of selective cytochrome P-450 inhibition has been successfully proven also in the area of endocrine disorders and hormonal induced tumors in man (TRACHTENBERG and PONT, 1984; STEVENS, 1985; BRODIE and SANTEN, 1986; SONINO, 1987; STEELE et al., 1987; SANTEN et al., 1989). Moreover, azoles may be used as highly valuable probes of the structures, mechanisms and biological roles of particular isozymes of cytochromes P-450 (MUKHTAR et al., 1984; HECHT et al., 1989).

4.3.3.1. Interference with drug metabolizing enzymes

Antifungal therapy requires long-term treatment in most cases and secondary medication during this time is common. Severe clinical consequences may arise from an inhibition of drug metabolism since increased and longer-lasting drug levels will frequently provoke adverse responses. On the other hand induction of drug metabolism may take place and cause the loss of efficacy of the second medication.

Azoles have the potential to fit into the active domains of a host of different cytochromes P-450 but they might also prefer one or few subtypes only. The minimum fraction of hepatic enzymes—comprising a multitude of subtypes—involved in azole binding can be estimated from the impaired or entirely blocked carbon monoxide binding: A minimum of 60–70% of the hepatic cytochromes P-450 in different species including man were assessed to bind ketoconazole, clotrimazole and miconazole with moderate to high affinity (SCHUSTER, 1985; Table 4). Itraconazole, fluconazole and SDZ 89-485 showed also strong binding to human hepatic enzymes with Ki's up to $4.10^6 \cdot M^{-1}$, however, a lower minimum fraction of binding enzymes was assessed (fewer subtypes?) with 35% for SDZ 89-485 and fluconazole and some 10% for itraconazole (Table 4; SCHUSTER and MEINGASSNER, 1988; SCHUSTER, unpublished). These azoles therefore bear the potential to inhibit several different subtypes of cytochrome P-450 and thereby the metabolism of different classes of drugs on concomitant therapy. However, whether inhibition will take place and to what extent will depend on the impaired subtype(s), on the dose and the stationary (hepatic) levels of concomitantly administered drugs.

Ketoconazole was shown to affect in vivo human plasma levels of cyclosporin A, chlordiazepoxide, methylprednisolone, antipyrine and S-mephenytoin whereas debrisoquine levels were less altered (FERGUSON et al., 1982; CLISSOLD, 1987; DANESHMEND and WARNOCK, 1988; ATIBA et al., 1989). These data indicate selectivity, with some cytochrome P-450 subtypes more affected than others. For fluconazole and itraconazole, the two other recently marketed oral drugs, few anecdotal clinical incompatibilities have been reported yet with respect to cyclosporin and warfarin (VAN CAUTEREM et al., 1989; GRANT and CLISSOLD, 1989). However, from the strong inhibitory effect which low doses of fluconazole exert on antipyrine metabolism in mice (LA DELFA et al., 1989) a substantial effect on human cytochrome P-450 subforms is plausible.

Studies on inhibition of human drug metabolism are clearly mandatory for azole drugs. For the well-established molecular basis of azole-caused drug–drug interactions investigations on the in vitro level using human tissues and purified enzymes will have a high predictive value for the in vivo situation. Systematic and quantitative in vitro screening with substrates and inhibitors for particular families/subtypes of cytochrome P-450 (MURRAY and REIDY, 1990) will not only reveal the selectivity of azoles but will also yield a comprehensive synopsis on the involvement of individual subtypes in the metabolism of a host of drugs. Studies of this type have started and will become milestones in predicting and avoiding incompatibilities due to drug–drug interactions.

Using cyclosporin A as substrate PICHARD et al. (1991) have recently screened 59 compounds for inhibiting/inducing effects on P450IIIA3 in human hepatocytes. The fundamental data obtained by this group and summarized in Table 6 will apply also to ketoconazole interference with drug metabolism since this drug exerts strongest interactions with P450IIIA3 (FABRE et al., 1988).

BALLARD et al. (1988) have approached the selectivity problem by investigating the influence of ketoconazole, miconazole, tioconazole on the site specific hydroxylation of testosterone in mouse hepatic microsomes. Several forms of cytochrome P-450—members of family II and III—participate in the hydroxylation of testosterone (MURRAY and REIDY, 1990) and azoles will reduce the extent of different metabolites according to their inhibitory potential towards the different subtypes. These investigations point to a weak selectivity of tioconazole which inhibits the individual activities in the range from 0.4–3.3μM, to an increased

Table 6. Effects of diverse drugs on P450IIIA3 activity in human hepatocytes measured by their effects on cyclosporin A metabolism. Data from PICHARD et al., 1991. A substantial inhibition of the metabolism of drugs in the first row by ketoconazole is likely

	Drugs exerting	
Inhibition	**Induction**	**No effect**
triacetyloleandomycin	rifampin	cefaperazone
erythromycin	sulfadimidine	cefotaxime
josamycin	phenobarbital	ceftazidime
midecamycin	phenytoin	isoniazide
ketoconazol	phenylbutazone	pefloxacin
miconazol	dexamethasone	vancocin
midazolam	sulfinpyrazone	trimethoprim
nifedipin	carbamazepine	amphotericin B
diltiazem		valproic acid
verapamil		quinidine
nicardipine		cimetidine
ergotamine		ranitidine
glibenclamide		omeprazole
bromocryptine		diclofenac
ethinylestradiol		aspirin
progesteron		paracetamole
cortisol		debrisoquin
		guanoxan
		captopril
		furosemide
		azetazolamide
		spartein
		gliclazide
		imipramine

selectivity of miconazole with inhibitory concentrations in the range of 0.2–10μM and a high selectivity of ketoconazole which inhibits 6β-hydroxylation at 0.08μM whereas the other activities require concentrations of 13– > 100μM (BALLARD et al., 1988). 6β-Hydroxylation is exerted by subforms of the P-450III-family and these findings are in line with the strong impairment of cyclosporine A metabolism by ketoconazole (HENRICSSON et al., 1990). Fluconazole interferes only weakly with all the subtypes involved in testosterone hydroxylation with Ki-values > 100μM (SHAW et al., 1987); obviously, strong binding of this drug (Table 4) to different subtypes takes place.

The strong impairment of the substrate binding of cyclosporin A and bromocriptine (which is competitive to other P-450III substrates) by different azoles was also shown in the author's laboratory (SCHUSTER and MEINGASSNER, 1988; SCHUSTER, unpublished). Using azoles at presumptive therapeutic levels ketoconazole and itraconazole abolished the binding of both substrates. Fluconazole had a substantial effect on cyclosporin A binding whereas SDZ 89-485 lacked an inhibitory interaction (Table 7).

Recent data on the novel macrolide immunosuppressive FK-506 showed competition with cyclosporin A for the same cytochrome P-450 and strong inhibition by ketoconazole but not by SDZ 89-485 (SCHUSTER, unpublished; OMAR et al., 1991). SDZ 89-485 might therefore be regarded as drug of choice in patients treated with substrates of P450AIII as comedication.

Besides inhibiting cytochrome P-450 activities, azoles have been found to induce these enzymes also. Clotrimazole, miconazole and tioconazole increase

Table 7. Effect of different azoles on the substrate (TYPE I) binding of cyclosporine A and bromocryptin to cytochrome P-450 in human liver microsomes. Spectrophotometric titrations at pH 7.4 and 25°C and calculations were performed as described in (Schuster, 1985). Two different liver samples were used; from a male kidney transplant donor (40 yrs, m) and a female (24 yrs, f). The binding constant Ks for cyclosporin and bromocryptin was determined in the absence and presence of 10 μM ketoconazole, itraconazole, SDZ 89-485 and 5 μM fluconazole. Data from SCHUSTER and MEINGASSNER, 1988 and SCHUSTER unpublished.

Azole	Ks [$10^6 \times M^{-1}$]			
	Cyclosporin A		Bromocryptin	
	m	f	m	f
none	0.086	0.089	1.20	0.84
SDZ 89-485	0.086	0.067	1.10	0.71
ketoconazole	*	n.t.	*	*
itraconazole	*	*	0.22	0.14
fluconazole	0.035	n.t.	1.10	n.t.

* means: up to a substrate concentration of 40–50 μM no TYPE I spectrum detectable; i.e. a strong inhibition of substrate binding.
n.t. not tested.

in vivo cytochrome P-450 content and monooxygenase activities in animal models (KAHL et al., 1980; RITTER and FRANKLIN, 1987a,b). Moreover, miconazole and clotrimazole were found to have an enzyme-inducing effect in humans (BURGESS and BODEY, 1972; HEINEMYER and JAECK, 1983) and these characteristics limited their use mostly to topical administration.

Ketoconazole exerts a biphasic effect on liver enzymes in mice, primarily acting as a strong inhibitor but changing to induction at later times (WHITEHOUSE et al., 1990). Superposition of inhibiting and inducing properties (as a response to chronic diminished enzyme levels?) may obscure the increased levels of subtypes (which may be small fractions of total cytochromes P-450) and the assessment of inducing properties should therefore prefer a direct measurement of enzyme and m-RNA levels over activity measurements. Itraconazole appears to have a low inducing effect in rodents with effects on hexobarbital hypnosis at doses > 40 mg/kg and there is no evidence for induction in man (VAN CAUTEREM et al., 1989; HEYKANTS et al., 1989). Fluconazole is a polar drug which is eliminated from the body mostly in unchanged form in all species investigated including man (SHAW et al., 1987); its high inducing potency in rodents even at low doses of 10 mg/kg·d (HEYKANTS et al., 1987) indicates an inducing potential also in man.

4.3.3.2. Inhibition of cytochromes P-450 involved in steroidal hormone synthesis

Most of the steps in the cascade of cholesterol to the individual steroidal hormones, including calcitriol, are catalyzed by specific subtypes of cytochrome P-450 and azoles interact to varying degrees with each of them (Fig. 10; Table 4). The high affinity ($Ki < \mu M$) which e.g. ketoconazole exhibits in vitro towards several of these enzymes, such as 17α-hydroxylase, 17,20-lyase (P450XVII), 11β-hydroxylase (P450XIB), desmolase (P450XIA), 1- and 24 vitamin D3-hydroxylases (e.g. SCHUSTER, 1985; SIKKA et al., 1985; NAGAI et al., 1986; LAMBERTS et al., 1986; REINHARDT and HORST, 1989; HOLLIS, 1990); has attracted high interest for the potential of undesired side effects as well as for a beneficial administration in endocrine disorders.

In fact, the high affinities for these cytochromes P-450 are reflected in the endocrine effects observed in patients during therapy. Ketoconazole at therapeutic and fairly higher doses transiently reduced human testosterone levels (PONT et al., 1982, 1984; SANTEN et al., 1983). These effects were clearly dose-dependent and accompanied by an incidence of gynecomastia of 0.2, 2 and 14% at doses of 200, 400 and 800 mg/day respectively (CLISSOLD, 1987). Impotence, reduced libido and sperm counts were observed in some cases (PONT et al., 1984). A decrease of testosterone levels was observed in females also with former elevated levels; high doses of ketoconazole gave rise to frequent menstrual irregularities (CLISSOLD, 1987). In these cases estrogen synthesis might have been impaired at two subsequent steps, at the level of androgen formation and additionally at the level of

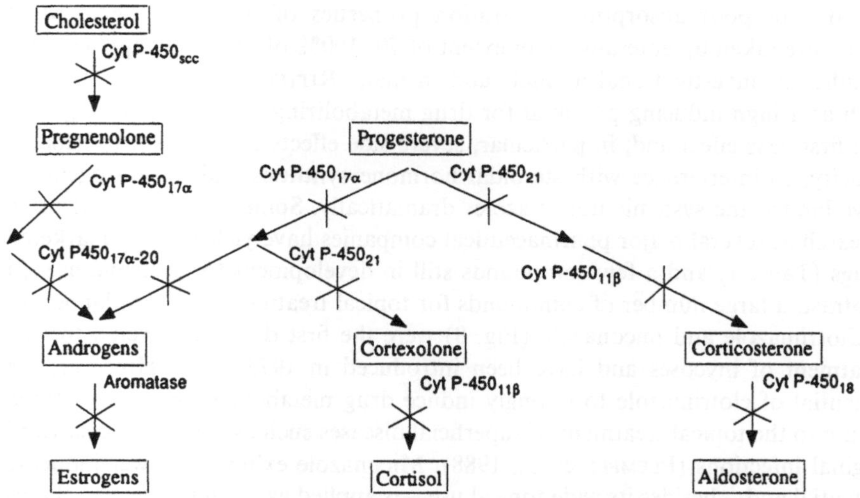

Fig. 10. Azole-vulnerable steps in the synthesis of steroidal hormones. A simplified scheme showing the cytochrome P-450 catalyzed steps which can be inhibited by azole derivatives.

aromatase for which ketoconazole exhibits lower but still substantial affinity (Table 4).

In addition to its effect on 17,20-lyase, ketoconazole showed a pronounced transient decrease in serum cortisol and a depressed response to ACTH-stimulation in patients (Pont et al., 1982; Graybill and Craven, 1983), effects explicable from the very strong inhibition of the 11β-hydroxylase (Nagai et al., 1986). Adrenals are in general the site at which highest lipophilic drug levels are detected in the body; this holds true for ketoconazole and even more pronounced for itraconazole (Grant and Clissold, 1989). High local inhibitor concentrations may outweigh a weaker inhibitory capacity and adrenal cytochromes P-450 should therefore be suspected as potential targets for lipophilic systemic azoles. In contrast, the polar fluconazole distributes even throughout the body (Brammer and Tarbit, 1987) and adrenal levels of this weak inhibitor will be too low to impair the local cytochromes P-450. However, fluconazole binds to the aromatase at μM concentrations (Table 4) and comparable levels prevail in placental tissue (Brammer and Tarbit, 1987).

4.4. Azoles in antifungal therapy

In vitro activities against pathogenic strains (see 4.3.2.) may indicate efficacy after topical treatment but will not warrant successful systemic therapy. This is clearly not due to poor absorption/penetration properties of the compounds. Most azoles are taken up enterally to an extent of 70–100% of the orally applied dose in different investigational animals and in man (RITTER, 1988). Other factors, such as a high inducing potential for drug metabolizing enzymes, a strong hepatic first pass effect and, in particular, severe side effects with respect to hepatotoxicity, to interference with steroidal hormone synthesis and to teratogenicity have limited the systemic use of azoles dramatically. Some 25 years of intensive research at several major pharmaceutical companies have yielded 3 (4) marketed drugs (Table 1) and a few compounds still in development for systemic use. In contrast, a large number of compounds for topical treatment has been launched.

Clotrimazole and miconazole (Fig. 8) were the first drugs discovered for the treatment of mycoses and have been introduced in 1972, 1973. However, the potential of clotrimazole to strongly induce drug metabolizing enzymes limited its use to the topical treatment of superficial diseases such as dermal mycoses and vaginal infections (PLEMPEL et al., 1988). Miconazole exhibits a lower inducing potential and—besides its wide topical use—is applied as a drug of second choice intravenously in severe systemic mycoses.

Ketoconazole (Fig. 8)—introduced in 1981 as Nizoral^{R}—heralded a new era of systemically active antimycotics. It is well absorbed after oral administration and achieves plasma and organ levels that indicated high efficacy against a wide spectrum of different fungi (CAUWENBERGH, 1987). As a consequence it was widely prescribed for all types of mycoses. Rare but serious side effects, in particular, hepatotoxicity and endocrine disorders due to the interference with testosterone and corticoid synthesis (CLISSOLD, 1987) have restricted its use now to a more controlled treatment mostly of severe systemic mycoses not responsive to other antimycotics.

Systemic infections caused by opportunistic fungi are an increasing problem in highly immunocompromised patients and contribute to their death (PERFECT, 1989). Prophylactic and therapeutic treatment of these severely ill patients requests antifungal drugs lacking the serious side effects of a therapy with amphotericin B a still-used standard (HÖFFKEN, 1989). Treatment with ketoconazole is also limited due to its impairment of steroidal hormone levels and moreover, due to its strong inhibition of cyclosporin A-metabolism (FERGUSON et al., 1982; BERTHAULT-PERES et al., 1987; WARNOCK, 1989). Treatment with cyclosporin A—the leading immunosuppressive agent administered after organ transplantation—bears the potential of severe side effects, in particular of dose-dependent nephrotoxicity (CALNE et al., 1978), which may become especially pronounced at the ketoconazole-provoked increased drug levels (see 4.3.3.1.).

The unsatisfying situation in the systemic treatment has directed the efforts to the development of active and less toxic, better tolerated drugs. The new generation of azole-drugs for systemic use—belonging to the triazoles—have highly improved properties with respect to a weak or lacking interference with steroidal hormone levels and also with the metabolism of concomitantly administered drugs (see 4.3.3.). Recently, two of these new drugs have been introduced to the market: fluconazole (RICHARDSON et al., 1985) and itraconazole (VAN CUTSEM et al., 1987).

In spite of moderate antifungal activities in vitro (MARRIOTT and RICHARDSON, 1987) exceptional pharmacokinetic properties have led to the breakthrough of fluconazole. Unlike the other azoles fluconazole is a polar molecule (Fig. 8); it is nevertheless well absorbed, shows a low binding to plasma proteins and an equal distribution throughout the body, including the central nervous system; the drug is present in plasma and in tissues mostly in its unbound, antifungally active form (BRAMMER and TARBIT, 1987). It shows no first pass effect and is eliminated mostly in an unchanged form; the stable high and (entirely) bioavailable plasma and tissue levels of some 4μM during oral therapy (BRAMMER and TARBIT, 1987) may outweigh the far higher antifungal activities of other far less bioavailable drugs such as ketoconazole and itraconazole. Fluconazole is successfully used in superficial candidosis such as vaginal candidosis and oral thrush of immunocompromised patients (WARNOCK, 1990; HAY, 1990). For systemic mycoses initial data on the effectivity in cryptococcal meningitis indicates a clear advantage over the standard amphotericin B with respect to a more convenient treatment and increased safety (WARNOCK, 1990). In spite of these highly favourable results recent reports on the emergence of resistant Candida strains may endanger and limit its range of applications (DUPONT, 1991; DUPOUY-CAMET et al., 1991).

Itraconazole (Fig. 8) shows two orders of magnitude higher lipophilia than ketoconazole and is highly protein-bound; plasma concentrations are several times lower than tissue levels and CSF levels are even three orders of magnitude lower (GRANT and CLISSOLD, 1989). Its absorption is variable, it undergoes a substantial first pass effect and becomes extensively metabolized (HEYKANTS et al., 1987). As a consequence of these events plasma levels may be below some 0.5μM and these low levels are reported to be associated with treatment failure (WARNOCK, 1990). Itraconazole is effective in the treatment of dermatomycoses, oral and vaginal candidoses and in deep-seated mycoses. A significant benefit was reported in various forms of aspergillosis (DUPONT, 1990), cryptococcosis, histoplasmosis and sporo-trichosis (GRAYBILL, 1989; WARNOCK, 1990).

A third compound, SDZ 89-485 (Fig. 8), combining optimal properties of the latter drugs shows promising antimycotic efficacy and a low toxicological potential (SCHUSTER and MEINGASSNER, 1988; BECHTER et al., 1988; GRASSBERGER, 1989; SCHAUDE et al., 1990; RYDER, 1990a). Besides a lacking potential to interfere with steroidal hormone synthesis (Table 4) it will also not impair the degradation of immune suppressives such as cyclosporine A (Table 7) and FK 506 (SCHUSTER and MEINGASSNER, 1988 and unpublished) and thus may be regarded as drug of choice in patients undergoing organ transplantation. It is highly lipophilic, becomes well absorbed after oral administration, is highly tissue-distributed and shows good metabolic stability in animal models (SCHUSTER, unpublished; FIERER et al., 1990).

SCH 39304 has proved successfully in the treatment of aspergillosis, cryptococcosis and other deep mycoses and shows an exceptional long half-life in humans of more than 50 hours (WARNOCK, 1990). Recently, the development of this drug was discontinued for toxicological reasons. Saperconazole (Fig. 8) is a closely related analogue of itraconazole which obviously has entered development and some of its preclinical data have been recently reported. It resembles other potent azoles in their antifungal activity and interferes only weakly with most mammalian cytochromes P-450 in steroidal hormone synthesis and drug metabolism (VANDEN BOSSCHE et al., 1989b).

5. Future prospects

A spectrum of drugs is used to treat efficiently a wide range of superficial mycoses and inhibition of ergosterol synthesis is the predominant mode of their action. However, it is not more than a handful antifungals which can be used for oral administration required in the therapy of deep-seated mycoses, wide-spread superficial mycoses and nail infections. All have short-comings in terms of efficacy and/or safety and therefore, efforts in the search for novel superior drugs should not be reduced.

Allylamines (section 3) interfere with ergosterol synthesis at the step of squalene epoxidase. They exert a highly selective and fungicidal action on filamentous and dimorphic fungi with an almost lacking potential for side effects. These properties will make the allylamines to new standards in the successful management of dermatophyte infections, especially involving the nails. Compared to the extremely susceptible dermatophytes fungi causing deep mycoses are less sensitive to terbinafine and with regard to the results in animal model systems its use seems to be limited. Attempts to augment the applicability of the powerful mechanistic principle are further requested. This principle applies also to cholesterol synthesis and the recently described potent inhibitors of mammalian squalene epoxidase will certainly stimulate an avalanche of pharmaceutic work and enzymologic research on this area.

The development of azole antimycotics (section 4) had been a milestone in the successful treatment of mycoses ranging from mucocutaneous infections to severe systemic mycoses. However, their mode of action—inhibition of 14α-lanosterol demethylase which is a fungal cytochrome P-450—implicates also a wide potential for side effects due to the inhibition of other members of the cytochrome P-450 superfamily in the host, especially members involved in steroidal hormone synthesis and drug metabolism. The search for new antifungal azoles aims therefore not only at fungicidal efficacy and improved pharmacokinetic properties but also at the increase of selectivity towards the fungal enzyme. A problem also encountered with this group of compounds is their type of activity which is fungistatic only and will therefore bear the potential of relapses when treatment is discontinued.

The side effects evoked with the azoles—especially with ketoconazole—have started a new story, namely the development of such compounds as endocrine drugs. Whereas ketoconazole will be of limited value e.g. in reducing androgen levels in man since it also blocks there corticoid synthesis and drug metabolism (STEVENS, 1985; FELDMAN, 1986) more selective analogs can be developed for steroid-blocking effects. The aromatase inhibitor CGS 16949A which is considered to be used in breast cancer and other estrogen induced disorders is one of the first examples (SANTEN et al., 1989). Considering the wide range of physiological processes catalyzed by different types of cytochrome P-450 the exceptional large therapeutic potential for highly specific inhibitors becomes evident which might even exceed the therapeutic volume of antibiotics.

Moreover, studies with azoles as cytochrome P-450 inhibitors can be used to gain a host of informations such as a mapping of active sites—also with respect to species variations of a particular enzyme—understanding of individual steps in the catalytic cycle of cytochrome P-450 and elucidating the role of particular cytochromes P-450 in physiological processes, in the biotransformation of environmentals and other xenobiotics and in toxic interactions.

Acknowledgement

The author is grateful to M. GRASSBERGER, G. HERZIG, A. P. JONES, H. MIETH, N. S. RYDER, and A. STUETZ for their help in preparing this chapter.

6. References

ATIBA, J. O., T. F. BLASCHKE, G. R. WILKINSON, (1989), Br. J. Clin. Pharmac. **28**, 161–165.

AOYAMA, Y., Y. YOSHIDA, Y. SONODA, Y. SATO, (1987), J. Biol. Chem. **262**, 1239–1243.

BACK, D. J., P. STEVENSON, J. F. TIJA, (1989), Br. J. Clin. Pharmacol. **28**, 166–170.

BALLARD, S. A., A. LODOLA, M. H. TARBIT, (1988), Biochem. Pharmacol. **37**, 4643–4651.

BALOCH, R. I., E. I. MERCER, T. E. WIGGINS, B. C. BALDWIN, (1984), Phytochemistry **23**, 2219–2226.

BATTIG, F. A., M. NEFZGER, G. SCHULZ, (1987), in: Recent Trends in the Discovery, Development and Evaluation of Antifungal Agents, (R. A. FROMTLING, ed.), J. R. Prous Science Publishers, Barcelona pp. 479–495.

BECHTER, R., M. HAMADA, T. NAKASHIMA, (1988), Revista Iberica de Micologia **5** (Suppl. 1), 88 (Abstract).

BERG, D., W. DRABER, H. VON HUGO, W. HUMMEL, D. MAYER, (1981), Z. Naturforsch. **36c** 798–803.

BERG, D., K. H. BÜCHEL, M. PLEMPEL, E. REGEL, (1986), TIPS, 233–238.

BERTHAULT-PERES, P., C. BONFILS, G. FABRE, S. JUST, J.-P. CANO, P. MAUREL, (1987), Drug Metab. Dispos. **15**, 391–398.

Bikle, D. D., M. K. Nemanic, E. Gee, P. Elias, (1986), J. Clin. Invest. **78**, 557–566.
Brodie, A. M., R. J. Santen, (1986), in: CRC critical reviews in oncology/hematology (S. Davis, ed.), Boca raton CRC Press, pp. 361–396.
Burgess, M. A., G. P. Bodey, (1972), Antimicrob. Agents Chemother. **2**, 423–426.
Brammer, K. W., M. H. Tarbit, (1987), in: Recent Trends in the Discovery, Development and Evaluation of Antifungal Agents, (R. A. Fromtling, ed.), J. R. Prous Science Publishers, Barcelona, pp. 141–149.
Calne, R. Y., D. J. G. White, S. Thiru, (1978), Lancet **2**, 1323–1327.
Cauwenbergh, G., (1987), in: Ketoconazole Today. A review of Clinical Experience, (H. E. Jones, ed.), ADIS Press Ltd., Manchester, UK, pp. 92–99.
Clayton, Y. M., (1987), in: Recent Trends in the Discovery, Development and Evaluation of Antifungal Agents, (R. A. Fromtling, ed.), J. R. Prous Science Publishers, Barcelona, pp. 433–439.
Clissold, S. P., (1987), in: Ketoconazole Today. A review of Clinical Experience, (H. E. Jones, ed.), ADIS Press Ltd., Manchester, UK, pp. 77–79.
Daneshmend, T. K., D. W. Warnock, (1988), Clin. Pharmacokin. **14**, 13–34.
Dupont, B., (1990), J. Am. Acad. Dermatol. **23**, 607–614.
Dupont, B., (1991), XI Congr. Int. Soc. Hum. Anim. Mycology. Abstracts **S1.1**, p. 23.
Dupouy-Camet, J., C. DiDonato, C. Viguie, A. Paugam, J. Vicens, P.-J. Valle, C. Tourte-Schaefer, (1991), XI Congr. Int. Soc. Hum. Anim. Mycology. Poster PS **3.27**, p. 127.
Eilenberg, H., I. Shechter, (1987), J. Lipid Res. **28**, 1398–1404.
Fabre, I., G. Fabre, P. Maurel, P. Berthault-Peres, J.-P. Cano, (1988), Drug Metabolism & Disposition **16**, 296–301.
Faergemann, J., H. Zehender, T. Jones, I. Maibach, (1990), Acta Derm. Venereol. **71**, 322–326.
Feldman, D., (1986), Endocrine Reviews **7**, 409–420.
Ferguson, R. M., D. E. Sutherland, R. L. Simmons, J. S. Najarian, (1982), Lancet **ii**, 882–883.
Fierer, J., T. Kirkland, F. Finley, (1990), Antimicrob. Agents Chemother. **34**, 13–16.
Fischer, R. T., J. M. Trzaskos, R. L. Magolda, S. S. Ko, C. S. Brosz, B. Larsen, (1991), J. Biol. Chem. **286**, 6124–6132.
Georgopoulos, A. G., G. Petranyi, H. Mieth, J. Drews, (1981), Antimicrob. Agents Chemother. **19**, 386–389.
Gibbons, G. F., K. A. Mitropoulos, (1973), Eur. J. Biochem. **40**, 267–273.
Grant, S. M., S. P. Clissold, (1989), Drugs **37**, 310–344.
Grassberger, M., (1989), Drugs of the Future **14**, 772–774.
Graybill, J. R., (1989), Eur. J. Clin. Microbiol. Infect. Dis. **8**, 402–412.
Grimus, R. C., I. Schuster, (1984), Xenobiotica **14**, 287–294.
Guengerich, F. P., (1983), Biochemistry **22**, 2811–2820.
Guengerich, F. P., (1991), J. Biol. Chem. **286**, 10019–10022.
Hall, S. D., F. P. Guengerich, R. A. Branch, G. R. Wilkinson, (1987), J. Pharmac. exp. Ther. **240**, 216–222.
Hay, R. J. (1990), J. Dermatol. Treatment **1**, Suppl. 2, 1–3.
Hecht, P., H. Berner, A. Stuetz, H. Egger, I. Schuster, (1989), in: Cytochrome P-450: Biochemistry and Biophysics. Proceedings from the 6th Int. Conf. on Biochemistry and Biophysics of Cytochrome P-450, Vienna, (I. Schuster, ed.), Taylor & Francis London, pp. 446–449.
Heinemeyer, G., G. Jaeck, (1983), Naunyn-Schmiedeberg's Arch. Phamacol. **322** (Suppl.) R128.
Henricsson, S., A. Lindholm, M. Aravoglu, (1990), Pharmacology & Toxicology **66**, 49–52.

HEYKANTS, J., A. VAN PEER, V. VAN DE VELDE, P. VAN ROOY, W. MEULDERMANS, K. LAVRIJSEN, R. WOESTENBORGHS, J. VAN CUTSEM, G. CAUWENBERGH, (1989), Mycoses **32**, (Suppl. 1), 67–87.
HEYL, B. L., D. J. TYRELL, J. D. LAMBETH, (1986), J. Biol. Chem. **261**, 2743–2749.
HITCHCOCK, C. A., K. DICKINSON, S. B. BROWN, E. G. V. EVANS, D. J. ADAMS, (1989), Biochem. J. **263**, 573–579.
HITCHCOCK, C. A., K. DICKINSON, S. B. BROWN, E. G. V. EVANS, D. J. ADAMS, (1990), Biochem. J. **266**, 475–480.
HÖFFKEN, G., (1989), Mycoses **32** (Suppl. 1), 88–95.
HOLLIS, B. W., (1990), Proc. Natl. Acad. Sci. USA **87**, 6009–6013.
HORIE, M., Y. TSUCHIYA, M. HAYASHI, Y. IIDA, Y. IWASAWA, Y. NAGATA, Y. SAWASAKI, H. FUKUZUMI, K. KITANI, T. KAMEI, (1990), J. Biol. Chem. **265**, 18075–18078.
IGLESIAS, J., G. F. GIBBONS, (1989), Biochem. J. **264**, 495–502.
IKUSHIRO, S., S. KOMONAMI, S. TAKEMORI, (1989), Biochim. Biophys. Acta **984**, 50–56.
JEFCOATE, C. R. (1978), Methods Enzymol. **52**, 258–279.
JENSEN, J. C., (1990), J. Dermatol. Treatment **1**, Suppl. 2, 15–18.
JONES, T. C., (1990), J. Dermatol. Treatment **1**, Suppl. 2, 29–32.
JONES, T. C., V. VILLARS, (1990), in: Handbook of Experimental Pharmacology, **96**: Chemotherapy of Fungal Diseases, (RYLEY, ed.), Springer-Verl. Berlin, pp. 483–503.
KAHL, F., D. E. FRIEDERICI, G. F. KAHL, W. RITTER, R. KREBS, (1980), Drug Metab. Dispos. **8**, 191–196.
KALB, V. F., C. W. WOODS, T. G. TURI, C. R. DEY, T. R. SUTTER, J. C. LOPER, (1987), DNA (NY) **6**, 529–537.
KATO, T., (1986), in: Sterol Biosynthesis Inhibitors and Anti-Feeding Compounds, (G. HANG, H. HOFFMAN, eds.), Springer Verl. Berlin, pp. 2–24.
KERKENAAR, A., J. M. VAN ROSSUM, G. G. VERSLUIS, J. W. MARSMAN, (1984), Pestic. Sci. **15**, 177–187.
KIRSCHNER, K., H. BISSWANGER, (1974), Annu. Rev. Biochem. **45**, 143–166.
KOENIG, K. H., E. H. POMMER, W. SANNE, (1965), Angew. Chemie **77**, 327–333.
KUCK, K. H., H. SCHEINPFLUG, (1986), in: Sterol Biosynthesis Inhibitors and Anti-Feeding Compounds, (G. HANG, H. HOFFMANN, eds.), Springer Verl. Berlin, p. 66.
KUEHN-VELTEN, W. N., T. BUNSE, M. E. C. FOERSTER, (1991), J. Biol. Chem. **286**, 6291–6301.
KUNZE, K. L., B. L. K. MANGOLD, C. WHEELER, H. S. BEILAN, P. R. ORTIZ DE MONTELLANO (1983), J. Biol. Chem. **258**, 4202–4207.
LA DELFA, I., Q. M. ZHU, Z. MO, T. F. BLASCHKE, (1989), Drug Metabolism and Disposition **17**, 49–53.
LAI, M. H., D. R. KIRSCH, (1989), Nucleic Acids Res. **17**, 804.
LAMBERT, A., R. MITCHELL, W. R. ROBERTSON, (1986), Biochem. Pharmacol. **35**, 3999–4004.
LAMBERTS, S. W. J., E. G. BONS, H. A. BRUINING, F. H. DE JONG, (1987), J. Pharm. Exper. Ther. **240**, 259–264.
LEES, N. D., F. W. KLEINHANS, M. C. BROUGHTON, D. E. PENNINGTON, V. A. RICKERS, H. BARD, (1989), Steroids **53**, 567–578.
LEES, N. D., M. C. BROUGHTON, D. SANGLARD, M. BARD, (1990), Antimicrob. Agents Chemother. **34(5)**, 831–836.
MARRIOTT, M. S., K. RICHARDSON, (1987), in: Recent Trends in the Discovery, Development and Evaluation of Antifungal Agents, (R. A FROMTLING, ed.), J. R. Prous Science Publishers, Barcelona, pp. 81–92.
M'BAYA, B., F. KARST, (1987), Biochem. Biophys. Res. Commun. **147**, 556–564.
MEINGASSNER, J. G., U. SLEYTR, G. PETRANYI, (1981), J. Invest. Dermatol. **77**, 444–451.
MERCER, E. I., (1987), in: Systemic Fungicides and Antifungal Compounds; Int. Symp. Reinhardsbrunn, 1986, H. LYR, C. POLTER, eds.) Akademie Verl. Berlin **253**, 115–120.

MERCER, E. I., (1988), in: Sterol Biosynthesis Inhibitors, (D. BERG, M. PLEMPEL, eds.), Ellis Horwood Ltd., Chichester UK, p. 120–150.
MIETH, H., G. PETRANYI, (1989), Clin. Exp. Dermatol. **14**, 104–107.
MIETH, H., (1990), J. Dermatol. Treatment **1**, Suppl. 2, 5–6.
MORITA, T., Y. NOZAWA, (1985), J. Invest. Dermatol. **85**, 434–437.
MORPURGO, G., G. SERLUPI-CRESCENZI, G. TECCE, F. VALENTE, D. VENETACCI, (1964), Nature **201**, 897–899.
MUKHTAR, H., B. J. DEL TITO, M. DAS, E. P. CHERNIAK, D. R. BICKERS, (1984), Cancer Res. **44**, 4233–4240.
MURRAY, M. G. F. REIDY, (1990), Pharmacol. Reviews **42**, 85–101.
NAGAI, K., I. MIYAMORI, M. IKEDA, H. KOSHIDA, R. TAKEDA, K. SUHARA, M. KATAGIRI, (1986), J. Steroid. Biochem. **24**, 321–323.
NAKAJIN, S., P. F. HALL, (1981), J. Biol. Chem. **256**, 3871–3876.
NUSSBAUMER, P., G. PETRANYI, A. STUETZ, (1991a), J. Med. Chem. **34**, 65–73.
NUSSBAUMER, P., N. S. RYDER, A. STUETZ, (1991b), Pestic. Sci. **31**, 437–455.
OHBA, M., R. SATO, Y. YOSHIDA, T. NOSHINO, H. KATSUKI, (1978), Biochem. Biophys. Res. Commun. **85**, 21–27.
OMAR, G., I. ALI SHAW, A. W. THOMSON, P. H. WHITING, M. D. BURKE, (1991), Transplantation Proc. **23**, 934–935.
ONO, T., K. BLOCH, (1975), J. Biol. Chem. **250**, 1571–1579.
ONO, T., S. OZASA, F. HASEGAWA, Y. IMAI, (1977), Biochim. Biophys. Acta **486**, 401–407.
ONO, T., K. NAKAZONO, H. KOSAKA, (1982), Biochim. Biophys. Acta **709**, 84–90.
ORTIZ DE MONTELLANO, P. R., N. O. REICH, (1986), in: Cytochrome P-450. Structure, Mechanism and Biochemistry, (P. R. ORTIZ DE MONTELLANO, ed.), Plenum Press, NY, p. 273–314.
PALTAUF, F., G. DAUM, G. ZUDER, G HOEGENAUER, G. SCHULZ, G. SEIDL, (1982), Biochim. Biophys. Acta **712**, 268–273.
PERFECT, J. P., (1989), Curr. Opinion Infect. Dis. **2**, 568–571.
PERFECT, J. P., (1990), Curr. Opinion Infect. Dis. **3**, 221–224.
PETRANYI, G., N. S. RYDER, A. STUETZ, (1984), Science **224**, 1239–1241.
PETRANYI, G., (1985), Mykosen (Suppl.), 37–43.
PETRANYI, G., J. G. MEINGASSNER, H. MIETH, (1987), Antimicrob. Agents Chemother. **31**, 1558–1561.
PICHARD, L., J. M. FABRE, J. DOMERGUE, G. FABRE, B. SAINT-AUBERT, G. MOURAD, P. MAUREL, (1991), Transplantation Proc. **23**, 978–979.
PINTO, W. J., R. LOZANO, B. C. SEKULA, W. R. NES, (1983), Biochem. Biophys. Res. Commun. **112**, 47–54.
PLEMPEL, M., D. BERG, K.-H. BÜCHEL, W. RITTER, (1988), in: Sterol Biosynthesis Inhibitors, (D. BERG, M. PLEMPEL, eds.), Ellis Horwood Ltd., Chichester, UK, pp. 350–396.
POLAK, A., D. DIXON, (1988), in Recent Trends in the Discovery, Development and Evaluation of Antifungal Agents, (R. A. FROMTLING, ed.), J. R. Prous Science Publishers, Barcelona, pp. 555–573.
POLAK, A., M. ZAUG, (1990), in: Handbook of Experimental Pharmacology, **96**: Chemotherapy of Fungal Diseases, (RYLEY, ed.), Springer Verl. Berlin, pp. 505–521.
PONT, A., P. L. WILLIAMS, S. AHZAR, R. E. REITZ, C., BOCHRA, (1982), Arch. Int. Med. **142**, 2137–2140.
PONT, A., J. R. GRAYBILL, P. C. CRAVEN, J. N. GALGANI, W. E. DISMUKES, (1984), Arch. Int. Med. **144**, 2150–2153.
PYE, G. W., M. S. MARRIOTT, (1982), Sabouraudia **20**, 325–329.
REINHARDT, T. A., R. L. HORST, (1989), Arch. Biochem. Biophys. **272**, 459–465.
RICHARDSON, K., K. W. BRAMMER, M. S. MARRIOTT, P. F. TROKE, (1985), Antimicrob. Agents Chemother. **27**, 832–835.

RITTER, J. K., M. R. FRANKLIN, (1987a), Biochem. Pharmacol. **36**, 2783–2787.
RITTER, J. K., M. R. FRANKLIN, (1987b), Drug Metab. Dispos. **15**, 335–343.
RITTER, W., (1988), in: Sterol Biosynthesis Inhibitors, (D. BERG, M. PLEMPEL, eds.), Ellis Horwood Ltd., Chichester UK, pp. 397–429.
ROGERSON, T. D., C. F. WILKINSON, K. HETARSKI, (1977), Biochem. Pharmacol. **26**, 1039–1042.
RODRIGUEZ, R. W., C. LOW, C. D. K. BOTTEMA, L. W. PARKS, (1985), Biochim. Biophys. Acta **837**, 336–343.
ROTTEM, S., V. C. CIRILLO, B. DEKRUYFF, M. SHIMIBKY, S. RAZIN, (1973), Biochim. Biophys. Acta **323**, 509–538.
RYDER, N. S., P. F. TROKE, (1982), in: Current Chemotherapy and Immunotherapy. American Society for Microbiology, (PERITI, P., GRASSI, G. G., eds.), Washington, p. 1016–1017.
RYDER, N. S., G. SEIDL, P. F. TROKE, (1984), Antimicrob. Agents Chemother. **25**, 483–487.
RYDER, N. S., M. C. DUPONT, (1984), Biochim. Biophys. Acta **794**, 466–471.
RYDER, N. S., M. C. DUPONT, (1985), Biochem. J. **230**, 765–770.
RYDER, N. S., (1985), Antimicrob. Agents and Chemother. **27**, 252–256.
RYDER, N. S., I. FRANK, M. C. DUPONT, (1986), Antimicrob. Agents Chemother. **29**, 858–860.
RYDER, N. S., (1987), in: Recent Trends in the Discovery, Development and Evaluation of Antifungal Agents, (R. A. FROMTLING, ed.), J. R. Prous Science Publishers, Barcelona, pp. 451–459.
RYDER, N. S., (1988), in: Sterol Biosynthesis Inhibitors, (D. BERG, M. PLEMPEL, eds.), Ellis Horwood Ltd., Chichester UK, p. 151–167.
RYDER, N. S., (1990a), J. Med. Vet. Mycology **28**, 385–394.
RYDER, N. S., (1990b), in: Biochemistry of Cell Walls and Membranes in Fungi, (P. J. JUNG, A. P. J. TRINCI, M. J. JUNG, M. W. GOOSEY, L. G. COPPING, eds.), Springer Verlag Berlin, pp. 189–203.
RYDER, N. S., H. MIETH, (1991), Current Topics in Medical Mycology **4**, 158–188.
SANTEN, R. J., H. VANDEN BOSSCHE, J. SIMOYENS, J. BRUGMANS, R. DE COSTER, (1983), J. Clin. Endocrin. Metabolism **57**, 732–736.
SANTEN, R. J., L. M. DEMERS, H. ADLERCREUTZ, H. HARVEY, S. SANTNER, S. SANDERS, A. LIPTON, (1989), J. Clin. Endocrin. Metab. **68**, 99–106.
SCHAUDE, M., H. ACKERBAUER, H. MIETH, (1987), Mykosen **30**, 281–287.
SCHAUDE, M., G. PETRANYI, H. ACKERBAUER, J. G. MEINGASSNER, H. MIETH, (1990), J. Med. Vet. Mycology **28**, 445–454.
SCHENKMAN, J. B., S. G. SLIGAR, D. L. CINTI, (1981), Pharmacol. Ther. **12**, 43–71.
SCHUSTER, I., (1985), Xenobiotica **15**, 529–546.
SCHUSTER, I., (1987a), in: Recent Trends in the Discovery, Development and Evaluation of Antifungal Agents, (R. A. FROMTLING, ed.), J. R. Prous Science Publishers, Barcelona, pp. 461–470.
SCHUSTER, I., (1987b), in: Recent Trends in the Discovery, Development and Evaluation of Antifungal Agents, (R. A. FROMTLING, ed.), J. R. Prous Science Publishers, Barcelona, pp. 471–478.
SCHUSTER, I., M. SCHAUDE, F. SCHATZ, H. MIETH, (1988), in: Sterol Biosynthesis Inhibitors, (D. BERG, M. PLEMPEL, eds.), Ellis Horwood Ltd., Chichester UK, pp. 449–470.
SCHUSTER, I., SCHNEEBERGER, D., (1989), in: Cytochrome P-450: Biochemistry and Biophysics. Proceedings from the 6th Int. Conf. on Biochemistry and Biophysics of Cytochrome P-450, Vienna, (I. SCHUSTER, ed.), Taylor & Francis London, pp. 204–207.
SCHUSTER, I., J. G. MEINGASSNER, (1988), Revista Iberica de Micologia **5** (Suppl. 1), p. 89 (Abstr.).
SCHUSTER, I., (1990), J. Dermatol. Treatment **1** (Suppl. 2), 7–9.

SEKIGAWA, Y., Y. SONODA, Y. SATO, (1988), Chem. Pharm. Bull. **36**, 3049–3054.
SHADOMY, S., A. ESPINEL-INGROFF, R. J. GEBHART, (1985), J. Med. Vet. Mycol. **23**, 125–132.
SHAFIEE, A., J. M. TRZASKOS, Y. K. PAIK, J. L. GAYLOR, (1986), J. Lipid Res. **27**, 1–10.
SHAW, J. T. B., M. H. TARBIT, P. F. TROKE, (1987), in: Recent Trends in the Discovery, Development and Evaluation of Antifungal Agents, (R. A. FROMTLING, ed.), J. R. Prous Science Publishers, Barcelona, pp. 125–139.
SHEETS, J. J., J. J. MASON, (1984), Drug Metabolism Dispos. **12**, 603–606.
SIKKA, S. C., R. S. SWERDLOFF, J. RAJFER, (1985), Endocrinology **116**, 1920–1925.
SONINO, N., (1987), New England J. Med. **317**, 812–818.
SONODA, Y., Y. SEKIGAWA, Y. SATO, (1989), Chem. Pharm. Bull. **37**, 718–722.
SOWDEN, J. M., B. R. ALLEN, (1990), J. Dermatol. Treatment **1**, Suppl. 2, 27–28.
STEELE, R. E., L. B. MELLOR, W. K. SAWYER, J. M. WASVARY, L. J. BROWNE, (1987), Steroids, **50**, 147–161.
STEVENS, D. A., (1985), Arch. Intern. Med. **145**, 913–815.
STEVENSON, D. E., J. W. WRIGHT, M. AKTHAR, (1988), J. Chem. Soc. Perkin Trans. **I**, 2043–2052.
STUETZ, A., (1987), Angew. Chemie, Int. Chem. Ed. Engl. **26**, 320–328.
STUETZ, A., (1988), Annals NY Acad. Sci. **544**, 46–62.
STUETZ, A., G. PETRANYI, (1984), J. Med. Chem. **27**, 1539–1543.
STUETZ, A., A. GEORGOPOULOS, W. GRANITZER, G. PETRANYI, D. BERNEY, (1986), J. Med. Chem. **29**, 112–125.
SUTTER, T. R., J. C. LOPER, (1989), Biochem. Biophys. Res. Commun. **160**, 1257–1266.
TESTA, B., P. JENNER, (1981), Drug Metab. Reviews **12**, 1–117.
TRACHTENBERG, J., A. PONT, (1984), Lancet **2**, 433–435.
TRZASKOS, J. M. S. KAWATA, J. L. GAYLOR, (1986), J. Biol. Chem. **261**, 14651–14657.
VAN CAUTEREM, H., A. LAMPO, J. VANDENBERGHE, P. VANPARYS, W. COUSSEMENT, R. DE COSTER, R. MARSBOOM, (1989), Mycoses (Suppl. 1), 60–66.
VAN CUTSEM, J., F. VAN GERVEN, P. A. J. JANSSEN, (1987), Rev. Infect. Dis. **9**, 15–32.
VANDEN BOSSCHE, H., (1985), in: Current Topics in Medical Mycology, (M. R. GINNIS, ed.), Springer-Verl. N. Y., pp. 313–351.
VANDEN BOSSCHE, H., (1988), in: Sterol Biosynthesis Inhibitors, (D. BERG, M. PLEMPEL, eds.), Ellis Horwood Ltd., Chichester UK, pp. 79–119.
VANDEN BOSSCHE, H., P. MARICHAL, J. GORRENS, M. C. COENE, G. WILLEMSENS, D. BELLENS, I. ROELS, H. MOEREELS, P. A. J. JANSSEN, (1989a), Mycoses **32**, 35–52.
VANDEN BOSSCHE, H., P. MARICHAL, G. WILLEMSENS, D. BELLENS, J. GORRENS, I. ROELS, M. C. COENE, L. LEJEUNE, P. A. J. JANSSEN, (1989b), Mycoses **33**, 335–352.
VAN TAMELEN, E. E., J. D. WILLETT, R. B. CLAYTON, K. E. LORD, (1966), J. Amer. Chem. Soc. **88**, 4752–4754.
VILLARS, V., JONES, T. C., (1989), Clin. Exper. Dermatol. **14**, 124–127.
VILLARS, V., T. C. JONES, (1990), J. Dermatol. Treatment **1**, Suppl. 2, 33–38.
WAKIL, S. J., (1989), Biochemistry **281**, 4523–4530.
WALLACE, P. G., M. HUANG, A. W. LINNANE, (1968), J. Cell Biol. **37**, 207.
WARNOCK, D. W., (1989), Curr. Opinion Infect. Dis. **2**, 362–366.
WARNOCK, D. W., (1990), Curr. Opinion Infect. Dis. **3**, 765–769.
WATSO, P. F., M. E. ROSE, S. W. ELLIS, H. ENGLAND, S. L. KELLY, (1989), Biochem. Biophys. Res. Commun. 164, 1170–1175.
WHITE, R. E., M. J. COON, (1980), Ann. Rev. Biochem. **49**, 315–356.
WHITEHOUSE, L. W., A. P. PAKUTS, C. J. PAUL, R. W. MUELLER, B. H. THOMAS, (1990), Can. J. Physiol. Pharmacol. **68**, 1136–1142.
YOSHIDA, Y., Y. AOYAMA, (1984), J. Biol. Chem. **259**, 1655–1660.
YOSHIDA, Y., Y. AOYAMA, (1986), in: in Vitro and in Vivo Evaluation of Antifungal Agents, (K. IWATA, H. VANDEN BOSSCHE, eds.), Elsevier Amsterdam, p. 123–134.

YOSHIDA, Y., Y. AOYAMA, (1987), Biochem. Pharmacol. **36**, 229–235.
YOSHIDA, Y., Y. AOYAMA, Y. SONODA, Y. SATO, (1989), in: Cytochrome P-450: Biochemistry and Biophysics. Proceedings from the 6th Int. Conf. on Biochemistry and Biophysics of Cytochrome P-450, Vienna, (I. SCHUSTER, ed.), Taylor & Francis London, 162–165.
YOSHIDA, Y., Y. AOYAMA, (1990), Chirality **2**, 10–15.

Chapter 5
Isozyme-Specificity of Cytochrome P450-mediated Hepatotoxicity

Y. Terelius, K. O. Lindros, E. Albano, and M. Ingelman-Sundberg

1. Introduction

Man is continuously exposed to a huge amount of foreign chemicals, many of which exert toxic effects, particularly after metabolic activation. This activating process is largely catalyzed by the cytochrome P450 system in the endoplasmic reticulum of the cell. The process is dependent on the amount of hepatic cytochrome P450, which after induction can rise to 2–4% of the total liver protein. The requirement of the liver, in particular, to metabolize thousands of different chemicals of diverse structures, is met by the existence of multiple forms of P450, each having a unique, but partially overlapping substrate specificity. Many toxicologically important drugs, precarcinogens and solvents are metabolically activated by essentially one or a limited number of P450 isozymes. In this review we will discuss recent advances focusing on the relationship between various isozymes of cytochrome P450 and specific metabolic activation of various hepatotoxins. This might have implications for the future development of isozyme specific inhibitors, possibly useful for therapy under special conditions.

The cytochrome P450 supergene family and its evolution

Cytochrome P450 constitutes a supergene family with presently 25 different members identified. As of today, more than 150 cytochromes P450 have been cloned. In mammals, having in total at least ten different gene families, mainly P450s belonging to families 1–4 catalyze the metabolism of xenobiotics. Of those, gene family 2 is the largest with more than 55 different members (NEBERT et al., 1991).

All the different cytochromes P450 have probably evolved from one single ancestor gene. The broad spectrum of various P450s is the result of gene duplications, gene conversions, gene insertions and mutations. GONZALEZ and NEBERT (1990) have proposed that the diversity in the P450 genes is a consequence of a continuous molecularly driven coevolution of plants producing phytoalexins and of animals responding to this by developing new enzymes able to detoxify these chemicals. In favour of this hypothesis there is evidence for numerous gene duplications about 400 million years ago, i.e. at the time when animals became terrestrial.

The number of P450 genes present in a single species is presently unknown but one could imagine that between 50 and 100 P450 genes might be present in man. In general, each P450 gene produces a single protein (enzyme). To date, there is only one known exception to this rule, where functional alternative splicing might occur. That is, differential processing of the P450 transcript such that the entire exons or portions of the exons are exchanged in order to produce an enzyme with a new catalytic activity (KIMURA et al., 1989).

The expression of various P450 forms is mainly controlled at the level of transcription. However, endogenous and exogenous factors can influence the amount of enzyme expressed also on e.g. the level of the stability of the transcribed mRNA, the rate of translation of the transcript and the stability of the enzyme. It appears that a particular form of cytochrome P450 is regulated at one or several of these levels by different xenobiotics. Thus, the steady state level of a specific form of P450 is governed by several different factors and is accordingly under a very complex control. These aspects are covered in more detail in recent reviews (GONZALEZ, 1989; PORTER and COON, 1991).

Although the principal task of the liver is to convert xenobiotics into forms that lead to detoxication, the initial activation step often results in a toxic intermediate. The ensuing toxicity, mediated by one of several forms of P450, can be direct or indirect. An indirect, often drastically enhanced toxicity, results from the induction by a xenobiotic of particular forms of P450. This is exemplified by the synergistic effect of ethanol on the toxicity of several different substrates for ethanol inducible CYP2E1. However, we will focus this review mainly on the direct metabolic activation by specific forms of P450. Presently, reliable information of direct roles for a particular form of P450 in the metabolic activation of a compound is rather limited. Hence, we will also discuss isozyme specific metabolism of hepatotoxins in general. Only a few studies based on use of characterized purified P450 forms or with specific inhibitory antibodies have been carried out. Therefore, we felt it necessary also to consider some results from experiments where the link to isozyme specificity has been obtained in indirect ways, e.g. by using specific inducers.

We focus our review on compounds with documented hepatotoxicity at relevant levels as opposed to chemicals that are mainly considered as carcinogens.

2. Chemicals and hepatotoxicity

The metabolic action of the P450 system can transform chemicals into intermediates displaying electrophilic properties and potentials to interact with nucleophilic targets within the cells. Intermediates with these properties cause alkylation of protein and DNA and depletion of hepatocyte glutathione (GSH), which protects against these reactive species. Furthermore, the metabolic activation of a number of compounds including halogenated alkanes, hydrazines and alcohols can cause the production of free radical intermediates, which not only are capable to bind to macromolecules, but also initiate the peroxidative breakdown of unsaturated fatty acid in the membranes and oxidize GSH and protein thiol groups (POLI et al., 1987; ORRENIUS et al., 1989; FARBER et al., 1990). As outlined in Figure 1 oxidative damage and alkylation of macromolecules are thought to result in the inactivation of many critical enzymatic functions including blockage of lipoprotein secretion and steatosis (POLI et al., 1987) as well as to affect critical cell functions such as the control of calcium homeostasis (ORRENIUS et al., 1989), the mitochondrial energy production and cytosceletal organization (NICOTERA

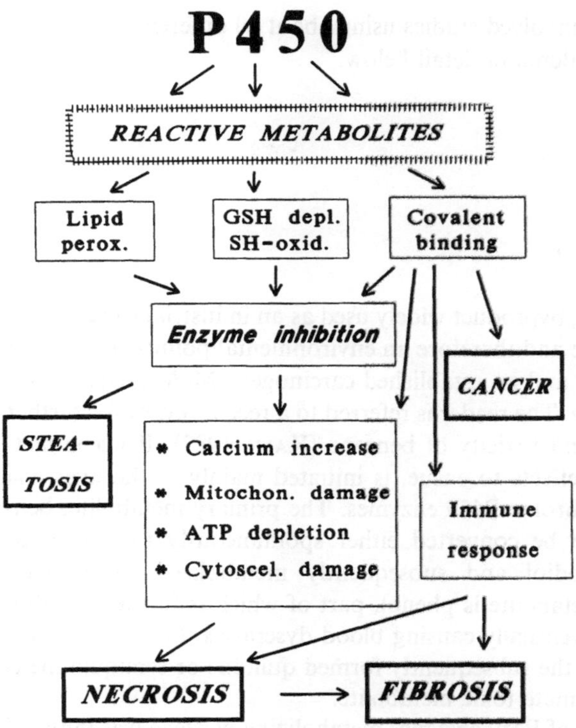

Fig. 1. Mechanisms involved in the hepatotoxic action of reactive intermediates formed by the catalytic action of various isozymes of cytochrome P450.

et al., 1990; FARBER et al., 1990; GORES et al., 1990). Extensive studies performed in the recent years have demonstrated that these alterations contribute to cause irreversible cell injury leading to liver necrosis (POLI et al., 1987; NICOTERA et al., 1990; FARBER et al., 1990; GORES et al., 1990). An additional mechanism by which reactive intermediates might contribute to liver damage is by stimulation of an immunological response (Fig. 1). This mechanism has recently received more attention, since several forms of cytochrome P450, including CYP2E1 (CEDERBAUM, 1991), have been found to be also localized in the plasma membrane of the hepatocyte (LOEPER et al., 1990). Metabolic activation in this compartment, leading to the formation of covalent protein adducts, can cause stimulation of the immune system to form autoantibodies against P450 (KIFFEL et al., 1989) or to stimulate Kupffer cells and Ito cells leading to inflammatory lesion of the liver and possibly to fibrosis.

It is clear that cytochrome P450 plays an important role in the hepatotoxic action of many different chemicals and that different isozymes exert separate roles in this regard. The research concerning metabolic activation by a particular form of P450 has hitherto mainly involved studies using about 20 different representative agents that will be considered in detail below.

2.1. Solvents

2.1.1. Benzene

Benzene is a gasoline refining byproduct widely used as an industrial solvent. It is also a component of gasoline and therefore an environmental pollutant. Benzene is a well known myelotoxin, and an established carcinogen (McMichael, 1988) but also exerts hepatotoxicity. The reader is referred to a recent review for further details on the metabolism and toxicity of benzene (Kalf, 1987). Benzene metabolism, required for toxic effects to ensue, is initiated mainly in the liver, via reactions catalyzed by cytochrome P450 enzymes. The primary metabolite, benzene epoxide, is thought to be converted either spontaneously to phenol or hydroxylated to a dihydrodiol and subsequently metabolized to catechol (Kalf, 1987). The major metabolite is phenol, part of which is further metabolized in the bone marrow, eventually causing blood dyscrasias. Phenol is oxygenated to hydroquinone, and the subsequently formed quinone or semiquinone is thought to represent the ultimate toxic metabolite.

Identification of the forms of P450 enzymes metabolizing benzene was initiated by using microsomes isolated from animals treated with the classical inducers phenobarbital (PB) and β-naphthoflavone (BNF) (Post and Snyder, 1983). In these studies, evidence for a low- and a high affinity benzene hydroxylase activity was obtained. The high-affinity form was increased by pretreatment with benzene and was also seen in BNF-treated microsomes. The low-affinity activity was observed in PB-induced microsomes. Beaune et al. (1983) separated two P450 isozymes, that were induced by both benzene and ethanol pretreatment. One of these forms was also induced by the classical P450 inducers PB, BNF and 3-methylcholanthrene, but at that time the respective isozymes were only partially characterized and therefore not identified.

Evidence for the involvement of an isolated purified P450 isozyme in benzene oxidation was presented by Ingelman-Sundberg and Hagbjörk (1982), who used reconstituted microsomal membrane vesicles and benzene-inducible P450 LM_{eb} (CYP2E1) isolated from rabbit liver. Similarly, Griffiths et al. (1986) reported that a P450 isozyme purified from PB-induced rat liver actively converted benzene to phenol in a reconstituted microsomal system. In addition, a mitochondrial form of P450 capable of oxidizing benzene has been described (Karaszkiewicz et al., 1986).

More recent studies on the role of different P450 isozymes in hepatic benzene metabolism are based on immunoinhibition studies combined with the use of selective inducers. In rat liver, inducers of CYP2E1 such as ethanol, acetone, pyrazole, benzene and fasting, all markedly activate microsomal benzene metabolism (INGELMAN-SUNDBERG and HAGBJÖRK, 1982; JOHANSSON and INGELMAN-SUNDBERG, 1988; NAKAJIMA et al., 1989; KOOP et al., 1989). The extent of increase in microsomal benzene metabolism by different inducers was similar to the increase in immunoreactive CYP2E1. Furthermore, poly- or monoclonal antibodies against CYP2E1 efficiently inhibit microsomal benzene metabolism at low substrate concentrations (JOHANSSON et al., 1988; JOHANSSON and INGELMAN-SUNDBERG, 1988; NAKAJIMA et al., 1989; KOOP et al., 1989). These studies confirm that CYP2E1 is a high-affinity benzene hydroxylase. Treatment with phenobarbital, that induces CYP2B1/2, resulted in increased microsomal conversion of benzene at high (mM) concentrations (NAKAJIMA et al., 1987) and this metabolism was inhibited by monoclonal antibodies to 2B1/2 (NAKAJIMA et al., 1989), suggesting that P450 2B acts as a low-affinity benzene hydroxylase. In addition to these isozymes, some high-K_m activity was observed with P450 1A and 2C11 isozymes. The pattern of contribution of isozymes appears to be similar in rabbit liver. Of six purified P450 isozymes, only the 2E1 form was active at low benzene concentrations, whereas at 2 mM benzene concentrations P450 2, presumably identical to the phenobarbital-inducible CYP2B4, exhibited almost as high activity as the 2E1 form and 7–10 times higher activity than the four other forms. The active P450 forms also efficiently oxidized phenol (KOOP et al., 1989).

Evidence for a different pattern in mice comes from the finding that pretreatment with methylcholanthrene of DBA/2 mice causes a greater response to benzene (HARPER and LEGATOR, 1987). To what extent species-specific patterns of benzene hydroxylation is associated with species differences in benzene hepatotoxicity is not clear. In a study comparing benzene metabolism in human, rat and mouse liver slices and microsomes, BRODFUERER et al. (1990) observed a rather similar pattern of metabolism and binding of benzene and concluded that metabolism is unlikely to explain species differences in toxicity.

Considering the central role of CYP2E1 in the conversion of benzene at the low concentrations, most likely to be seen in the liver after benzene exposure, conditions leading to the induction of this enzyme will increase the sensitivity to benzene toxicity. One obvious confounding factor is chronic exposure to benzene, which by itself induces 2E1 (INGELMAN-SUNDBERG and HAGBJÖRK, 1982; ARINC et al., 1991). It is also obvious that alcoholics form a high-risk group. Certain inducers may, however, also activate phase II conjugation pathways, thereby alleviating toxicity by accelerating the conversion of benzene into less toxic metabolites (SCHRENK and BOCK, 1990).

Acetone treatment increases CYP2E1 severely in bone marrow, an increase which parallels the rate of benzene oxidation in this tissue (SCHNIER et al., 1989). This implies an important role of CYP2E1 in hematopoetic toxicity exerted by benzene. Thus, taken together, it appears that CYP2E1 is the isozyme of highest importance for benzene metabolism and toxicity and accordingly, conditions known to increase the amount of this isozyme will also be synergistically influencing its toxic and carcinogenic effects.

2.1.2. Carbon tetrachloride

Carbon tetrachloride (CCl_4) is a well known hepatotoxin, that causes fatty degeneration and centrilobular necrosis in the liver. In the late 19th century and in the beginning of this century, CCl_4 was frequently used in treating hookworms and even as an anesthetic. Industrially it was used in fire extinguishers and in dry cleaning. Its use has been restricted after the toxicity was established, but CCl_4 is still a frequently used industrial solvent. Toxicity of carbon tetrachloride is generally accepted to result from reductive metabolism by cytochrome P450 to the trichloromethyl free radical ($CCl_3\cdot$) which rapidly reacts with oxygen to form the trichloromethylperoxi radical ($CCl_3OO\cdot$) (Packer et al., 1978). Although the use of ESR spectroscopy in combination with the spin trapping agent phenyl-t-butyl nitrone (PBN) has confirmed the formation of $CCl_3\cdot$ radical during both in vitro and in vivo metabolism of CCl_4 (Poyer et al., 1978; Albano et al., 1982), the presence of $CCl_3OO\cdot$ so far has not received the direct demonstration, because of the instability of this PBN adduct at temperatures above −50°C (Symons et al., 1982). Nonetheless, the recent detection of carbon dioxide anion free radical in the bile of rats receiving CCl_4, is interpreted as an indirect evidence supporting the formation of $CCl_3OO\cdot$ in vivo (Knecht and Mason, 1988). Once generated, CCl_4 free radical intermediates are able to alkylate cellular macromolecules and initiate lipid peroxidation, leading to a series of disturbances in hepatocyte functions, responsible for causing steatosis and cell necrosis (Slater, 1982; Poli et al., 1987; Recknagel et al., 1989). See Figure 2 for an overview of the metabolism and toxic effects of this solvent.

Carbon tetrachloride is commonly used for experimental studies on the mechanisms of liver injury and as a model compound characteristically damaging the centrilobular region of the liver. In spite of the abundant literature on CCl_4 (Recknagel et al., 1989; Williams and Burk, 1990), the cytochrome P450-mediated process initiating its activation has only recently started to be unravelled. Besides the numerous reports showing an increase in CCl_4 toxicity following PB pretreatment (Williams and Burk, 1990), the role of CYP2B isoenzyme family in catalyzing the metabolic activation of CCl_4 is indicated by the marked destruction of cytochrome P450 occurring in the liver of PB-treated animals (Lindström and Anders, 1978). Subsequently, studies using polyacrylamide gel electrophoresis of liver microsomal proteins confirmed the selective loss of the 52,000 dalton form of P450 induced by PB or Arochlor 1254 (CYP2B1-2) without significant modification of 56,000 daltons form stimulated by Arochlor 1254 itself and β-naphthoflavone or 3-MC (CYP1A1) (Noguchi et al., 1982a; Head et al., 1981). Moreover, reconstituted vesicles containing cytochrome P450 reductase

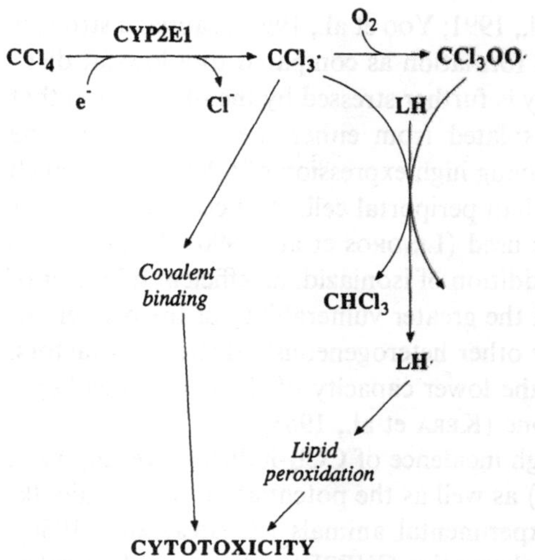

Fig. 2. Mechanisms of activation of carbon tetrachloride and its hepatotoxic effects. There is a pronounced connection between the reductive metabolism catalyzed by CYP2E1 and its hepatotoxic effects.

and the 52,000 dalton P450 isoenzyme have been shown to produce $CCl_3\cdot$ free radicals (NOGUCHI et al., 1982b) and PB induction is able to double the free radical in liver microsomes (CHEESEMAN et al., 1985).

CCl_4 metabolism and toxicity has been reported to be enhanced by rat pretreatment with acetone or other ketones (HEWITT et al., 1983), by ethanol (STRUBELT et al., 1978) and by starvation (SIEGERS et al., 1982) or uncontrolled diabetes (HEWITT et al., 1983), conditions known to increase the liver content of CYP2E1.

More recently, it was reported that in reconstituted microsomal membranes the purified ethanol-inducible form (CYP2E1) converted CCl_4 to chloroform vastly more efficiently as compared to two other isozymes, CYP1A2 and CYP2B4 (JOHANSSON and INGELMAN-SUNDBERG, 1985). The possibility that CYP2E1 might be specifically active in catalyzing the activation of CCl_4 is supported by the observation that substrates of these isoenzymes such as dimethylnitrosamine, isoniazid, cimetidine, tryptamine and diethyldithiocarbamate efficiently inhibited the reduction of CCl_4 in either liver microsomes and reconstituted system containing purified 2E1 (ENGLISH and ANDERS, 1985; PERSSON et al., 1990; GUENGERICH et al., 1991). Furthermore, antibodies against CYP2E1 are effective in inhibiting the haloalkane metabolism by both rat and human CYP2E1 (JOHANSSON et al., 1988; GUENGERICH et al., 1991). REINKE et al. (1988) have shown that ethanol feeding accelerated the rate of trichloromethylradical formation from CCl_4 in the liver. These authors have also observed that ethanol together with a high-fat diet, a regimen recently shown to potentiate ethanol induction of CYP2E1 (TAKAHASHI et al., 1991; YOO et al., 1991), caused a stronger stimulation of trichloromethyl radical formation as compared to a low-fat diet.

The role of CYP2E1 in CCl_4 toxicity is further stressed by the observation that in primary cultures of hepatocytes isolated from either the periportal or the perivenous region, perivenous cells, having high expression of CYP2E1 are much more vulnerable to damage by CCl_4 than periportal cells particularly when cells from ethanol-pretreated animals were used (LINDROS et al., 1990). In these cells the toxic effects were blocked by the addition of isoniazid, an efficient inhibitor of CYP2E1. It should be mentioned that the greater vulnerability of the perivenous region may be further accentuated by other heterogeneously distributed factors, such as the lower oxygen tension or the lower capacity of the perivenous hepatocytes for replenishment of glutathione (KERA et al., 1988).

These findings might explain the high incidence of CCl_4-induced liver injury in alcoholics (WILLIAMS and BURK, 1990) as well as the potentiation of the haloalkane toxicity by ethanol feeding in experimental animals (ZIMMERMANN, 1986). Not unexpectedly, pretreatment with alternative CYP2E1 inducers such as other aliphatic alcohols (RAY and MEHENDALE, 1990) or pyrazole (EBEL, 1989) also potentiated CCl_4 hepatotoxicity.

In conclusion, the data so far available indicated that both CYP2B and CYP2E families are involved in catalysing reductive activation of CCl_4. However, results obtained using reconstituted systems indicated that the rate of CCl_4 reduction by CYP2E1 is at least 100 fold higher than that of CYP2B4 (JOHANSSON and INGELMAN-SUNDBERG, 1985). A possible explanation for such a difference might be due to the fact that CYP2E1 is constitutively in its high spin state and has a high capacity for one electron reduction of dioxygen. Since the reduction potentials of oxygen and CCl_4 are quite close (HANZLIK, 1981), the latter may efficiently compete with oxygen in the univalent reduction process. Results presented by PERSSON et al. (1990) indeed show that this is the case.

Evidence for the importance of CYP2E1 in the activation process has further been given by BURK et al. (1988) who have reported that following rat pretreatment with isoniazid, an inducer of CYP2E1, the microsomal metabolism of CCl_4 is still five times more efficient than in PB-induced microsomes. In addition, diethyldithiocarbamate (DEDC) a fairly specific inhibitor of CYP2E1 (GUENGERICH et al., 1991) protect mice in vivo against the toxic effect of CCl_4 (DEN TONKELAAR and VAN LOGTEN, 1974; SIEGERS et al., 1978).

2.1.3. Chloroform and other trihalogenated methanes

Chloroform has primarily central nervous effects giving inebriation, anesthesia and narcosis. At low levels of exposure, the primary toxic response is hepatotoxicity leading to fatty liver and centrilobular necrosis. Kidney damage may also occur, primarily in the convoluted tubules and the epithelium of HENLE's loops (ZIMMERMANN, 1978).

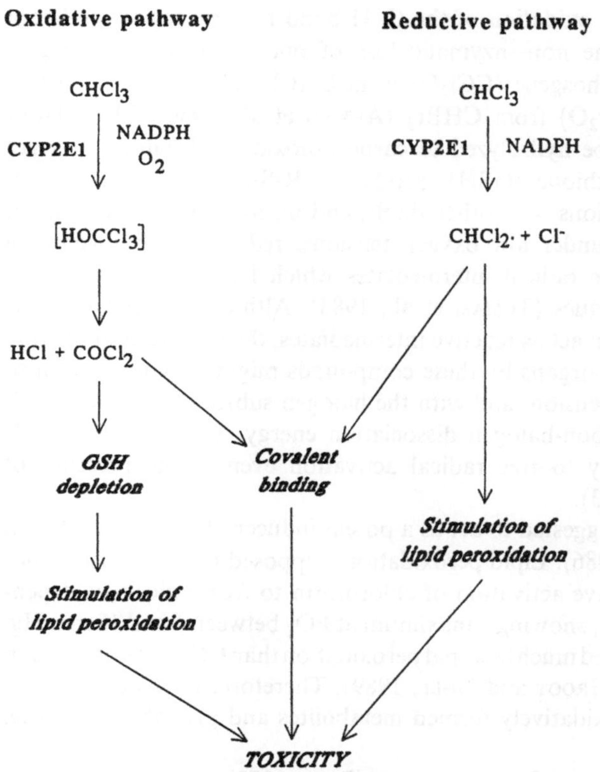

Fig. 3. Two different pathways for metabolic activation of chloroform in the liver. CYP2E1 appears to be the major catalyst of the activation.

Chloroform ($CHCl_3$) is the most widely used among the trihalogenated methane derivatives and has found application as an industrial solvent and intermediate in chemical synthesis. Until 1976, when it was banned by the Food and Drug Administration because suspected carcinogenicity (IARC, 1978), $CHCl_3$ was also employed as a preservative and a flavour enhancer in pharmaceutical products such as cough syrups, mouth washes and dentifrices. Moreover, $CHCl_3$ along with other halogenated compounds, is a common contaminant of drinking water potabilized by chlorination (Natl. Acad. Sci. Rep., 1968).

Chloroform is metabolized, by the cytochrome P-450 system, mainly to carbon dioxide probably via trichloromethanol and phosgene reacting with water (MANSUY et al., 1977). Free radicals are also formed in a P450-dependent reductive metabolism under anaerobic conditions (TOMASI et al., 1985). The toxicity of chloroform as well as that of other trihalogenated methanes has been demonstrated to require activation by the microsomal monoxygenase system. Two main metabolic pathways have been characterized (Fig. 3). The first consists of the P450-catalyzed oxidation of the C–H bond to form trihalomethanol (CCl_3OH) followed by the non-enzymatic loss of one halogen atom to give dihalocarbonyls such as phosgene (CCl_2O) from $CHCl_3$ (MANSUY et al., 1977) or dibromocarbonyl (CBr_2O) from $CHBr_3$ (AHMED et al., 1980). These latter compounds might either be hydrolyzed to carbon dioxide and halides or react with thiols, mainly glutathione (GSH), producing R-S-formyl halides which, following a series of reactions with other thiols, end up forming CO (AHMED et al., 1980). Alternatively, under low oxygen tensions, reductive dehalogenation takes place producing free radical intermediates which have been detected by ESR spin trapping techniques (TOMASI et al., 1983). Although both dihalocarbonyls and free radicals can act as reactive intermediates, the relative contribution to the toxicity in different organs by these compounds might vary in relation to the differences in oxygen tension, and with the halogen substitutions in the molecule. Decreasing the carbon-halogen dissociation energy (C-Cl > C-Br > C-I) enhances the susceptibility to free radical activation even in the presence of oxygen (TOMASI et al. 1983).

Chloroform has been suggested to act as a potent inducer of lipid peroxidation in vivo (EKSTRÖM et al., 1986). Lipid peroxidation, supposed to be a consequence of P450-dependent reductive activation of chloroform to free radicals, is dependent on the oxygen tension, showing a maximum at PO_2 between 1 and 10 mm Hg. However, chloroform caused much less lipid peroxidation than CCl_4, even in relation to its hepatotoxicity (DE GROOT and NOLL, 1989). Therefore, for haloforms like chloroform, mainly the oxidatively formed metabolites and probably phosgene, contribute to toxicity.

Following PB-pretreatment of rats, phosgene (POHL, 1979) and chloroform free radical production (TOMASI et al., 1985) are about twice those in control microsomes. The amount of covalent binding of reactive metabolites to liver protein parallels the hepatotoxic effect (ILET et al., 1973). Nonetheless, MCMARTIN et al. (1981) have reported that, while PB increases the hepatotoxicity of $CHCl_3$ this treatment does not alter its nephrotoxicity. This is consistent with an earlier study showing that, after chloroform intoxication of PB-induced rats, GSH depletion and phosgene formation, was evident in the liver, but not in kidney (POHL, 1979). Conversely, fasting (MCMARTIN et al., 1981) or acetone pretreatment (HEWITT et al., 1983) are able to stimulate both liver and kidney toxicity of trihalomethanes, suggesting that induction of CYP2E1 might be relevant for the expression of chloroform damage.

In vitro-studies have shown that microsomes from isoniazid-treated rats metabolize chloroform 9-fold better than control microsomes (TERELIUS and INGELMAN-SUNDBERG, 1988). This metabolism was inhibited to a large extent by anti-CYP2E1 IgG.

Chloroform metabolism was reconstituted in a membranous system with NADPH-cytochrome P450 reductase, rat ethanol-inducible CYP2E1 and cytochrome b_5 (TERELIUS and INGELMAN-SUNDBERG, 1991). It was found that the apparent affinity of CYP2E1 was strikingly high ($K_m = 4\,\mu M$) and the apparent V_{max} obtained was 15 nmol/nmol CYP2E1/min. By contrast, no detectable metabolism was seen using reconstituted membranes containing phenobarbital-inducible rabbit CYP2B4. BRADY et al. (1989) have also shown that chloroform metabolism in microsomes from acetone-treated rats was 3-fold that in microsomes from control rats. The chloroform metabolism was reconstituted using purified CYP2E1 plus b_5 but no activity was registered with phenobarbital-inducible rat CYP2B1. Ketones, like 2-hexanone, 2-butanone and acetone, are known to potentiate the hepatotoxicity of chloroform (HEWITT et al., 1987) and have been shown to induce both CYP2E1 and CYP2B1 (BRADY et al., 1989). Since CYP2B1 did not metabolize chloroform, these authors suggested that the potentiation of hepatotoxicity was mainly due to induction of CYP2E1.

Taken together, there is sufficient evidence to affirm that CYP2E1 is the principal catalyst responsible for the metabolism of $CHCl_3$ and that PB-inducible forms have only a minor role in this process. These results have been further supported by GUENGERICH et al. (1991), who found an almost complete inhibition of chloroform metabolism by anti-human CYP2E1 IgG in human liver microsomes and specific inhibition by diethyldithiocarbamate.

2.1.4. Dichloromethane

Dichloromethane (DCM) is considerably less hepatotoxic than chloroform or carbon tetrachloride and has replaced them as a general purpose solvent, degreaser in textile industries and cleaning fluid. Early reports showed elevated amounts of carboxyhaemoglobin in the blood as a result of human exposure to dichloromethane vapors (RATNEY et al., 1974; STEWART et al., 1972) and subsequent investigations demonstrated that carbon monoxide was formed from the halomethane (KUBIC and ANDERS, 1975, 1978). According to the metabolic pathway proposed by KUBIC and ANDERS, cytochrome P450 catalyzes oxygen insertion in dihalomethane molecules to yield the hydroxy derivatives, which subsequently rearrange to formyl halides. These compounds have been shown to decompose yielding CO as products. The observed formyl halide intermediates may react as alkylating agents towards macromolecules, explaining the covalent binding of radioactive dichloromethane residues to microsomal protein and lipids (ANDERS et al., 1977; KUBIC and ANDERS 1978).

Pretreatment of rats with phenobarbital or 3-methylcholanthrene has been demonstrated to increase CO formation by liver microsomes incubated with dichloromethane by 360 and 146% respectively (KUBIC and ANDERS, 1975). However, phenobarbital pretreatment does not influence CO production in vivo (ROTH et al., 1975), although ANDERS and coworkers (1977) reported that PB induction increased dichloromethane covalent binding to proteins by 4 fold. More recently, PANKOW (1988) and PANKOW and HOFFMAN (1989) have reported that rats pretreated with isoniazid, acetone or fasting metabolize dichloromethane to CO twice more efficiently than control animals. Consistently, microsomes from isoniazid treated rats catalyzed oxidation of dichloromethane at a rate much higher compared to control (TERELIUS and INGELMAN-SUNDBERG, 1988). This reaction is inhibited by antibodies against CYP2E1 and can be reconstituted in a membranous system with P450 reductase, cytochrome b_5 and CYP2E1 with the apparent V_{max} of 15 nmol/min/nmol and the K_m of 10 μM (TERELIUS and INGELMAN-SUNDBERG, 1991). By contrast CYP2B4 was virtually ineffective. These findings allow us to conclude that, as other halomethanes, DCM is mainly metabolized by CYP2E1 and that the induction of this P450 form might be regarded as a potential risk factor in the case of an otherwise relatively safe compound.

2.1.5. Ethers

Exposure to methyl t-butyl ether (MTBE) has been shown to alter various muscle, kidney and liver metabolic activities. In liver microsomes from acetone- or phenobarbital-induced rats, t-butanol and formaldehyde was formed (BRADY et al., 1990). Metabolism was inhibited 35% by monoclonal anti-CYP2E1, suggesting a partial contribution by this isozyme. MTBE-pretreatment induced the pentoxyresorufin dealkylase activity 50-fold without affecting N-nitrosodimethylamine demethylase activity. Immunoblot analysis showed an elevation in CYP2B1 but not in CYP2E1.

Quite conclusive evidence has been presented for the metabolism of diethyl ether by CYP2E1 (BRADY et al., 1987) and for the role of diethyl ether as an inducer of the isozyme. Acute exposure of fasted rats to ether anesthesia has recently been shown to cause marked oxidative hepatotoxicity as determined by increased amount of ethane exhalation, enhanced chemiluminescence and increased rate of malondialdehyde formation (PIETAN et al., 1991). The toxic effects were accompanied by acute depletion of cytochrome P450 and in particular CYP2E1-dependent catalytic activities were rapidly lost. The data indicate that CYP2E1 might form a very reactive radical of diethylether that causes heme loss and could be able to initiate lipid peroxidation. Further ESR-studies might give an answer to the existence and the nature of such a radical.

2.1.6. Trichlorethylene

Trichlorethylene (TRI) is considered to be of low acute toxicity to humans and the principle effect is a depression of the central nervous system. At high concentrations TRI is hepatotoxic and, to a lesser extent, nephrotoxic (TAKAMATSU, 1962; BROWN et al., 1990). In rats, CHAKRABARTI and TUCHWEBER (1988) found an involvement of a capacity-limited metabolism in the expression of acute nephrotoxicity due to TRI. TRI has been shown to cause cancer in mice and rats and metabolism is believed to play an important role in cytotoxicity and carcinogenicity. The metabolic pathways for TRI appear to be qualitatively similar in mice, rats and humans. TRI is metabolized mainly to chloral (2,2,2-trichloroacetaldehyde), possibly via an epoxide, and further to trichloroethanol or trichloroacetic acid. Minor microsomal metabolites include CO, CO_2, glyoxalic acid and trichloroethylene oxide. In hepatocytes, covalent adducts are formed with proteins as well as nucleic acids. MILLER and GUENGERICH (1983) have presented evidence that the epoxide is not an obligate intermediate in the formation of chloral. They postulated the initial formation of an oxygenated TRI-P450 intermediate. Trichloroethylene-derived chloral, trichloroethanol, trichloroacetic acid, monochloroacetic acid and N-hydroxyacetylaminoethanol have been detected in humans in vivo.

Toxic species that have been suggested include TRI-epoxide (HENSCHLER, 1977; BOLT et al., 1982; DEKANT et al., 1984), dichloroacetic acid (HERREN-FREUND et al. 1987; NELSON and BULL, 1988), unstable intermediates formed from an oxygenated TRI-P450 complex (MILLER and GUENGERICH, 1982; MILLER and GUENGERICH, 1983) and trichloroethanol (TCE) or a free radical derived from TCE (GONTHIER and BARRET, 1989). Recently it has been shown that TRI cause lipid peroxidation in rat liver (OGINO et al., 1991).

MILLER and GUENGERICH (1983) have shown that PB-treatment of rats, but not β-NF, increased the metabolism of trichloroethylene to chloral and trichlorethanol in hepatocytes and increased protein- and nucleic acid-adduct formation. Phenobarbital pretreatment of mice and rats increased TRI-epoxide formation 4-fold in liver microsomes but gave only a 25% increase of protein and DNA adducts. Levels of protein- and DNA-adducts in human microsomes were comparable to levels in microsomes from untreated rats, whereas mice had higher levels of DNA-adducts and chloral formation. Trichloroethylene metabolism was studied in reconstituted systems with eight different rat P-450 isozymes. CYP2B was the isozyme that metabolized trichloroethylene to the greatest extent, based on chloral formation, but P-450IA was also efficient and only $P\text{-}450_{UT\text{-}F}$, $P450_{PB\text{-}D}$ and $P450_{PB/PCN\text{-}E}$ gave no measurable chloral. All eight different P450 isozymes gave TRI-protein adducts when incubated with TRI, but only one isozyme, CYP2B, produced detectable levels of epoxide. Anti-CYP2B IgG inhibited about two-thirds of the chloral formation in microsomes from PB-treated animals but did not inhibit the formation of protein- or DNA-adducts. Suicidal heme destruction occurs with trichloroethylene to the extent of 40% of the heme in a microsomal system in 30 min at 37°C. The destruction probably occurs by attack of pyrrolic nitrogen of the heme on the intermediate, as demonstrated by ORTIZ DE MONTELLANO et al. (1982) for other olefins.

Microsomes from isoniazid-treated rats have been shown to exhibit a 3-fold higher metabolism of trichloroethylene as compared to control microsomes

(TERELIUS and INGELMAN-SUNDBERG, 1988) indicating an important role taken by CYP2E1. However, NAKAJIMA et al. (1990), recently showed that there are three forms of trichloroethylene-metabolizing enzymes in rat liver: the low K_m type, which was present in untreated rats, increased with ethanol-induction, a high-K_m form was induced by phenobarbital, while 3-methylcholanthrene treatment caused induction of a form with intermediary K_m. Phenobarbital enhanced TRI metabolism mainly at high TRI concentrations, whereas ethanol enhanced the metabolism predominantly at low trichloroethylene concentrations and seemed to be substrate-inhibited at higher TRI concentrations. Both ethanol and phenobarbital also increased the metabolism of trichloroethanol to chloral hyrate, with the ethanol-inducible enzyme being the low-K_m form. Their studies correlated well with a study in vivo by NAKAJIMA et al. (1988), where ethanol-pretreatment of rats enhanced TRI metabolism more than PB at TRI concentrations in the inhaled air below 2000 ppm, whereas PB-induced enhancement was more pronounced than that of ethanol at TRI concentrations above 4000 ppm.

Purified ethanol-inducible CYP2E1, in the presence of cytochrome b_5, was recently shown to metabolize TRI at about 20-fold higher rates than phenobarbital-inducible CYP2B4 at low TRI concentrations (2 μM) (TERELIUS and INGELMAN-SUNDBERG, 1991). Cytochrome b_5 enhanced TRI metabolism using both enzymes, but there was a significant metabolism also in the absence of cytochrome b_5. With CYP2E1, the apparent Km was as low as 3 μM and the Vmax about 18 nmol/nmol P450 and minute. These results are in line with recent inhibition studies in microsomes with anti human CYP2E1 IgG by GUENGERICH et al. (1991) and together they indicate an important role of CYP2E1 in TRI metabolism at smaller concentrations.

2.1.7. Tri- and tetrachloroethanes

Although tri- and tetrachloroethanes are less hepatotoxic than CCl_4 and CCl_3 and have largely replaced them as general purpose solvents, the exposure to 1,1,2- and 1,1,2,2-tetrachloroethane can produce liver damage in both humans and experimental animals (PARKER et al., 1979). Moreover, polychloroethanes have been reported to cause hepatocellular carcinomas in rodents (PARKER et al., 1979). Microsomal monoxygenase system is mainly involved in the metabolism of chloroethanes producing reactive intermediates capable to alkylate liver macromolecules (SIPES and GANDOLFI, 1982). As other haloalkanes, tri- and tetrachloroethanes might undergo to both oxidative and reductive metabolism, the latter resulting in the formation of free radical species (THOMPSON et al., 1984; ALBANO et al., 1985). The involvement of different CYP isoenzymes in the metabolism of polychloromethanes has not been studied in detail, however, PB, but not 3-MC, induction has been shown to increase the toxicity of both 1,1,1- and 1,1,2-trichloroethanes (CARLSON, 1973). MACDONALD and coworkers (1982) have demonstrated that following acetone pretreatment non-hepatotoxic doses of 1,1,2-trichloroethane became capable to cause substantial liver injury. This suggests that, beside PB-inducible forms, other cytochrome P-450 isoenzymes might activate polychloroethane and potentially increase the hepatotoxic risk. According to this interpretation recently it has been reported that the oxidation of 1,1,1,2-tetrafluoroethane is primarily catalyzed by CYP2E1 (OLSON et al., 1991). Furthermore, in human microsomes, activation of 1,1,1-trichloroethane is inhibited by diethyldithiocarbamate, a relative isozyme specific inhibitor of CYP2E1 and, furthermore, by antibodies against this isozyme (GUENGERICH et al., 1991). In addition, also the oxidation of 1,1,1,2-tetrafluoroethane is primarily catalyzed by CYP2E1 (OLSON et al., 1991). These data again indicate the unique versatility of CYP2E1 in the metabolism of these kinds of compounds.

2.2. Drugs

2.2.1. Acetaminophen

Acetaminophen (paracetamol) is widely used as an analgesic, especially in the U.S. and U.K., where acetaminophen overdose has become a leading cause of hepatic injury (CLARK et al., 1973; RUMACK et al., 1981). The drug is relatively safe when used in prescribed doses (DAVIDSON and EASTHAN, 1966; BOYD and BERECZKY, 1966). The main reason for its reputation as a hepatotoxin stems from its frequent misuse in large doses, in attempted suicide. In the United States more than one thousand annual cases of suspected acetaminophen overdoses occur (SMILKSTEIN et al., 1988). The dose dependence is further complicated by the large individual variability in the rate of disposition of acetaminophen. Alcoholics are forming a particular high-risk group and frequently develop hepatic lesions after prolonged drug use (SEEFF et al., 1986). A detailed review on the vast literature describing the hepatotoxicity of acetaminophen has recently appeared (NELSON, 1990).

Hepatic metabolism of acetaminophen (AAP) involves mainly conjugation by glucuronidation and sulfation (Fig. 4). It has been estimated that in man these pathways account for about 80% of the total conversion (PRESCOTT, 1983; BLACK, 1984). In addition to these major excretion products, 3-hydroxy- and 3-methoxy-acetaminophen and glutathione-derived conjugates are found in the urine (SLATTERY et al., 1987). The thioethers are detoxication products of glutathione and N-acetyl-p-benzoquinone imine (NAPQI), which is the established toxic metabolite formed via oxidative activation of AAP in a process catalyzed by the cytochrome P-450 system (MINER and KISSINGER, 1979; DAHLIN et al., 1984). Although this oxidative pathway only accounts for a minor fraction of AAP disposal, MITCHELL and associates convincingly demonstrated (MITCHELL et al., 1973; POTTER et al., 1973) that this reaction was responsible for the toxicity

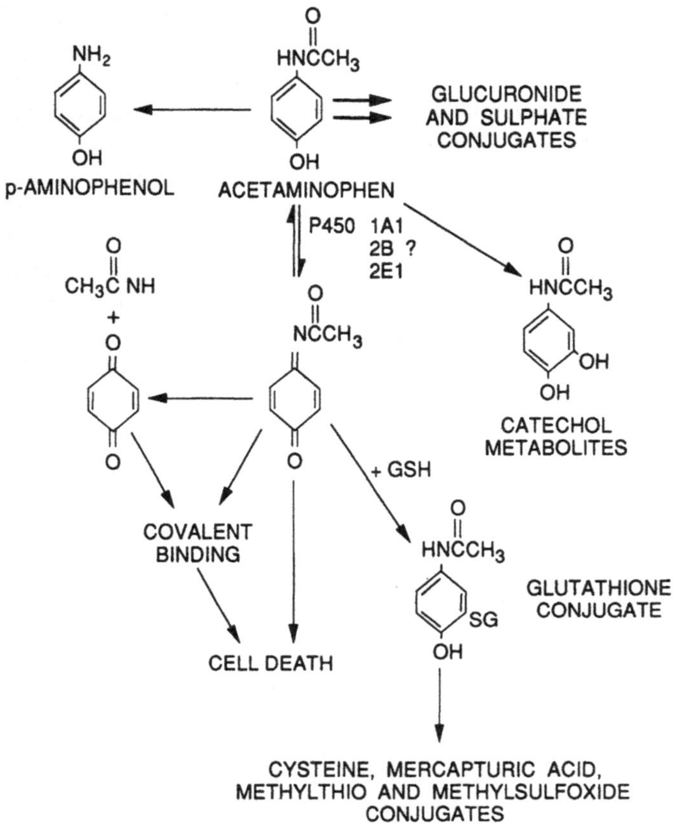

Fig. 4. Metabolic pathways in the conversion of acetaminophen to toxic intermediates. Oxidative activation leading to NAPQI formation is a minor pathway. The isozymes of highest importance in NAPQI formation appears to belong to the CYP1A and CYP2E1 families.

of acetaminophen. The exact mechanisms of NAPQI formation are not yet clear. If several mechanisms are involved, these may be catalyzed by separate P450 isozymes and it therefore becomes important to elucidate the quantitative role of these pathways both in the normal and induced state. Several studies suggest that P450 associated activation of AAP is initiated by abstraction of a hydrogen radical either from the phenolic oxygen or from the nitrogen group. By subsequent recombination and dehydration two alternative two-electron oxidation products, 3′-hydroxyacetaminophen and NAPQI, are formed (Dahlin et al., 1984; Potter et al., 1987; Harvison et al., 1988; van de Straat, 1988) (cf. Fig. 4).

Availability of purified P450 isozymes have enabled NMR studies on the theoretical catalytic capacity of active sites to catalyze the formation of NAPQI.

Based on the measurement of distances between the heme iron of cytochrome P450 isozyme and the hydrogen position on the structure of AAP, it was suggested that at the active site of the CYP1A1 form the heme iron is closer to the phenolic group as compared to the amide nitrogen, while the structure of CYP2B1 would better favor interaction between the heme iron and the nitrogen group. Although such studies are performed under artificial conditions, and therefore should be treated with caution, they tentatively suggest that the structure of CYP1A1 would favor NAPQI formation more efficiently than CYP2B1 (VAN DE STRAAT et al., 1987).

A clue to the role of specific isozymes in the activation of AAP can be obtained by the use of isozyme-specific inducers. Phenobarbital induction does not seem to enhance AAP toxicity (POULSEN et al., 1985) or microsomal AAP oxidation (VAN DE STRAAT et al., 1986) in rats. By contrast, pretreatment with 3-methylcholanthrene enhanced oxidation of AAP in rat liver microsomes (VAN DE STRAAT et al., 1986) and increased toxicity both in isolated rat hepatocytes and in the rat in vivo (ALBANO et al., 1984; NELSON, 1990). Moreover, treatment with both this compound and with 2,3,7,8-tetrachlorodibenzo-p-dioxin caused a 3-fold elevation of the excretion of the AAP-GSH conjugate in bile in hamsters (MADHU et al., 1989) suggesting a role for the 3-methylcholanthrene-inducible isozyme CYP1A1 or 1A2 in AAP activation, at least in these species. Since neither ethanol nor isoniazid pretreatment increased the biliary excretion of AAP-GSH in hamsters (MADHU et al., 1989), CYP2E1 may play a lesser role in this species.

A number of studies based on either a comparison of the catalytic capacity of different isozymes in reconstituted systems (STEELE et al., 1983; MORGAN et al., 1983; RAUCY et al., 1989) or on enhanced AAP toxicity after chronic ethanol pretreatment (MCCLAIN et al., 1980; MOLDEUS et al., 1980; SATO et al., 1981; WALKER et al., 1983) do, however, suggest a major involvement of the ethanol-inducible CYP2E1 in AAP activation in rats and mice.

As mentioned above, AAP toxicity is common among alcoholics. However, several animal studies do suggest that potentiation of AAP by ethanol may be mediated also by other factors than the induced CYP2E1 isozyme, factors that are associated with the ethanol-induced shift in the NAD/NADH redox state of the liver or with an effect on the hepatic GSH status (THUMMEL et al., 1988). A lowered glutathione level has also been incriminated as a risk factor for AAP toxicity in alcoholics (LAUTERBURG and VELEZ, 1988). Accordingly, the observation of an enhanced AAP toxicity after chronic ethanol exposure need not necessarily to be a consequence of an enhanced activation of APAP to NAPQI mediated by ethanol-induced CYP2E1 induction, but could also be due to a reduced defence capacity of liver cells after ethanol ingestion. Increased AAP hepatotoxicity has, on the other hand, been observed to persist in chronically ethanol-treated rats even after withdrawal of ethanol (SATO et al., 1981), suggesting an involvement of ethanol-induced accelerated AAP activation. Also speaking in favor of a role of induced CYP2E1 is the finding that the acute presence of ethanol reduces the hepatotoxicity of AAP (SATO et al., 1981; ALTOMARE et al., 1984).

The role of P450 isozymes has also recently been studied by comparing AAP toxicity in primary cultures of hepatocytes isolated from the periportal or perivenous region (ANUNDI et al., 1991). The perivenous hepatocytes, with high expression of CYP2E1, were somewhat more vulnerable to AAP than periportal cells, but the difference was smaller than that previously observed for CCl_4 (LINDROS et al., 1990) and only moderately enhanced when cells from ethanol-pretreated animals were used. This indicates a less dominant role for CYP2E1 in AAP activation as compared to CCl_4. This is also supported by a comparison between six different rabbit P450s, showing that both CYP2E1 and CYP1A1 were equally efficient in acetaminophen metabolism (KOOP and COON, 1986).

Also in humans, there is evidence for the involvement of CYP2E1, where a patient receiving the CYP2E1 inducer isoniazid developed unexpectedly severe liver and renal injury after ingestion of acetaminophen (MURPHY et al., 1990).

In conclusion, the present evidence suggest a central role for CYP2E1 in AAP activation into NAPQI both in man and rat; however the involvement of the CYP1A isozymes, which is crucial for AAP toxicity in hamsters and rats, needs to be elucidated. Notably, the CYP1A2 form has been found to efficiently catalyze the deethylation of phenacetin to acetaminophen in rat liver (SESARDIC et al., 1990).

2.2.2. Halothane and other fluorinated anaesthetics

The administration of halothane (1,1,1-trifluoro-2-bromo-2-chloroethane) as a general anaesthetic has been implicated as a causative factor in several cases of post-anaesthetic jaundice and hepatitis. Although, unexplained hepatitis after halothane anaesthesia is rare (1 in 10,445 patients), abnormal liver enzyme levels in the serum and jaundice are relative frequent, occurring, respectively, in 1 of 700 and 1 of 1,900 people in the general population (SCHEMEL, 1976; ZIMMERMAN, 1978). A molecular basis for halothane-induced hepatitis, involving immune mediated hepatotoxicity has recently been presented by KENNA (1991).

Studies on halothane metabolism have demonstrated that cytochrome P450 catalyzes both reductive and oxidative biotransformation pathways and the prevalence of one over the other depends upon the oxygen tension (NASTAINCZYK et al., 1978). Trifluoroacetate is the main oxidation product of halothane, while reductive metabolism leads to the formation of both 1,1,1-trifluoro-2-chloroethane (CTE) and 1,1-difluoro-2-chloroethylene (DCE) which are thought to result from, respectively, one and two electron transfers (SIPES et al., 1981). See Figure 5 for a schematic representation of halothane metabolism and cytotoxicity.

Both halothane metabolic pathways have been implicated in the pathogenesis

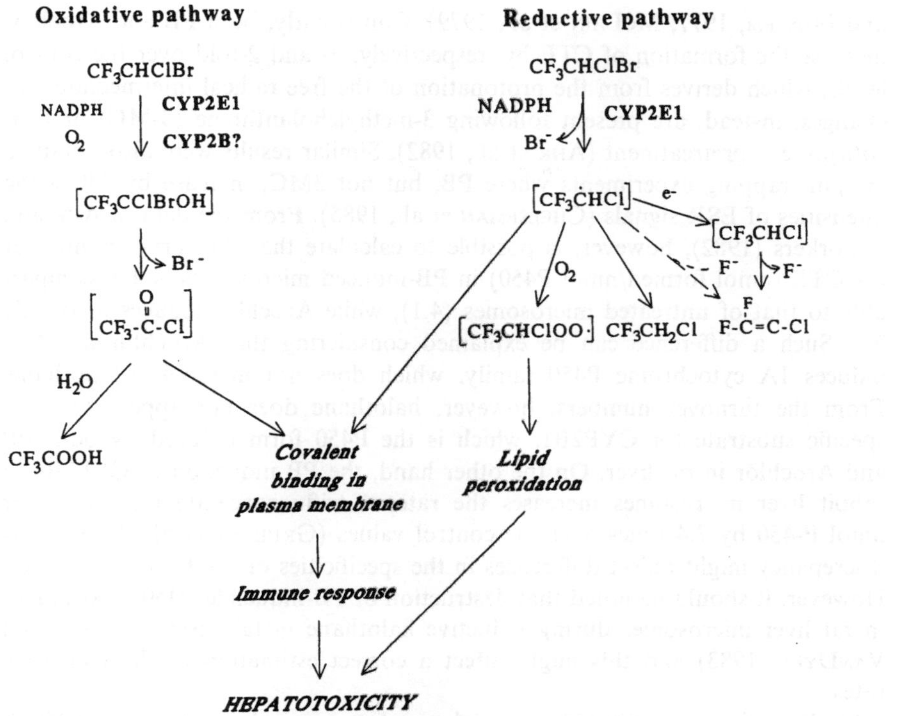

Fig. 5. Reductive P450-dependent activation of halothane to toxic intermediates. CYP2E1 plays an important role in this pathway.

of liver injury (POHL and GILLETE, 1982). Nonetheless, several reports suggest that the reductive pathway is more important for the development of hepatotoxicity (BROWN and SIPES, 1977) and indeed, in vitro and in vivo binding of halothane metabolites to lipids and proteins is enhanced by low oxygen tensions (BROWN and SIPES, 1977; SIPES et al., 1981). Similarly, hypoxic conditions are essential for the development of liver damage in experimental animals (MCLAIN et al., 1979). A free radical intermediate is produced by one electron reduction of halothane (TOMASI et al., 1983) and this reactive species is responsible for alkylation of macromolecules, stimulation of lipid peroxidation and probably cellular injury (TRUDELL et al., 1982; DE GROOT and NOLL, 1983; TOMASI et al., 1983). The production of such radicals in vivo has recently been shown by trapping halothane radicals in bile with the spin trap PBN (HUGHES et al., 1991).

Toxicological studies have demonstrated that, beside hypoxia, pretreatment of rats with polychlorophenols (Arochlor 1254) or PB is needed to mimic in experimental animals liver lesions induced by halothane in humans (REYNOLDS and MOSLEM, 1977; MCLAIN et al., 1979). Consistently, both PB and Arochlor increase the formation of CTE by, respectively, 3- and 2-fold over the control levels, which derives from the protonation of the free radical intermediate. No changes, instead, are present following 3-methylcholanthrene (3-MC) or benzo(a)pyrene pretreatment (AHR et al., 1982). Similar results were also obtained in spin trapping experiments where PB, but not 3MC, increase by 240% the intensities of ESR signals (CHEESEMAN et al., 1985). From the data of AHR and coworkers (1982), however, is possible to calculate that the turnover number for CTE (nmol formed/nmol P450) in PB-induced microsomes is 4.4, comparable to that of untreated microsomes (4.1), while Arochlor reduces it to only 2.4. Such a difference can be explained considering that Arochlor as 3MC induces 1A cytochrome P450 family, which does not metabolize halothane. From the turnover numbers, however, halothane does not appear to be a specific substrate for CYP2B1, which is the P450 form induced by both PB and Arochlor in rat liver. On the other hand, the PB induction of CYP2B4 in rabbit liver microsomes increases the rate of trifluoroacetate formation per nmol P-450 by 7.4 times over the control values (GRUENKE et al., 1988). This discrepancy might reflect differences in the specificities of the P450 isoenzymes. However, it should be noted that destruction of PB-inducible P450s takes place in rat liver microsomes during reductive halothane metabolism (KRIETER and VANDYKE, 1983) and this might affect a correct estimation of the metabolic rates.

Studies using reconstituted systems have also shown that among the P4502B family, the isoenzyme 2B1 (RLM5) is three times more efficient than 2B2 (RLM4) in performing a two electron reduction of halothane to CDE (VANDYKE et al., 1988).

Results by TAKAGI et al. (1983) have demonstrated that, upon halothane exposure, chronically ethanol fed rats displayed centrilobular liver damages, further potentiated by lowering the oxygen tension. Similarly, isoniazid pretreatment of guinea pigs also greatly enhances the toxicity of halothane administered in the presence of oxygen (LIND et al., 1990). In the latter study, animals receiving 25 mg/kg of isoniazid for seven days displayed a 150% elevation in the plasma levels of both oxidative and reductive metabolites of the anaesthetic, suggesting that CYP2E1 may be specifically involved in halothane activation. Consistently, induction by imidazole of CYP2E1 in rabbit microsomes increase the rate of halothane oxidation 18-fold, as expressed per nmol P450 and antibodies against CYP2E1 caused 90% inhibition of the same reaction (GRUENEKE et al., 1988). This evidence suggests that halothane might be preferentially activated by CYP2E1 and this might be important in explaining the occurrence of halothane toxicity in the general population. On this respect it is noteworthy that chronic ethanol feeding also promotes the metabolism and toxicity of enfluorane, another fluorinated general anaesthetic (TSUTSUMI et al., 1990).

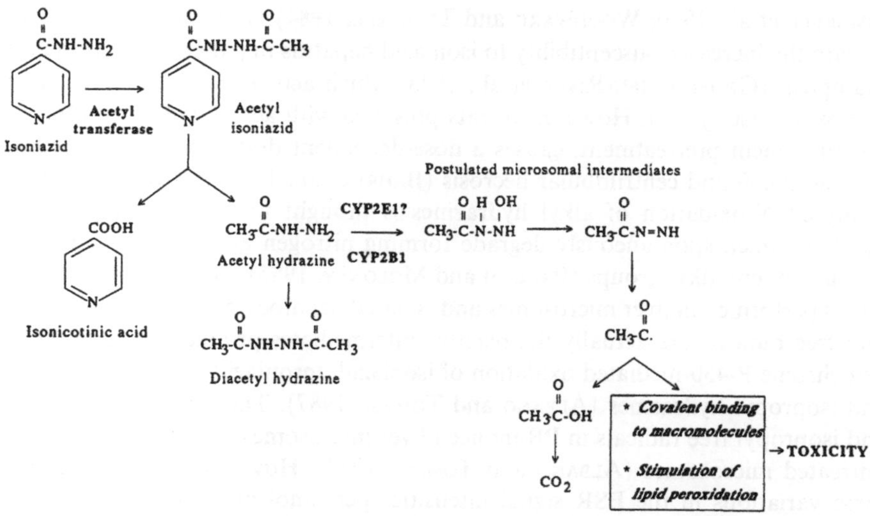

Fig. 6. Activation of isoniazid requires an initial acetylation step. Although isoniazid is a CYP2E1 inhibitor and inducer, studies are lacking showing the involvement of this isozyme in the activation of the alkylhydrazine.

2.2.3. Isoniazid

The antituberculoic agent isoniazid (isonicotinic acid hydrazide) has been reported to cause clinically overt hepatic injury in about 1% and abnormal serum enzyme levels in 10–20% of patients taking the drug for chemoprophylaxis for tuberculosis (MITCHELL et al., 1976; ZIMMERMANN, 1978). The toxicity of isoniazid as well as that of the parent drug iproniazid (isonicotinic acid 2-isopropyl hydrazide) depends upon metabolic activation and acetylation is the initial step in isoniazid biotransformation. Acetylisoniazide and iproniazid are then hydrolyzed respectively, to the acetyl- and isopropyl hydrazines, respectively, which undergo further oxidation by cytochrome P450 (TIMBRELL, 1979) (see Fig. 6).

The acetylation step is the subject of a polymorphic distribution among the caucasian population. Between 40 and 70% of the population are considered as slow acetylators and this has recently been shown to be associated with two differently mutated alleles of a specific N-acetyltransferase (NAT2). Base substitutions at T341 or G590 results in gene products which are either slowly translated or enzymes that are rapidly degraded (BLUM et al., 1991).

Studies performed in experimental animals to clarify the mechanisms of isoniazid, acetylisoniazid and iproniazid toxicity indicated that pretreatment of rats with phenobarbital (PB) increases the toxicity of these compounds. This effect was accompanied by 1.5–1.8 fold stimulation in the covalent binding of the acetyl and isopropyl portions of the molecules to hepatic protein (TIMBRELL, 1979;

Timbrell et al., 1980; Woodward and Timbrell, 1984). The effect of PB might explain the increased susceptibility to isoniazid hepatitis in patients receiving also rifampicin (Groenhagen-Riska et al., 1978) which acts as a PB-like inducer of monoxygenase system. However, in rats poisoned with acetyl hydrazine PB, but not rifampicin pretreatment, causes a dose-dependent destruction of liver cytochrome P-450 and centrilobular necrosis (Bahri et al., 1981). Cytochrome P450 catalyzed N-oxidation of alkyl hydrazines is thought to give rise to unstable diazenes which spontaneously degrade forming nitrogen and free radicals from the substituent alkyl groups (Prough and Moloney, 1985). Spin trapping experiments performed in liver microsomes and isolated hepatocytes have demonstrated that free radicals are actually the reactive intermediates produced as a result of cytochrome P-450-mediated oxidation of isoniazid, iproniazid as well as of acetyl and isopropyl hydrazines (Albano and Tomasi, 1987). The formation of acetyl and isopropyl free radicals in PB-induced liver microsomes is about twice that of untreated microsomes (Albano and Tomasi, 1987). However, when expressing these variations in the ESR signal intensities per nmol of P450, no significant changes over the controls are apparent. Thus PB-inducible isoenzymes do not appear to be specifically active in metabolizing these hydrazines and the more efficient activation is likely to depend upon the high level of cytochrome P450. According to this interpretation, PB pretreatment has been found to enhance the V_{max} of acetylhydrazine covalent binding to microsomal protein without changing the apparent Km values (Timbrell, 1979).

A doubling in the levels of acetylhydrazine covalent binding has been observed in liver microsomes from rats pretreated for ten days with isoniazid, despite the absence of any modification in the cytochrome P450 content (Powell-Jackson et al., 1982). Subsequent studies have shown that isoniazid is an effective inducer of CYP2E1, but do not change the overall amount of the cytochrome (Ryan et al., 1985, 1986). Although epidemiological reports show that chronic alcohol consumption increases the risk of liver damage by isoniazid (Dickinson et al., 1981), the possibility that CYP2E1 might be specifically responsible of alkylhydrazine activation has not been thoroughly investigated. On this respect, preliminary experiments in our laboratories indicate that induction of CYP2E1 by the combination of acetone administration and starvation leads to a 3–4 fold elevation in the spin trapping of free radical from methyl, ethyl and isopropyl hydrazines.

2.3. Others

2.3.1. Aflatoxin

Aflatoxin has long been known to be hepatotoxic, causing fatty liver and necrosis, as well as being hepatocarcinogenic. Aflatoxin B1 (AFB1) is detoxified to aflatoxin M1 (AFM1) via AFB1-4-hydroxylase (FALETTO et al., 1988). The activation is primarily caused by the P450-dependent formation of the 8,9-epoxide (SWENSON et al., 1977. FUKUHARA and colleagues have identified a 3-methylcholanthrene-inducible P450 in Syrian Golden hamsters which has a higher potency towards AFB1 activation than P450s from other species (FUKUHARA et al., 1989 and 1990). After induction, this isozyme constituted 40% of the total P450. SHAYIQ and AVADHANI (1990) recently purified a phenobarbital-inducible male-predominant P450 isozyme, named P450mt4, from rat liver mitochondria. This isozyme seems to be a close relative to CYP2B1 and metabolized aflatoxin exclusively in the presence of mitochondrial-specific ferredoxin and ferredoxin-reductase.

Results by SHIMADA et al. (1989) indicated that in the human, CYP3A4 is the pedominant form of P450 involved in the genotoxic activation of AFB1 in the liver. Other authors (RAMSDELL et al., 1991) failed to show any high specificity between the association of AFB1 metabolism and CYP3A4 at relevant concentrations of AFB1, although these authors concluded that CYP3A enzymes might be responsible for 8,9-epoxide formation at higher substrate concentrations.

A fetal form of P450, P450HFLa, structurally related to, but not identical to CYP3A4, has also been shown to activate AFB1 (KITADA et al., 1990).

Also other human P-450 isozymes have been shown to be involved in the metabolic activation of AFB1 to toxic or mutagenic compounds (FORRESTER et al., 1990). CYP3A3/4 proteins were important in the metabolism and mutagenicitiy of AFB1 and constituted the major P450s in the human liver samples, but also antibodies to CYP1A2 and CYP2A1 inhibited 8,9-diol formation. In addition, the expression of a CYP2C protein correlated with AFB1 metabolism and mutagenicity.

The development of P450 expression systems in hepatoma cells by means of recombinant vaccinia virus has allowed studies about the contribution of several different human P450s to metabolic activation of AFB1 (AOYAMA et al., 1990). These investigators found that five various P450s, CYP1A2, 2A3, 2B7, 3A3 and 3A4, could contribute to metabolic activation of aflatoxin as assessed by production of His revertants of *Salmonella typhimurium* in the Ames test. The same P450s also catalyzed conversion of AFB1 to DNA-bound derivatives, whereas six other human P450s (CYP2C8, 2C9, 2D6, 2E1, 3A5 and 4B1) were ineffective in both tests.

In conclusion, no overwhelming evidence has been provided regarding any P450 isozyme responsible for AFB1 activation, although isozymes of the CYP3A family belong to those with highest activity.

2.3.2. Bromobenzene

Bromobenzene metabolism and toxicity has been studied for over 100 years. The centrilobular hepatic necrosis seen in rats or mice treated with bromobenzene is associated with covalent binding (Reid and Krishna, 1973). The relative reactivity of related chemicals in vivo, expressed as covalent binding to rat liver DNA, is: 1,2-dibromoethane > bromobenzene > 1,2-dichloroethane > chlorobenzene > epichlorohydrin > benzene (Prodi et al., 1986).

Bromobenzene is metabolized by hepatic microsomes to two different epoxide intermediates, which then rearrange to form either ortho- or para-bromophenol. Humans preferentially metabolize halobenzenes through the hepatotoxic 3,4-epoxide pathway and at greater rates than mice, suggesting that humans are susceptible to halobenzene-induced hepatotoxicity.

Phenobarbital induces both 2,3- and 3,4-epoxide pathways in rats and enhance hepatotoxicity, while 3-methylcholanthrene or β-naphthoflavone cause a selective increase in the 2,3-epoxide pathway only (Lau and Zannoni, 1979). In fact, 3-methylcholanthrene pretreatment of rats can protect against bromobenzene hepatotoxicity by inducing the comparatively nontoxic pathway to 2-bromophenol (Zampaglione et al., 1973). This principle has been shown to be relevant also using purified enzymes. Thus, Lau and Zannoni (1981) have shown that purified phenobarbital-inducible rabbit liver LM_2 (CYP2B4) specifically catalyzes the 3,4-epoxidation with rearrangement to *p*-bromophenol, whereas β-naphthoflavone-inducible rabbit LM_4 (CYP1A2) specifically catalyzes the 2,3-epoxidation yielding o-bromophenol.

Diabetes potentiates the hepatotoxicity of bromobenzene and carbon tetrachloride in rats without ketonemia (Watkins et al., 1988). This potentiation was, however, less pronounced with bromobenzene as compared to carbon tetrachloride. It was therefore speculated that bromobenzene was not a preferred substrate for CYP2E1 which was induced by the diabetes.

In conclusion, it appears that bromobenzene is metabolically activated primarily by phenobarbital-inducible P450 species.

2.3.3. Ethanol and acetaldehyde

Ethanol is unique in being both a drug and a nutrient. The hepatotoxicity of alcohol is low, compared to most other drugs, but nevertheless becomes significant in many countries where the daily ingestion of large amounts of alcoholic beverages is common. The mechanisms responsible for ethanol-induced liver damage are still largely obscured. (Lieber, 1988, 1990). Acetaldehyde, the potentially toxic proximate metabolite from ethanol, has been incriminated in many of the effects of alcohol (Lindros, 1978), including liver damage (Lieber, 1988). Alcohol dehydrogenase mediated ethanol oxidation results in a dramatic shift in the NAD/NADH redox ratio (Forsander et al., 1965) and this, in possible conjunction with a relative hypoxia in the centrilobular liver region (Jauhonen et al., 1982), might be a pathogenic factor. In addition, ethanol treatment is known to cause an oxidative stress by induction of microsomal oxidase activity and NADPH-dependent lipid peroxidation mediated by CYP2E1 (EKSTRÖM and INGELMAN-SUNDBERG, 1989).

Based on the original observation by ORME-JOHNSON and ZIEGLER (1965), LIEBER and DE CARLI demonstrated the existence of a cytochrome P450-dependent Microsomal Ethanol Oxidizing System (MEOS) (LIEBER and DE CARLI, 1968). In control livers, the contribution of this sytem to ethanol metabolism seems to be of minor importance (NORSTEN et al., 1989) although the adaptive increase seen after chronic ethanol ingestion might make its contribution quantitatively somewhat more significant (HANDLER et al., 1988). It has therefore been speculated that the oxygen requiring conversion of ethanol to acetaldehyde via this pathway might contribute to ethanol-induced hepatotoxicity. The major part of the microsomal ethanol oxidation is carried out by the ethanol inducible CYP2E1 (KOOP et al., 1982; INGELMAN-SUNDBERG and JOHANSSON, 1984; WRIGHTON et al., 1986) and a major effect of ethanol on the hepatotoxicity of more than 50 different chemicals now identified as substrates for CYP2E1 (Table 2), is connected by the property of ethanol as a CYP2E1 inducer.

A marked increase of CYP2E1 after alcohol exposure is seen both in rats (about 10-fold) (JOHANSSON et al., 1988) and in humans (WRIGHTON et al., 1986; EKSTRÖM et al., 1989). The expression of CYP2E1, both constitutively and after induction with ethanol, is restricted to the centrilobular region of the liver and, in particular, to the three to four layers of hepatocytes most proximal to the central vein (INGELMAN-SUNDBERG et al., 1988). The concentration of CYP2E1 in these induced hepatocytes may approach 0.1 mM. The molecular basis for this heterogeneous distribution has been shown to be inherent in a regioselective expression of the CYP2E1 gene in the perivenous hepatocytes (JOHANSSON et al., 1990). This is of interest, since ethanol causes selective destruction of this acinar region.

As mentioned, one possible mechanism for ethanol mediated hepatotoxicity might be connected with the ability of CYP2E1 to produce reactive oxygen radicals (INGELMAN-SUNDBERG and HAGBJÖRK, 1982; INGELMAN-SUNDBERG and JOHANSSON, 1984). CYP2E1 effectively reduces dioxygen (INGELMAN-SUNDBERG and JOHANSSON, 1984; GORSKY et al., 1984). Similarly, the enzyme can reduce carbon tetrachloride, chloroform and dioxygen to reactive species (JOHANSSON and INGELMAN-SUNDBERG, 1985). Dioxygen and carbon tetrachloride compete for CYP2E1 electrons (PERSSON et al., 1990). The CYP2E1 dependent reduction of dioxygen takes place at physiological concentrations of dioxygen and is inhibited in an isozyme-specific manner by isoniazid, octylamine and tryptamine (PERSSON et al., 1990), all compounds interacting directly with the P450 heme.

Oxygen radicals generated by P450 can initiate membrane lipid peroxidation (EKSTRÖM and INGELMAN-SUNDBERG, 1986). CYP2E1 was found to be the most efficient isozyme out of eight by initiating NADPH dependent lipid peroxidation in reconstituted membranes. Furthermore, anti CYP2E1 IgG inhibited P450-dependent but not NADPH-cytochrome P450 reductase-dependent lipid peroxidation in microsomes (Ekström and Ingelman-Sundberg, 1989). The rate of NADPH consumption, and formation of either hydrogen peroxide or superoxide anions correlated well to the amount of CYP2E1 in various membranes. In addition, CYP2E1 was induced in both liver and lung microsomes by oxygen treatment of animals (Tindberg and Ingelman-Sundberg, 1989) and pretreatment of the rats with acetone, in order to induce the CYP2E1 level, reduced the survival time of the rats in an oxygen atmosphere.

Studies using ESR spectroscopy in combination with the spin trapping agent 4-pyridyl-1-oxide-t-butyl nitrone (4-POBN) have demonstrated the formation of a free radical, identified as 1-hydroxyethyl radical, in rat liver microsomes incubated in the presence of ethanol and of an NADPH regenerating system (Albano et al., 1987, 1988a). We also found that alcohol free radicals are similarly produced by liver microsomes during the metabolism of various other aliphatic alcohols, indicating the existence of a common activation pathway for these compounds (Albano et al., 1987, 1991). Spin trapping experiments also show that a non-enzymatic mechanism involving the presence of hydroxyl (OH·) radicals, originating from iron-catalyzed degradation of hydrogen peroxide and an enzymatic pathway involving cytochrome P450 are responsible for the formation of hydroxyethyl radicals (Albano et al., 1988a; Reinke et al., 1990). The role of cytochrome P450, and particularly of CYP2E1, in catalyzing the free radical activation of alcohol has been recently established using reconstituted membrane systems and anti CYP2E1 IgG (Albano et al., 1991). Moreover, Knecht et al. (1990) have reported the spin trapping in vivo of hydroxyethyl free radicals during ethanol metabolism by ADH-negative deer-mice. This finding is of great importance since it confirms that the observations in vitro concerning hydroxyethyl radicals generation have relevance in explaining ethanol toxicity in the living animals. The possible involvement of reactive oxygen species or hydroxyethyl radicals in stimulating lipid peroxidation is suggested by the observation that hepatocytes isolated from livers of rats chronically fed with ethanol exhibit a higher sensitivity to the pro-oxidant effects of alcohol as compared to cells obtained from untreated animals (Albano et al., 1988b). Moreover, the addition of 4-methylpyrazole, an inhibitor of alcohol dehydrogenase to hepatocytes obtained from ethanol pretreated rats do not significantly interfere with the stimulation of lipid peroxidation by ethanol (Albano et al., 1988b).

Although the mechanisms of free radical mediated activation of alcohols by cytochrome P450 has not yet been completely elucidated, we have observed that in both liver microsomes and in reconstituted membranes containing CYP2B1 or CYP2E1, the formation of alcohol radicals correlates with the capacity of different isozymes to produce reactive oxygen radicals. The alcohol radical formation is inhibited by isoniazid, tryptamine and octylamine (Albano et al., 1991), agents that specifically inhibit the CYP2E1-dependent formation of hydrogen peroxide and P450-dependent NADPH-oxidase activity (Persson et al., 1990). This indicates that the generation of alcohol radicals should occur during the process of

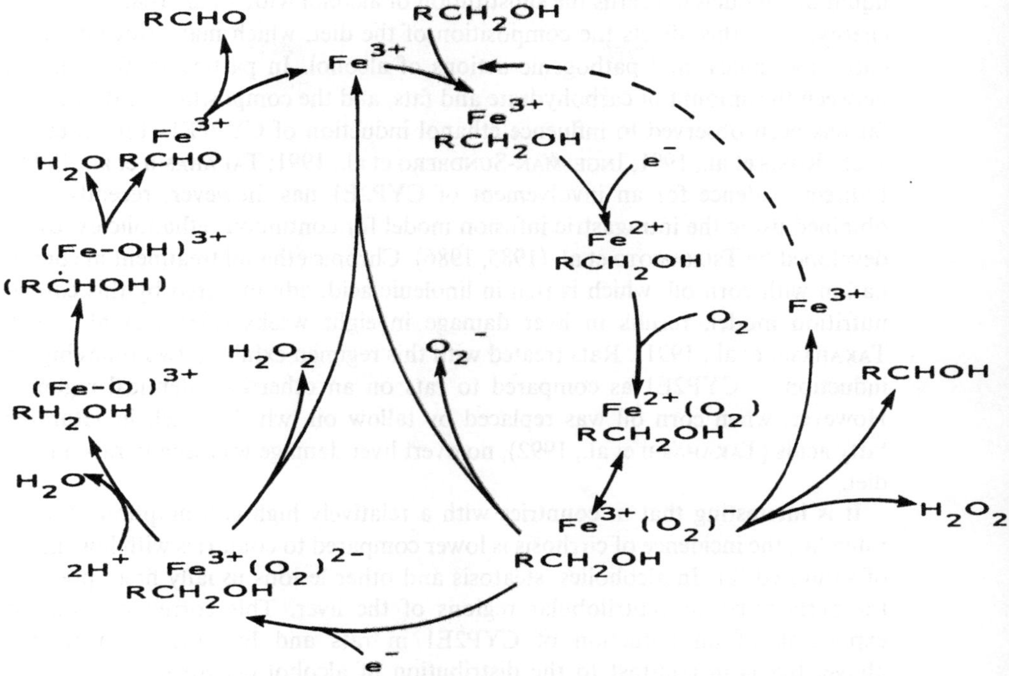

Fig. 7. Proposed mechanism for the CYP2E1-dependent formation of reactive hydroxyethyl radicals from ethanol involving primarily the oxidase activity of the hemoprotein yielding a ferric P450 oxy complex with sufficient energy to abstract a proton from the α-carbon.

univalent reduction of dioxygen and possibly be carried out by the ferric cytochrome P450 oxy-complex (P450-$Fe^{3+}O_2^-$) formed during the reduction of heme-oxygen (Fig. 7). In such a state the cytochrome should be sufficiently reactive to abstract a proton from the α-carbon of the alcohol yielding a carbon centered radical and H_2O_2 (Fig. 7).

These studies all suggest that CYP2E1 generates reactive oxygen species as well as hydroxyethylradicals that can initiate microsomal lipid peroxidation. Induction of this isozyme by ethanol might contribute to an enhanced oxidative stress in the cells and an increased amount of dioxygen utilization selectively in the centrilobular region of the liver acinus. Future work in vivo, utilizing CYP2E1-specific inhibitors, might provide us with an answer to the toxicological importance of CYP2E1-dependent formation of oxygen radicals under various conditions.

However, experimentally, it has proved difficult to convincingly demonstrate a direct role of CYP2E1 in hepatotoxicity induced only by ethanol. A major obstacle has been to find a nutritionally adequate efficiently working animal model to induce liver damage. A serious problem associated with the use of liquid diet models concerns the substitution of alcohol with an alternate source of energy, since this affects the composition of the diet, which may critically influence biochemical and pathogenic actions of alcohol. In particular, the relation between the amount of carbohydrate and fats, and the composition of the dietary fat has been observed to influence ethanol induction of CYP2E1 (LIEBER et al., 1988; RONIS et al., 1991; INGELMAN-SUNDBERG et al., 1991; TAKAHASHI et al., 1992). Indirect evidence for an involvement of CYP2E1 has, however, recently been obtained using the intragastric infusion model for continuous ethanolic exposure developed by TSUKAMOTO et al. (1985, 1986). Chronic ethanol treatment in combination with corn oil, which is rich in linolenic acid, administered by this enteral nutrition model, results in liver damage in eight weeks (NANJI et al., 1989; TAKAHASHI et al., 1991). Rats treated with this regimen exhibit a two times higher induction of CYP2E1 as compared to rats on an otherwise identical regimen. However, when corn oil was replaced by tallow oil, which is rich in saturated fatty acids (TAKAHASHI et al., 1992), no overt liver damage was seen in rats on this diet.

It is interesting that in countries with a relatively high consumption of saturated fat, the incidence of cirrhosis is lower compared to countries with low intake of saturated fat. In alcoholics, steatosis and other lesions usually first appear in the perivenular or centrilobular regions of the liver. This correlates with the expression of an induction of CYP2E1 in rats and humans as mentioned above, but is in contrast to the distribution of alcohol dehydrogenase activity which is rather evenly distributed in the liver acinus (VÄÄNÄNEN et al., 1984; BENGTSON et al., 1988).

In conclusion, a major hepatotoxic effect of alcohol is attributed to its induction on CYP2E1, enhancing metabolic activation and subsequent toxic actions of carbon tetrachloride, acetaminophen, nitrosoamines, bromobenzene and other liver toxic CYP2E1 substrates.

Acetaldehyde, the proximal metabolite in ethanol breakdown, has been suggested to explain many of the actions of alcohol (LINDROS, 1978; LIEBER, 1990). A possible mechanism leading to injury involves the formation of protein adducts. These covalent modifications might result in the formation of autoantibodies. The covalent binding might also cause inactivation of enzymes, decreased rate of DNA repair, alteration of the microtubules, plasma membranes and mitochondria and also impairment of the mitochondrial oxygen utilization. Experimentally, acetaldehyde causes glutathione depletion and lipid peroxidation and stimulates hepatic collagen synthesis, promoting fibrosis (cf. LIEBER, 1990).

Acetaldehyde protein adducts are also demonstrated in conjunction with ethanol metabolism. Interestingly, a single major 52 kD protein in rat liver microsomes, that reacted with anti-acetaldehyde adduct-specific immunoglobulin, also was recognized by anti-CYP2E1 IgG (BEHRENS et al., 1988). The observation suggests that CYP2E1 might be a target protein for acetaldehyde binding in liver microsomes in vivo. This suggestion is supported by recent data by NIEMELÄ et al. (1991) who demonstrated immunoreactive acetaldehyde protein adducts in the centrilobular region of liver from alcoholics.

Recently, TERELIUS et al. (1991) showed that microsomes from starved and acetone treated rats with elevated CYP2E1 levels metabolizd acetaldehyde about eight times faster than control microsomes. The affinity of CYP2E1 for acetaldehyde was almost three orders of magnitude higher than for ethanol: an apparent K_m of 30 μM was obtained both for the CYP2E1-dependent metabolism in microsomes as well as in reconstituted membranous systems. No detectable metabolism was registered with phenobarbital-inducible rabbit CYP2B4 in the reconstituted membranes. Thus, ethanol-inducible CYP2E1 might contribute to acetaldehyde metabolism not only in liver but in other organs as well, e.g. in the brain (HANSSON et al., 1990) and might be responsible for adduct formation. Future studies will give an answer to a possible role of CYP2E1 in acetaldehyde toxicity and immune response. In any case it is interesting to note that CYP2E1 can effectively metabolize ethanol in two steps to yield acetate as the ultimate product.

2.3.4. Nitrosamines

Nitrosamines must be metabolically activated to exert their cytotoxic or carcinogenic actions. Activation is currently thought to occur via oxygenation of the α-carbon catalyzed by a microsomal cytochrome P450 dependent process (YANG et al., 1987). The best studied nitrosamine is N-nitrosodimethylamine (NDMA), the prototype of these carcinogens and a member of the N-nitrosodialkylamines. The metabolism of the N-nitrosodialkylamines is thought to occur either by dealkylation or by nitrosation. Of these processes, the nitrosation is generally believed to lead to formation of chemically less reactive intermediates, while the oxidation of the alkyl group is the process held responsible for the production of intermediates ultimately leading to toxic and carcinogenic effects (YANG et al., 1987).

The metabolism of the model compound N-nitrosodimethylamine (NDMA) has been estimated from the rate of demethylation. The NDMA demethylase activity was found to be enhanced in microsomes from animals pretreated with acetone, ethanol, pyrazole or isopropanol and was also enhanced in the diabetic state or after fasting (YANG et al., 1985; THOMAS et al., 1987). Subsequently, a P450 isozyme termed nitrosamine demethylase was isolated from ethanol-treated rabbits (KOOP et al., 1982; MORGAN et al., 1982). P450 isozymes with similar properties were then isolated from rat (RYAN et al., 1985) and human (WRIGHTON et al., 1986) liver and the enzyme was found to be identical with the ethanol-inducible CYP2E1. Although it is generally agreed that this enzyme plays a pivotal role in the metabolism by demethylation of NDMA, other enzymes are also involved.

YANG et al. (1985) studied the capacity of various P450 enzymes to metabolize nitrosamines by the use of reconstituted microsomal membranes and six purified P450 isozymes obtained from variously pretreated rabbits. The highest activity was observed using CYP2E1, while the isozymes obtained after induction with phenobarbital, isosafrole or imidazole only were active at high NDMA concentrations and two other isozymes were inactive.

LEVIN et al. (1986) made a similar study, comparing eleven P450 isozymes purified from rat liver. Compared to all other isozymes, CYP2E1, isolated after induction with ethanol or isoniazid, exhibited an at least 15-fold higher demethylation rate. Furthermore, addition of cytochrome b_5 resulted in a several-fold increase in the affinity of CYP2E1 for NDMA, suggesting an essential role for this cytochrome in the catalytic process. Some contribution by other isozymes is indicated by the existence of several different K_m values for NDMA in isolated microsomes (YANG et al., 1985).

More direct evidence for the role of P450 CYP2E1 in the toxicity of NDMA was recently obtained by comparing NDMA toxicity in primary cultures of hepatocytes isolated from the periportal or perivenous region (ANUNDI et al., 1991). Perivenous hepatocytes with a high expression of CYP2E1 were clearly more damaged by NDMA than periportal cells. Furthermore, these effects were enhanced when cells from ethanol-pretreated animals were compared and the toxicity could be inhibited by inclusion of isoniazid, recently shown to be an efficient inhibitor of CYP2E1 (PERSSON et al., 1990).

The pattern of metabolism in human liver seems to be rather similar. Microsomal preparations from human liver catalyzed dealkylation of the diethyl-, dimethyl-, benzylmethyl-, and butylmethylnitrosamines much more efficiently than nitrosation at low substrate concentration (YOO et al., 1988). There was a large interindividual variation in the capacity for NDMA demethylation, but this capacity correlated with the amount of a P450 isozyme immunoreactive against the rat CYP2E1. This suggests a crucial role for the ethanol-inducible P450 form in NDMA demethylation also in human liver. The large interindividual differences suggest corresponding differences in vulnerability to nitrosamine carcinogenicity, differences that may be determined not only by genetic factors, but also by the nutritional status.

Alcoholics exhibit elevated levels of CYP2E1 (WRIGHTON et al., 1987; EKSTRÖM et al., 1989; TSUTSUMI et al., 1990). This is probably associated with enhanced rates of demethylation and denitrosation observed in liver microsomes from alcoholics (AMELIZAD et al., 1989), making these individuals more susceptible to nitrosamine-induced hepatocarcinogenicity. Experimentally, potentiation of the hepatotoxicity of NDMA has been reported after pretreatment with acetone or isopropanol, in diabetes and after fasting, all states known to cause induction of CYP2E1. This enhanced vulnerability may include extrahepatic organs, such as the lung, which exhibited enhanced NDMA metabolism after chronic ethanol exposure (CARLSON, 1990). On the contrary, ethanol pretreatment was reported to inhibit diethylnitrosamine (NDEA)-induced tumor development, suggesting that this dialkyl nitrosamine derivative is metabolized by a different pattern of P450 isozymes as compared to NDMA. In fact, in transfected human cell lines, NDEA was found to be metabolized more efficiently by CYP2A3 as compared to CYP2E1, which on the other hand converted NDMA more efficiently (CRESPI et al., 1990).

In conclusion, comparison of the catalytic capacity of purified P450 isozymes, in vivo induction experiments and immunological data all point to a dominant role of CYP2E1 in the demethylation of NDMA in the liver of several species including man. Much less is known about the isozymes catalyzing the denitrosation of the nitrosamines. Such studies are needed in order to evaluate whether the nitrosation has any role in the carcinogenic and toxic actions of the nitrosamines. In addition, the isozyme profiles involved in the metabolism of other nitrosamines need to be studied to evaluate the role of the other isozymes in the activation of the nitrosamine family to potentially carcinogenic and cytotoxic intermediates as a whole.

2.3.5. Thioacetamide and thiobenzamide

These agents are thiono-sulphur containing xenobiotics which cause necrosis in the centrilobular region. They can also cause formation of cirrhosis and, in some instances, hepatic tumours (HUNTER et al., 1977; ZIMMERMAN, 1978; MALVADI et al., 1984). Metabolic studies have indicated that thioacetamide and thiobenzamide are substrates for both the cytochrome P450-dependent monooxygenase system and the flavin-dependent monooxygenase (FADM) (HUNTER et al., 1977; CASHMAN and HANZLIK, 1981; TYNES and HODGSON, 1983). The proposed pathway of activation involves the S-oxidation to the respective S-oxides (PORTER and NEAL, 1978; CHIELI et al., 1980). The S-oxides formed are further oxidized by the same enzymatic systems to postulated dioxides which hydrolyze producing reactive species responsible for macromolecule alkylation. Experiments using differently radiolabelled thioacetamides have shown that such a reactive intermediate does not contain the sulphur atom, and suggest that acetamide actually binds to proteins by interacting with the ε-amino group of lysine (PORTER et al., 1979; DYROFF and NEAL, 1981). Thus, according to this metabolic pathway thioacetamide S-oxide is more effective as alkylating agent and more hepatotoxic than the parent compound (PORTER et al., 1979). Thioacetamide and thioacetamide S-oxide toxicity is enhanced by pretreatment of rats with phenobarbital, whereas destruction of cytochrome P450 by cobalt chloride has a strong inhibitory effect (HUNTER et al., 1977). Antibodies against P450 inhibited metabolism of thioacetamide S-oxide completely (VADI and NEAL, 1981). Conversely, in spite of the fact that cytochrome P450 was reported to account for 35% of thiobenzamide metabolism in liver microsomes (TYNES and HODGSON, 1983), a clear modulation of thiobenzamide toxicity by cytochrome P450 inducers such as PB has not been observed.

A number of reports have shown that thioacetamide toxicity is potentiated by administration of ethanol, by fasting and by experimental diabetes (MALING et al., 1975; STRUBELT et al., 1981; EL-HAWARI and PLAA, 1983; NICOLAEV et al., 1986), conditions all known to induce CYP2E1. Consistently, acute administration of pyrazole, a CYP2E1 substrate, has been found to decrease toxicity of thioacetamide in control rats. In support of this CHIELI et al. (1990) have observed that acetone treatment of rats doubles the rate of thiobenzamide oxidation. Apparently, in microsomes of acetone treated rats, P450-dependent metabolism accounts for 70% of the total thiobenzamide oxidation as compared to about 40% observed in microsomes from native rats. Pretreatment with acetone is also capable to stimulate thiobenzamide toxicity and such an effect is prevented by coadministration of acetone which acts as a competitive inhibitor of thioacetamide activation (CHIELI et al., 1990). Thus, these results indicate that CYP2E1 is specifically involved in the metabolism of thiono-sulfur containing compounds in agreement with the recent observation that carbon disulfide is similarly metabolized by this isozyme (SNYDERWEIN et al., 1988).

3. Conclusions

CASSARETT and DOULL (1975) presented at least 100 different chemicals which are considered to be hepatotoxic either by causing liver cirrhosis, fatty infiltration, or hepatitis like cholestasis. Many of those chemicals can be considered to require metabolic activation by the cytochrome P450 system before exerting their toxicity. When reviewing the connection between P450 isozymes and hepatotoxicity it is evident that this field is in its developing phase. The majority of research is based on the use of 10–15 different hepatotoxic model chemicals. Only in a limited amount of cases, isozyme specific antibodies or reconstituted systems containing various forms of purified P450s have been used. Table 1 summarizes the presently known associations between P450 isozymes and hepatotoxicity. Research related to isozyme-specific hepatotoxicity has hitherto been concerned mainly about effects of classical inducers like phenobarbital and 3-methylcholanthrene on the metabolism and toxicity of various agents. One must remember that the knowledge about the number and characteristics of the P450 isozymes is still very limited and that several more forms have to be characterized with regard to its role in hepatotoxicity. Another important obstacle in our understanding about isozyme specific hepatotoxicty is the great interspecies variation with respect to P450 isozymes responsible for metabolic activation. With the exception of CYP2E1 and perhaps some other isozymes, the substrate specificity of a particular isozyme is not very well conserved between species. The most important

Table 1. Examples of P450 isozyme involvement in activation and metabolism of hepatotoxic chemicals

Compound	P450-isozyme	Representative references
acetaminophen	CYP1A*	VAN DE STRAAT, 1986
	CYP2E1*	RAUCY et al., 1989
aflatoxin	CYP1A2, 2A3, 2B7*, CYP3A3, 3A4*	AYOMA et al., 1990
	CYP2C	FORRESTER et al., 1990
	CYP3A4*	SHIMADA et al., 1989
	HFLa*	KITADA et al., 1990
	P450mt4	SHAYIQ and AVADHANI, 1990
benzene	CYP2E1	JOHANSSON and INGELMAN-SUNDBERG, 1988
		NAKAJIMA et al., 1989
bromobenzene	CYP2B4	LAU et al., 1979
	CYP1A2	LAU et al., 1981
carbon tetrachloride	CYP2E1*	JOHANSSON and INGELMAN-SUNDBERG, 1985
		LINDROS et al., 1990
chloroform	CYP2E1* (oxidative)	TERELIUS and INGELMAN-SUNDBERG, 1988
	CYP2E1 & CYP2B1* (reductive)	TERELIUS and INGELMAN-SUNDBERG, 1991
		BRADY et al., 1989
dimethylnitrosamine	CYP2E1*	YANG et al., 1985; WRIGHTON et al., 1986
		LEVIN et al., 1986; WATKINS et al., 1988
diethylether	CYP2E1	PIETAN et al., 1991
halothane	CYP2E1*	TSUTSUMI et al., 1990
	CYP2B	GRUENKE et al., 1988
thioacetamide	CYP2E1	CHIELI et al., 1990
trichloroethylene	CYP2E1	TERELIUS and INGELMAN-SUNDBERG, 1988, 1991
		GUENGERICH et al., 1991
	CYP2B	NAKAJIMA et al., 1990
		MILLER and GUENGERICH, 1982
		MILLER and GUENGERICH, 1983

* Denotes isozyme involved in metabolic activation of the parent compound.

species is man and the development of expression systems with a panel of various human P450 isozymes stably expressed (CRESPI et al., 1990) will be a most important tool in the future for work on the roles of various human P450 isozymes in the hepatotoxic action of chemicals.

Evidence for the involvement of one or two isozymes of cytochrome P450 in the hepatotoxicity of a given compound is thus only slowly accumulating. However, one might consider that in many cases the toxicity of a xenobiotic will be strongly influenced by factors that influence the expression of these particular forms(s) of

Table 2. Inducers and substrates for CYP2E1. The list considers inducers and substrates as described in August 1991. Only selected references are given for each compound.

Compounds	Inducer	Substrate	References
alcohols			
—1,3-butanediol	§		§ LI et al., 1989
—n-butanol	§	#	# LUCAS et al., 1990; KOOP et al., 1984 § MORGAN et al., 1982
—ethanol	§	#	§ KOOP et al., 1982; INGELMAN-SUNDBERG and HAGBJÖRK, 1982; JOHANSSON et al., 1988 # MORGAN et al., 1982; KOOP et al., 1982; LUCAS et al., 1990; EKSTRÖM et al., 1987
—glycerol		#	# WINTERS and CEDERBAUM, 1990
—methanol		#	# MORGAN et al., 1982
—n-pentanol		#	# KOOP et al., 1984
—1-propanol		#	# MORGAN et al., 1982
—2-propanol	§		§ ELIASSON et al., 1988
aldehydes			
—acetaldehyde	§	#	# TERELIUS et al., 1991; § LUCAS et al., 1990
—p-nitrobenzaldehyde		#	# WERRINGLOER, unpublished
alkanes			
—ethane		#	# TERELIUS and INGELMAN-SUNDBERG, 1988
—hexane		#	# TERELIUS and INGELMAN-SUNDBERG, 1988, 1991
—pentane		#	# TERELIUS and INGELMAN-SUNDBERG, 1986
aromatic hydrocarbons			
—benzene	§	#	§ INGELMAN-SUNDBERG et al., 1982; JOHANSSON et al., 1988 # JOHANSSON et al., 1988; NAKAJIMA et al., 1990
—styrene		#	# TERELIUS and INGELMAN-SUNDBERG, unpubl.
—toluene		#	# TERELIUS and INGELMAN-SUNDBERG, 1988; 1991
—m-xylene	§	#	§ RAUNIO et al., 1990; # TERELIUS and INGELMAN-SUNDBERG, unpubl.
ethers			
—diethylether	§	#	§ BRADY et al., 1988 # BRADY et al., 1987
—methyl t-butyl ether		#	# BRADY et al., 1990
halogenated anesthetics			
—enfluorane		#	HOFFMAN et al., 1989
—halothane		#	GRUENKE et al., 1988
—Methoxyflurane		#	HOFFMAN et al., 1989
—Sevoflurane		#	HOFFMAN et al., 1989

Table 2. continued

Compounds	Inducer	Substrate	References
halogenated hydrocarbons			
—carbon tetrachloride		#	# JOHANSSON and INGELMAN-SUNDBERG, 1985; EKSTRÖM et al., 1989; PERSSON et al., 1990
—chloroform		#	# TERELIUS and INGELMAN-SUNDBERG, 1988, 1991; BRADY et al., 1989; GUENGERICH et al., 1991
—dichloromethane		#	# TERELIUS and INGELMAN-SUNDBERG, 1988, 1991; GUENGERICH et al., 1991
—dichloropropane		#	# GUENGERICH et al., 1991
—ethylene dichloride		#	# GUENGERICH et al., 1991
—methylenechloride		#	# GUENGERICH et al., 1991
—trichloroethylene	§	#	§ KOOP et al., 1982 # TERELIUS and INGELMAN-SUNDBERG, 1988, 1991; NAKAJIMA et al., 1990
—vinylchloride		#	# GUENGERICH et al., 1991
heterocyclic compounds			
—imidazole	§		§ KOOP and COON, 1984; INGELMAN-SUNDBERG and JÖRNVALL, 1984; PORTER et al., 1989
—isoniazid	§	#	§ RYAN et al., 1985, 1986 # RYAN et al., 1986
—ketoconazole	§		§ HOSTETLER et al., 1989
—4-methylpyrazole	§		§ THOMAS et al., 1987; PALAKODETY et al., 1988
—Pyrazole	§	#	§ INGELMAN-SUNDBERG and JÖRNVALL 1984, KOOP et al 1985; CLEJAN and CEDERBAUM, 1990 # CLEJAN et al., 1989; CLEJAN and CEDERBAUM, 1990
—pyridine	§	#	§ KIM et al., 1988 # KIM et al., 1988
ketones			
—acetone	§	#	§ KOOP and CASAZZA, 1985; JOHANSSON et al., 1986; RONIS and INGELMAN-SUNDBERG, 1989 # KOOP and CASSAZZA, 1985; JOHANSSON et al., 1986; BRADY et al., 1989
—2-butanone	§	#	# TERELIUS and INGELMAN-SUNDBERG, 1988, 1991; BRADY et al., 1989
—methyl t-butyl ketone		#	# TERELIUS and INGELMAN-SUNDBERG, 1991
Miscellaneous compounds			
acetaminophen		#	# MORGAN et al., 1983; RAUCY et al., 1989
acetoacetate		#	# KOOP and CASSAZZA, 1985
acetal		#	# KOOP et al., 1985

Table 2. continued

Compounds	Inducer	Substrate	References
miscellaneous compounds cont.			
acrylonitrile		#	# GUENGERICH et al., 1991
aniline		#	# MORGAN et al., 1982; KOOP et al., 1984; LUCAS et al., 1990
azoxymethane		#	# SOHN et al., 1991
bromobenzene		#	# WATKINS et al., 1988
carbon disulfide		#	# BRADY et al., 1987
chloroxazone		#	# PETER et al., 1990
clotrimazole	§		§ KHAN et al., 1989
dimethylsulfoxide	§		§ THOMAS et al., 1987; ELIASSON et al., 1988
ethylcarbamate		#	# GUENGERICH et al., 1991
methylazoxymethanol		#	# SOHN et al., 1991
p-nitrophenol		#	# KOOP, 1986; HÉTU and JOLY, 1988 LUCAS et al., 1990
N-nitrosodimethylamine		#	# HONG and YANG, 1985; YANG et al., 1985
N-nitroso-2,6-dimethylmorpholine		#	# KOKKINAKIS et al., 1985
N-nitrosomethyl-ethylamine		#	# YANG et al., 1985; YANG et al., 1987
N-nitrosopyrrolidine		#	# MCCOY and KOOP, 1988b
oxygen	§	#	§ TINDBERG and INGELMAN-SUNDBERG, 1989 # HÉTU and JOLY, 1988; TINDBERG and INGELMAN-SUNDBERG, 1989; PERSSON et al., 1990
thiobenzamide		#	# CHIELI et al., 1990
vinylcarbamate		#	# GUENGERICH et al., 1991
Other treatments			
alloxan- or streptozotocin-induced diabetes	§		§ SONG et al., 1987; THOMAS et al., 1987
birth	§		§ SONG et al., 1986; UMENO et al., 1988
high fat	§		§ LIEBER et al., 1988; YOO et al., 1991
hypophysectomy	§		§ WILLIAMS and SIMONET, 1988; WAXMAN et al., 1989
low caloric diet	§		§ LEAKEY et al., 1989
spontaneous diabetes	§		§ FAVREAU et al., 1987; BELLWARD et al., 1988; THUMMEL and SCHENKMAN, 1990
starvation or fasting	§		§ MILLER and YANG, 1984; JOHANSSON et al., 1988; IMAOKA et al., 1990; JOHANSSON et al., 1990

P450 activating this compound. This is exemplified by CYP2E1, which activates numerous compounds exerting hepatotoxicity in the centrilobular region of the liver acinus where this form of P450 is expressed and specifically induced by substrates, hormones and dietary factors (Table 2). This isozyme is involved in the activation of the by far highest number of cytotoxic chemicals among different P450 isozymes and over 60 different CYP2E1 substrates have been identified. Exogenous, environmental factors acting as inducer of CYP2E1, therefore will have a strong influence on the hepatotoxic action of several other kinds of compounds. In addition, reports start to appear describing the genetic polymorphism of several P450 genes, among them CYP2E1 (Uematsu et al., 1991) where a direct correlation between one particular genotype and an eliminated risk for receiving lung cancer was described. For instance, at the moment it cannot be excluded that the pronounced interindividual sensitivity for developing liver cirrhosis and necrosis is related to the P450 genotype we express. The coming decade of P450 research will give us an insight into this fascinating area.

Acknowledgement

We thank Mrs Kerstin Hall for valuable aid in the preparation of this manuscript. The work in the authors' laboratories are supported in part by grants from The Swedish Medical Research Council, The Swedish Alcohol Research Fund and from O. E. Edla Johanssons Stiftelse.

4. References

Ahmed, A. E., V. L. Kubic, J. L. Stevens and M. W. Anders, (1980), Fed. Proc. **39:** 3150–3155.

Ahr, H. R., L. J. King, W. Nastainczyk, and V. Ullrich, (1982), Biochem. Pharmacol. **31:** 383–390.

Albano, E., K. A. K. Lott, T. F. Slater, A. Stier, M. C. R. Symons, and A. Tomasi, (1982), Biochem. J. **204:** 593–603.

Albano, E., G. Poli, E. Chiarpotto, F. Biasi, and M. U. Dianzani, (1984), Chem. Biol. Interact. **47:** 249–263.

Albano, E., A. Tomasi, K. H. Cheeseman, V. Vannini, and M. U. Dianzani, (1985), in: Free Radicals in Liver Injury, (G. Poli, K. H. Cheesman, and M. U. Dianzani, T. F. Slater, eds.), pp. 7–16, IRL Press, Oxford.

Albano, E. and A. Tomasi, (1987), Biochem. Pharmacol. **36:** 2913–20.

Albano, E., A. Tomasi, L. Goria-Gatti, G. Poli, V. Vannini, and M. U. Dianzani, (1987), Free Rad. Res. Communs. **3:** 234–249.

Albano, E., G. Poli, A. Tomasi, L. Goria-Gatti, and M. U. Dianzani, (1988), in: Medical, Biochemical and Chemical Aspects of Free Radicals, (O. Hayaishi, E. Niki, M. Kondo and T. Yoshikawa, eds.), pp. 1389–1392, Elsevier, Amsterdam.

Albano, E., A. Tomasi, L. Goria-Gatti, and M. U. Dianzani, (1988), Chem. Biol. Interact **65:** 223–234.

Albano, E., A. Tomasi, J. O. Persson, Y. Terelius, L. Goria-Gatti, M. Ingelman-Sundberg, and M. U. Dianzani, (1991), Biochem. Pharmacol. **46:** 1895–1902.

Altomare, E., M. A. Leo, and C. S. Lieber, (1984), Alcoholism: Clin. Exp. Res. **8:** 405–408.

AMELIZAD, S., K. E. APPEL, M. SCHOEPKE, C. S. RÜHL, and F. OESCH, (1989), Cancer Lett. **46:** 43–49.
ANDERS, M. W., V. L. KUBIC, and A. E. AHMED, (1977), J. Environ. Pathol. Toxicol. **1:** 117–24.
ANUNDI, I., T. LÄHTEENMÄKI, and K. O. LINDROS, (1991), Alcohol Alcoholism. **26:** 234.
ARINC, E., O. ADALI, M. ISCAN, and T. GÜRAY, (1991), Arch. Toxicol. **65:** 186–190.
AYOMA, T., S. YAMANO, P. S. GUZELIAN, H. V. GELBOIN, and F. J. GONZALEZ, (1990), Proc. Natl. Acad. Sci. USA **87:** 4790–4793.
BAHRI, A. K., C. S. CHIANG, and J. A. TIMBRELL, (1981), Tox. Appl. Pharm. **60:** 561–568.
BEAUNE, P., J. FILINOIS, E. LEPREVOST, and J. LEROUX, (1983), Drug Metab. Dispos. **11:** 499–507.
BEHRENS, U. J., M. HOERNER, J. M. LASKER, and C. S. LIEBER, (1988), Biochem. Biophys. Res. Commun. **154:** 584–590.
BELLWARD, G. D., T. CHANG, B. RODRIGUES, J. H. MCNEILL, S. MINES, D. E. RYAN, W. LEVIN, and P. E. THOMAS, (1988), Mol. Pharmacol. **33:** 140–143.
BLACK, M., (1984), Ann. Rev. Med. **35:** 577–593.
BLUM, M., A. DEMIERRE, D. M. GRANT, M. HEIM, and U. A. MEYER, (1991), Proc. Natl. Acad. Sci. (USA) **88:** 5237–5241.
BOLT, H. M., R. J. LAIB, and J. G. FILSER, (1982), Biochem. Pharmacol. **31:** 1–4.
BOYD, E. M. and G. M. BRERECZKY, (1966), Br. J. Pharmacol. **26:** 606–614.
BRADY, J. F., M. J. LEE, M. LI, H. ISCHIZAKI, and C. S. YANG, (1987), Mol. Pharmacol. **33:** 148–154.
BRADY, J. F., D. LI, H. ISHIZAKI, M. LEE, S. M. NING, F. XIAO, and C. S. YANG, (1989), Toxicol. Appl. Pharmacol. **100:** 342–349.
BRADY, J. F., F. XIAO, S. M. NING, and C. S. YANG, (1990), Arch. Toxicol. **64:** 157–160.
BRODFUERER, J. I., D. E. CHAPMAN, T. J. WILKE, and G. POWIS, (1990), Drug Metab. Disp. **18:** 20–27.
BROWN, B. R. and I. G. SIPES, (1977), Biochem. Pharmacol. **26:** 2091–2094.
BROWN, B. R., D. G. FARRAR, and C. G. DE ROOIJ, (1990), Regul. Toxicol. Pharmacol. **11:** 24–41.
BURK, R. F., K. E. HILLAND, and J. M. LANE, (1988), Biochem. Biophys. Res. Commun. **152:** 1463–1467.
CARLSON, G. P., (1973), Life Sci. **13:** 67–73.
CARLSON, G. P., (1990), Cancer Lett. **54:** 153–156.
CASHMAN, J. R. and R. P. HANZLIK, (1981), Biochem. Biophys. Res. Commun. **98:** 147–153.
CASSARETT and DOULL's Toxicology, (1975), The Basic Science of Poisons, (J. DOULL, C. D. KLAASSEN, M. O. AMDUR, eds.), Macmillan Publishing Co. Inc., Second Edition.
CEDERBAUM, A. I., (1991), Alcoholism: Clin. Exp. Res., Abstract, RSA-meeting.
CHAKRABARTI, S. K. and B. TUCHWEBER, (1988), J. Toxicol. Environ. Health **23:** 147–158.
CHEESEMAN, K. H., E. ALBANO, A. TOMASI, and T. F. SLATER, (1985), Environ. Health Perspects. **64:** 85–101.
CHIELI, E., G. MALVALDI, and D. SEGNINI, (1980), Toxicol. Lett. **7:** 175–180.
CHIELI, E., M. SAVIOZZI, P. PUCCINI, V. LONGO,, and P. G. GERVASI, (1990), Arch. Toxicol. **64:** 122–127.
CLARK, R., V. BORIRAKCHANYAVAT, A. R. DAVIDSON, and R. P. THOMPSON, (1973), Lancet **i:** 66–70.
CLEJAN, L. A. and A. I. CEDERBAUM, (1990), Biochim. Biophys. Acta. **1034:** 233–237.
CLEJAN, L. A., D. R. KOOP, and A. I. CEDERBAUM, (1989), Drug Metab. Dispos. **17:** 694–698.
CRESPI, C. L., R. LANGENBACH, and B. W. PENMAN, (1990a), Prog. Clin. Biol. Res. **B:** 97–106.

Crespi, C. L., B. W. Penman, J. A. Leakey, M. P. Arlotto, M. A. Stark, T. Turner, D. T. Steimel, K. Rudo, R. L. Davies, and R. L. Langenbach, (1990b), Carcinogenesis **11:** 1293–1300.
Dahlin, D. C., G. T. Miwa, A. Y. H. Lu, and S. D. Nelson, (1984), Proc. Natl. Acad. Sci. USA. **81:** 1327–1331.
Davidson, D. G., and W. N. Easthan, (1966), Br. Med. J. **2:** 497–499.
de Groot, H. and T. Noll, (1983), Hepatology **3:** 601–606.
de Groot, H. and T. Noll, (1989), Toxicol. Appl. Pharmacol. **97:** 530–537.
Dekant, W., M. Metzler, and D. Henschler, (1984), Biochem. Pharmacol. **33:** 2021–2027.
Den Tonkelaar, E. M. and M. J. van Logten, (1974), Tox. Appl. Pharmacol. **30:** 96–106.
Dickinson, D. S., W. C. Bailey, B. I. Hirschowitz, S. J. Soong, L. Eidus, and M. M. Hoggkin, (1981), J. Clin. Gastroenterol. **3:** 271–279.
Dyroff, M. C. and R. A. Neal, (1981), Cancer Res. **41:** 3430–3435.
Ebel, R. E., (1989), Biochem. Biophys. Res. Commun. **161:** 615–618.
Ekström, G. and M. Ingelman-Sundberg, (1986), Eur. J. Biochem. **158,** 195–201.
Ekström, T., A. Ståhl, K. Sigvardsson, and J. Högberg, (1986), Acta Pharmacol. Toxicol. **58:** 289–296.
Ekström, G., C. Norsten, T. Cronholm, and M. Ingelman-Sundberg, (1987), Biochemistry **26:** 7348–7354.
Ekström, G. and M. Ingelman-Sundberg, (1989), Biochem. Pharmacol. **38:** 1313–1319.
Ekström, G., C. von Bahr, and M. Ingelman-Sundberg, (1989), Biochem. Pharmacol. **38,** 689–693.
El-Hawari, A. M. and G. L. Plaa, (1983), Toxicol. Lett. **17:** 293–300.
Eliasson, E., I. Johansson, and M. Ingelman-Sundberg, (1988), Biochem. Biophys. Res. Commun. **150:** 436–443.
English, R. and M. W. Anders, (1985), Drug Metab. Disp. **13:** 449–452.
Faletto, M. B., P. L. Koser, N. Battula, G. K. Townsend, A. E. Maccubbin, H. V. Gelboin, and H. L. Gurtoo, (1988), J. Biol. Chem. **263:** 12187–12189.
Farber, J. L., M. E. Kyle, and J. B. Coleman, (1990), Lab. Invest. **62,** 670–679.
Favreau, L. V., D. M. Malchoff, J. E. Mole, and J. B. Schenkman, (1987), J. Biol. Chem. **262:** 14319–14326.
Forrester, L. M., G. E. Neal, D. J. Judah, M. J. Glancey, and C. R. Wolf, (1990), Proc. Natl. Acad. Sci. (USA) **87:** 8306–8310.
Forsander, O. A., N. Räihä, M. Salaspuro, and P. H. Mäenpää, (1965), Biochem. J. **94:** 259–265.
Fukuhura, M., K. Nagata, K. Mizokami, Y. Yamazoe, A. Takanaka, and R. Kato, (1989), Biochem. Biophys. Res. Commun. **162:** 265–272.
Fukhura, M., K. Mizokami, M. Sakaguchi, Y. Niimura, K. Kato, S. Inouye, and A. Takanaka, (1990), Biochem. Pharmacol. **39:** 463–469.
Gonthier, B. P. and L. G. Barret, (1989), Toxicol. Lett. **47:** 225–234.
Gonzalez, F. J., (1989), Pharmacol. Ther. **45,** 1–38.
Gonzalez, F. J. and D. W. Nebert, (1990), Trends Genet. **6:** 182–186.
Gores, G. J., B. Herman, and J. J. Lemaster, (1990), Hepatology **11:** 690–698.
Gorsky, L. D., D. R. Koop, and M. J. Coon, (1984), J. Biol. Chem. **259:** 6812–6817.
Griffiths, J., G. Kalf, nd R. Snyder, (1986), in: Biological Reactive Intermediates III: Mechanisms of Action in Animal Models and Human Disease, (J. J. Kocsis, D. J. Jollow, C. M. Witmer, J. O. Nelson, and R. Snyder, eds.), Plenum Press, New York, 213–225.
Groenhagen-Riska, C., P. E. Hellström, and B. Froeseth, (1978), Am. Rev. Respir. Dis. **118:** 461–466.

GRUENKE, D. L., K. KONOPKA, D. R. KOOP, and L. A. WASKELL, (1988), J. Pharmacol. Exp. Ther. **246:** 454–459.
GUENGERICH, P. F., D.-H. KIM, and M. IWASAKI, (1991), Chem. Res. Tox. **4:** 168–179.
HANDLER, J. A., D. R. KOOP, M. J. COON, Y. TAKEI, and R. G. THURMAN, (1988), Arch. Biochem. Biophys. **264:** 114–124.
HANSSON, T., N. TINDBERG, M. INGELMAN-SUNDBERG, and C. KÖHLER, (1990), Neuroscience **34:** 451–463.
HANZLIK, R. P. , (1981), Biochem. Pharmacol. **30:** 3027–3030.
HARPER, B. and M. S. LEGATOR, (1987), Mutat. Res. **179:** 23–31.
HARVISON, P. J., F. P. GUENGERICH, M. S. RASHED, and S. D. NELSON, (1988), Chem. Res. Toxicol. **1:** 47–52.
HEAD, B., D. E. MOODY, C. H. WOO, and E. A. SMUCKELER, (1981), Tox. Appl. Pharmacol. **61:** 286–295.
HENSCHLER, D., (1977), Environ. Health Perspect. **21:** 61–64.
HERREN-FREUND, S. L., M. A. PEREIRA, M. D. KHOURY, and G. OLSON, (1987), Toxicol. Appl. Pharmacol. **90:** 183–189.
HETU, C. and J. G. JOLY, (1988), Biochem. Pharmcol. **37:** 421–426.
HEWITT, W. R., E. M. BROWN, and G. L. PLAA, (1983), Toxicol. Lett. **16:** 285–296.
HEWITT, L. A., C. VALIQUETTE, and G. L. PLAA, (1987), Can. J. Physiol. Pharmacol. **65:** 2313–2318.
HOFFMAN, J., K. KONOPKA, C. BUCKHORN, D. R. KOOP and L. WASKELL, (1989), Br. Anaesth. **63:** 103–108.
HONG, J. and S. YANG, (1985), Carcinogenesis **6:** 1805–1809.
HOSTETLER, K. A., S. A. WRIGHTON, D. T. MOLOWA, P. E. THOMAS, W. LEVIN, and P. S. GUZELIAN (1989) Mol. Pharmacol. **35:** 279–285.
HUGHES, H. M., I. M. GEORGE, J. C. EVANS, C. C. ROWLANDS, G. M. POWELL, and C. G. CURTIS, (1991), Biochem. J. **277:** 795–800.
HUNTER, A. L., M. A. HOLSHER, and R. A. NEAL, (1977), J. Pharmacol. Exp. Ther. **200:** 439–448.
ILET, F. K., W. D. REID, I. G. SIPES, and G. KRISHNA, (1973), Exp. Mol. Pathol. **19:** 215–221.
IMAOKA, S., Y. TERANO, and Y. FUNAE, (1990), Arch. Biochem. Biophys. **278:** 168–178.
INGELMAN-SUNDBERG, M. and A.-L. HAGBJÖRK, (1982), Xenobiotica **12:** 673–686.
INGELMAN-SUNDBERG, M. and I. JOHANSSON, (1984), J. Biol. Chem. **259:** 6447–6458.
INGELMAN-SUNDBERG, M. and H. JÖRNVALL, (1984), Biochem. Biophys. Res. Commun. **124:** 375–382.
INGELMAN-SUNDBERG, M., I. JOHANSSON, K. E. PENTTILÄ, H. GLAUMANN, and K. O. LINDROS, (1988), Biochem. Biophys. Res. Commun. **157:** 55-60.
INGELMAN-SUNDBERG, M., I. JOHANSSON, Y. TERELIUS, E. ELIASSON, G. EKSTRÖM, R. BÜHLER, and K. O. LINDROS, in: Nutrition and Toxicology, (D. V. PARKE, ed.), Taylor and Francis, in press.
JAUHONEN, E. BARAONA, M. MIYAKAWA, and C. S. LIEBER, (1982), Alcoholism: Clin. Exp. Res. **6:** 350–357.
JOHANSSON, I. and M. INGELMAN-SUNDBERG, (1985), FEBS Lett. **183:** 265–269.
JOHANSSON, I., E. ELIASSON, C. NORSTEN, and M. INGELMAN-SUNDBERG, (1986), FEBS Lett. **196:** 59–64.
JOHANSSON, I. and M. INGELMAN-SUNDBERG, (1988), Cancer Res. **48:** 5387–5390.
JOHANSSON, I., G. EKSTRÖM, B. SCHOLTE, D. PUZYCKI, H. JÖRNVALL, and M. INGELMAN-SUNDBERG, (1988), Biochemistry **27:** 1925–1934.
JOHANSSON, I., K. O. LINDROS, H. ERIKSSON, and M. INGELMAN-SUNDBERG, (1990), Biochem. Biophys. Res. Commun. **173:** 331–338.
KALF, G. F., (1987), CRC Critical Rev. Toxicol. **18:** 141–159.

KARASZKIEWICZ, J. W., R. SNYDER, and G. F. KALF, (1986), Fed. Proc. **45:** 1748–1752.
KENNA, J. G., (1991), Biochemical Society Transaction **19:** 191–195.
KERA, Y., K. E. PENTTILÄ, and K. O. LINDROS, (1988), Biochem. J. **254:** 411–417.
KERGER et al., (1988).
KHAN, W. A., C. KUHN, H. F. MERK, S. S. PARK, H. V. GELBOIN, D. R. BICKERS, and H. MUKHTAR, (1989), Drug Metab. Disp. **17:** 360–364.
KIFFEL, L., J. LOEPER, J. C. HOMBERG, and J. P. LEROUX, (1989), Biochem. Biophys. Res. Commun. **159:** 282–289.
KIM, S. G., D. E. WILLIAMS, E. G. SCHUETZ, P. S. GUZELIAN, and R. F. NOVAK, (1988), J. Pharmacol. Exp. Ther. **246:** 1175–1182.
KIMURA, H., K. SOGAWA, Y. SAKAI, and Y. FUJII-KURIYAMA, (1989), J. Biol. Chem. **264:** 2338–2342.
KITADA, M., M. TANEDA, K. OHTA, K. NAGASHIMA, K. ITAHASHIAND, and T. KAMATAKI, (1990), Cancer Res. **50:** 2641–2645.
KNECHT, K. T. and R. P. MASON, (1988), Drug Metab. Disp. **16:** 813–817.
KNECHT, K. T., B. U. BRADFORT, R. P. MASON, and G. R. THUMAN, (1990), Mol. Pharmacol. **38:** 26–30.
KOKKINAKIS, D. M., D. R. KOOP, D. G. SCARPELLI, M. J. COON and P. F. HOLLENBERG, (1985), Cancer Res. **45:** 619–624.
KOOP, D. R., E. T. MORGAN, G. E. TARR, and M. J. COON, (1982), J. Biol. Chem. **257:** 8472–8480.
KOOP, D. R. and M. J. COON, (1984), Mol. Pharmacol. **25:** 494–501.
KOOP, D. R., G. D. NORDBLOM, and M. J. COON, (1984), Arch. Biochem. Biophys. **235:** 228–238.
KOOP, D. R. and J. P. CASSAZZA, (1985), J. Biol. Chem. **260:** 13607–13612.
KOOP, D. R., B. L. CRUMP, G. D. NORDBLOM, and M. J. COON, (1985), Proc. Natl. Acac. Sci. USA **82:** 4065–4069.
KOOP, D. R., (1986), Mol. Pharmacol. **29:** 399–404.
KOOP, D. R. and M. J. COON, (1986), Alcoholism Clin. Exp. Res. **10** (suppl.): 44–49.
KOOP, D. R., C. L. LAETHEM, and G. G. SCHNIER, (1989), Toxicol. Appl. Pharmacol. **98:** 278–288.
KRIETER, A. P. and R. A. VAN DYKE, (1983), Chem. Biol. Interacts. **44:** 219–235.
KUBIC, V. L. and M. W. ANDERS, (1975), Drug Metab. Disp. **2:** 104–112.
KUBIV, V. L. and M. W. ANDERS, (1978), Biochem. Pharmacol. **27:** 2349–2355.
LAU, S. S. and V. G. ZANNONI, (1979), Toxicol. Appl. Pharmacol. **50:** 309–318.
LAU, S. S. and V. G. ZANNONI, (1981), Mol. Pharmacol. **20:** 234–235.
LAUTERBURG, B. H. and M. E. VELEZ, (1988), GUT **29:** 1153–1157.
LEAKEY, J. E., H. C. CUNNY, J. J. BAZARE, P. J. WEBB, R. J. FEUERS, P. H. DUFFY, and R. W. HART, (1989), Mech. Ageing Dev. **48:** 145–155.
LEVIN, W., P. E. THOMAS, N. OLDFILDAND, and D. E. RYAN, (1986), Arch. Biochem. Biophys. **248:** 158–165.
LI, D., J. F. BRADY, M. J. LEE, and C. S. YANG, (1989), Toxicol. Lett. **45:** 141–147.
LIEBER, C. S. and L. M. DECARLI, (1968), Science **162:** 917–918.
LIEBER, C. S. and W. LEVIN, (1985), J. Biol. Chem. **260:** 6385–6393.
LIEBER, C. S., (1988), New Engl. J. Med. **319:** 1639–1650.
LIEBER, C. S., J. M. LASKER, L. M. DECARLI, J. SAELI, and T. WOJTOWICZ, (1988), J. Pharm. Exp. Ther. **247:** 791–795.
LIEBER, C. S., (1990), Pharmacol. Ther. **46:** 1–41.
LIND, R. C., A. J. GANDOLFI, and P. M. HALL, (1990), J. Appl. Toxicol. **10:** 161–165.
LINDROS, K. O., (1978), in: Research advances in alcohol and drug problems, (Y. ISRAEL, F. B. GLASER, H. KALANT, R. E. POPHAM, W. SCHMIDT, R. G. SMART, eds.), Plenum, New York **4:** 111–176.

Lindros, K. O., C. Yuan, and K. E. Penttilä, (1990), Hepatology **12:** 1092–1097.
Lindström, T. D. and M. W. Anders, (1978), Toxicol. Lett. **1:** 307–14.
Loeper, J., V. Descatoire, M. Maurice, P. Beaune, G. Feldmann, D. Larrey, and D. Pessayre, (1990), Hepatology **11:** 850–858.
Lucas, D., F. Berthou, Y. Dreano, H. H. Floch, and J. F. Menez, (1990), Alcohol Clin. Exp. Res. **14:** 590–594.
MacDonald, J. R., A. J. Gandolfi, and I. G. Sipes, (1982), Toxicol. Lett. **13:** 57–69.
Madhu, C., Z. Gregus, and C. D. Claasen, (1989), J. Pharm. Exptl. Ther. **248:** 1069–1076.
Malvadi, G., E. Chieli, and M. Saviozzi, (1984), Arch. Toxicol. **55:** 34–38.
Mansuy, D., Ph. Beaune, T. Cresteil, M. Lange, and J.-P. Leroux, (1977), Biochem. Biophys. Res. Commun. **79:** 513–517.
McClain, C. J., J. P. Kromhult, F. J. Peterson, J. L. Holzman, (1980), JAMA **244:** 251–253.
McCoy, G. D. and D. R. Koop, (1988), Cancer Res. **48:** 3987–3992.
McLain, G. E., I. G. Sipes, and B. R. Brown, (1979), Anestesiol. **51:** 321–326.
McMartin, D. N., J. A. O'Connor, and L. S. Kaminsky, (1981), Res. Commun. Chem. Pathol. Pharmacol. **31:** 99–110.
McMichael, A. J., (1988), IARC Sci. Publ. **85:** 3–18.
Miller, R. E. and F. P Guengerich, (1982), Biochemistry **21:** 1090–1097.
Miller, R. E. and F. P. Guengerich, (1983), Cancer Res. **43:** 1145–1152.
Miller, K. W. and C. S. Yang, (1984), Arch. Biochem. Biophys. **229:** 483–491.
Miner, D. J. and P. T. Kissinger, (1979), Biochem. Pharmacol. **28:** 3285–3290.
Mitchell, J. R., D. J. Jollow, W. Z. Potter, D. C. Davis, J. R. Gillette, and B. B. Brodie, (1973), J. Pharmacol. Exp. Ther. **187:** 185–194.
Mitchell, J. R., H. J. Zimmerman, K. G. Ishak, S. P. Thorgeirsson, J. R. Timbrell, W. R. Snodgrass, and S. D. Nelson, (1976), Ann. Intern. Med. **94:** 181–92.
Moldéus, P., B. Andersson, A. Norling, and K. Ormstad, (1980), Biochem. Pharmacol. **29:** 1741–1745.
Morgan, E. T., D. R. Koop, and M. J. Coon, (1982), J. Biol. Chem. **257:** 13951–13957.
Morgan, E. T., D. R. Koop, and M. J. Coon, (1983), Biochem. Biophys. Res. Commun. **112:** 8–13.
Murphy, R., R. Schwarz, and P. B. Watkins, (1990), submitted.
Nakajima, T., T. Okino, and A. Sato, (1987), Biochem. Pharmacol. **36:** 2799–2804.
Nakajima, T., T. Okino, S. Okuyama, T. Kaneko, I. Yonekura, and A. Sato, (1988), Toxicol. Appl. Pharmacol. **94:** 227–237.
Nakajima, T., E. Elovaara, S. S. Park, H. V. Gelboin, E. Hietanen, and H. Vainio, (1989), Carcinogenesis **10:** 1713–1717.
Nakajima, T., R.-S. Wang, N. Murayama, and A. Sato, (1990), Toxicol. Appl. Pharmacol. **102:** 546–552.
Nanji, A. A., C. L. Mendenhall, and S. W. French, (1989), Alcoholism. Clin. Exptl. **13:** 15–19.
Nastainczyk, W., V. Ullrich, and H. Sies, (1978), Biochem. Pharmacol. **27:** 387–392.
Nebert, D. W., D. R. Nelson, M. J. Coon, R. W. Estabrook, R. Feyereisen, Y. Fujii-Kuriyma, F. J. Gonzalez, F. P. Guengerich, I. C. Gunsalus, E. F. Johnson, J. C. Loper, R. Sato, M. R. Waterman, and D. J. Waxman, (1991), DNA Cell Biol. **10:** 1–10.
Nelson, M. A. and R. J. Bull, (1988), Toxicol. Appl. Pharmacol. **94:** 45–54.
Nelson, S. D., (1990), Sem. Liver Dis. **10:** 267–278.
Nicotera, P., G. Bellomo, and S. Orrenius, (1990), Chem. Res. Toxicol. **3:** 484–494.
Niemelä, O., T. Juvonen, and S. Parkkila, (1991), J. Clin. Invest. **87:** 1367–1374.
Nikolaev, V., M. Kerinova, E. Naydenova, S. Dimov, G. Satov, and E. Ivanov, (1986), Toxicol. **38:** 203–208.

Noguchi, T., K. L. Fong, E. K. Lai, L. Olson, and P. B. McCay, (1982a), Biochem. Pharmacol. **31:** 609–614.
Noguchi, T., K. L. Fong, E. K. Lai, S. S. Alexander, M. M. King, L. Olson, J. L. Poyer, and P. B. McCay, (1982b), Biochem. Pharmacol. **31:** 615–24.
Norsten, C., T. Cronholm, G. Ekström, J. Handler, R. G. Thurman, and M. Ingelaman-Sundberg, (1989), J. Biol. Chem. **264:** 5593–5597.
Ogino, K., T. Hobara, and H. Kobayashi, (1991), Environ. Contam. Toxicol. **46:** 417–421.
Olson, M. J., S. G. Kim, C. A. Reidy, J. T. Johnson, and R. F. Novak, (1991), Drug Metabolism and Disposition **19:** 298–303.
Orme-Johnson, W. H. and D. M. Ziegler, (1965), Biochem. Biophys. Res. Commun. **21:** 78–82,
Orrenius, S., D. J. McConkey, G. Bellomo, and P. Nicotera, (1989), Trends. Pharmacol. Sci. **10:** 281–285.
Ortiz de Montellano, P. R. K., K. L. Kunze, H. S. Beilan, and C. Wheeler, (1982), Biochemistry **21:** 1331–1339.
Packer, J. E., T. F. Slater, and R. L. Willson, (1978), Life Sci. **32:** 2617–20.
Palakodety, R. B., L. A. Clejan, G. Krikun, D. E. Feierman, and A. I. Cederbaum, (1988), J. Biol. Chem. **263:** 878–884.
Pankow, D., (1988), Biomed. Biochim. Acta **47:** 293–295.
Pankow, D. and P. Hoffmann, (1989), Arch. Toxicol. Suppl. **13:** 202–203.
Parker, J. C., G. E. Casey, L. J. Bahlman, N. A. Leidel, D. Rose, H. P. Stein, A. W. Thomas, and J. M. Lane, (1979), Am. Ind. Hyg. Assoc. J. **40:** A46–60.
Persson, J. O., Y. Terelius, and M. Ingelman-Sundberg, (1990), Xenobiotica **20:** 887–900.
Peter, R., R. Böcker, Ph. Beaune, M. Iwasaki, F. P. Guengerich, and C. S. Yang, (1990), Chem. Res. Toxicol. **3:** 566–573.
Pohl, L. R., (1979), in: Reviews in Biochemical Toxicology, (E. Hodgson, J. R. Bend, and R. M. Philpot, eds.), Vol. **1:** pp. 79–107, Elsevier/North Holland, New York.
Pohl, L. R. and J. R. Gillette, (1982), Anestes. Anal. **61:** 809–811.
Poli, G., E. Albano, and M. U. Dianzani, (1987), Chem. Phys. Lipids **45:** 117–142.
Porter, W. R. and R. A. Neal, (1978), Drug Metab. Dispos. **6:** 379–388.
Porter, W. R., M. J. Gudzinowicz, and R. A. Neal, (1979), J. Pharmacol. Exp. Ther. **208:** 386–391.
Porter, T. D., S. C. Khani, and M. J. Coon, (1989), Mol. Pharmacol. **36:** 61–65.
Porter, T. D. and M. J. Coon, (1991), J. Biol. Chem. **266:** 13469–13472.
Post, G. B. and R. Snyder, (1983), J. Toxicol. Environ. Health **11:** 811–825.
Potter, W. Z., D. C. Davis, J. R. Mitchell, D. J. Jollow, J. R. Gillette, and B. B. Brodie, (1973), J. Pharmacol. Exp. Ther. **187:** 203–210.
Potter, D. W. and J. A. Hinson, (1987), J. Biol. Chem. **262:** 966–973.
Poulsen, H. E., A. Larche, and N. T. Pedersen, (1985), Pharmacol. **30:** 100–108.
Powell-Jackson, P. R., J. M. Tredger, H. M. Smith, M. Davis, and R. Williams, (1982), Biochem. Pharmacol. **31:** 4031–4034.
Poyer, J. L., R. A. Floyd, P. B. McCay, E. G. Janzen, and E. R. Davis, (1978), Biochem. Biophys. Acta **539:** 402–409.
Prescott, L. F., (1983), Drugs **25:** 290–314.
Prodi, G., G. Arfellini, A. Colacci, S. Grilliand, and M. Mazzullo, (1986), Toxicol. Pathol. **14:** 438–444.
Prough, R. A. and S. J. Moloney, (1985), in: Bioactivation of Foreign Compounds, (M. W. Anders, ed.), pp. 433–49, Academic Press, Orlando.
Ramsdell, H. S., A. Parkinson, A. C. Eddy, and D. L. Eaton, (1991), Toxicol. Appl. Pharmacol. **108:** 436–447.

RATNEY, R. S., D. H. WEGMAN, and H. B. ELKINS, (1974), Arch. Environ. Health **28:** 223–229.
RAUCY, J. L., J. M. LASKER, C. S. LIEBER, and M. BLACK, (1989), Arch. Biochem. Biophys. **271:** 279–283.
RAUNIO, H., J. LIIRA, E. ELOVAARA, V. RIIHIMÄKI, and O. PELKONEN, (1990), Toxicol. Appl. Pharmacol. **103:** 175–179.
RAY, S. D. and H. M. MEHENDALE, (1990), Fundam. Appl. Toxicol. **15:** 429–440.
RECKNAGEL, R. O., E. A. GLENDE, J. A. DOLAK, and R. L. WALLER,(1989), Pharmacol. Ther. **43:** 139–154.
REID, W. D. and G. KRISHNA, (1973), Exp. Mol. Pathol. **18:** 80–90.
REINKE, L. A., E. K. LAI, and P. B. MCCAY, (1988), Xenobiotica **18:** 1311–1318.
REYNOLDS, E. S. and M. J. MOSLEM, (1977), Anestesiol. **47:** 19–27.
RONIS, M. J. J. and M. INGELMAN-SUNDBERG, (1989), Xenobiotica **19:** 1161–1165.
RONIS, M. J. J., C. K. LUMPKIN, M. INGELMAN-SUNDBERG, and T. M. BADGER, (1991), Alcohol. Clin. Exp. Res. **15:** 693–699.
ROTH, R. P., R. T. DREW, R. J. LO, and J. R. FOUTS, (1975), Tox. Appl. Pharmacol. **33:** 427–437.
RUMACK, B. H., R. C. PETERSON, G. C. KOCH, and I. A. AMARA, (1981), Arch. Interna. Med. **141:** 380–385.
RYAN, D. E., L. RAMATHAN, S. IIDA, P. E. THOMAS, M. HANIU, J. E. SHIBELY, C. S. LIEBER, and W. LEVIN, (1985), J. Biol. Chem. **260:** 6385–6393.
RYAN, D. E., D. R. KOOP, P. E. THOMAS, M. J. COON and W. LEVIN, (1986), Arch. Biochem. Biophys. **246:** 633-644.
SATO, C., M. YOSHIRO, and C. S. LIEBER, (1981), Gastroenterology **80:** 140–148.
SCHEMEL, W. H., (1976), Anesth. Analg. **55:** 810–812.
SCHNIER, G. G., LAETHEM, C. L. and D. R. KOOP, (1989), J. Pharmacol. Exp. Ther. **251:** 790–796.
SCHRENK, D. and K. W. BOCK, (1990), Drug Metab. Disp. **18:** 720–725.
SEEFF, L. B., B. A. CUCCHERINI, H. J. ZIMMERMAN, E. ALDER, and S. B. BENJAMIN, (1986), Ann. Intern. Med. **104:** 399–404.
SESARDIC, D., K. J. COLE, R. J. EDWARDS, D. S. DAVIES, P. E. THOMAS, W. LEVIN, and A. R. BOOBIS, (1990), Biochem. Pharmacol. **39:** 499–506.
SHAYIQ, R. M. and N. G. AVADHANI, (1990), Biochemistry **29:** 866–873.
SHIMADA, T., M. V. MARTIN, S. D. PRUESS, L. J. MARNETT, and F. P. GUENGERICH, (1989), Cancer Res. **46:** 6304–6312.
SIEGERS, C.-P., J. G. FILSER, and H. M. BOLT, (1978), Tox. Appl. Pharmacol. **146:** 709–716.
SIEGERS, C. P., O. STRUBELT, and E. DOST-KEMPF, (1982), Toxicol. Lett. **10:** 423–426.
SIPES, I. G., A. J. GANDOLFI, and B. R. BROWN, (1981), in: Drug Reaction in the Liver, (M. DAVIS, J. M. TREDYER, and R. WILLIAMS, eds.), pp. 219–35, Pitman Medical, London.
SIPES, I. G. and A. J. GANDOLFI, (1982), in: Toxicology of the Liver, (G. PLAA and W. R. HEWITT, eds.), pp. 181–212, Raven Press, New York.
SYMONS, M. C. R., E. ALBANO, T. F. SLATER, and A. TOMASI, (1982), J. Chem. Soc. Paraday Trans. **78:** 2205–2214.
SLATER, T. F., (1982), in: Free Radicals, Lipid Peroxidation and Cancer, (D. C. H. M. MCBRIEN and T. F. SLATER, eds.), pp. 243–276, Academic Press, London.
SLATTERY, J. T., J. M. WILSON, T. F. KALHORN, and S. D. NELSON, (1987), Clin. Pharmacol. Ther. **41:** 413–418.
SMILKSTEIN, M. J., G. L. KNAPP, K. W. KULIG, and B. H. RUMACK, (1988), N. Engl. J. Med. **319:** 1557–1562.
SNYDERWINE, E. G., R. KROLL, and R. J. RUBIN, (1988), Toxicol. Appl. Pharmacol. **93:** 11–21.

Sohn, O. S., H. Ischizaki, C. S. Yang, and E. S. Fiala, (1991), Carcinogenesis **12:** 127–131.
Song, B.-J., H. V. Gelboin, S.-S. Park, C. S. Yang, and F. J. Gonzalez, (1986), J. Biol. Chem. **261:** 16689–16697.
Song, B.-J., T. Matsunaga, J. P. Hardwick, S. S. Park, R. L. Veech, C. S. Yang, H. V. Gelboin, and F. J. Gonzalez, (1987), Mol. Endocrinol. **1:** 542–547.
Steele, C. M., H. A. Masson, J. M. Battershill, G. G. Gibson, and C. Ioannides, (1983), Res. Commun. Chem. Pathol. Pharmacol. **40:** 109–119.
Stewart, R. D., T. N. Fisher, M. J. Hosko, J. E. Petersen, E. D. Baretta, and H. C. Dott, (1972), Science **176:** 295–296.
Strubelt, O., F. Obermeier, C. P. Siegers, and M. Volper, (1978), Toxicology **10:** 261–270.
Strubelt, O., E. Dost-Kempf, C. Siegers, M. Younes, M. Vopel, V. Preuss, and D. J. Dreekman, (1981), Tox. Appl. Pharmacol. **103:** 905–912.
Swenson, D. H., J.-K. Lin, E. C. Miller, and J. A. Miller, (1977), Cancer Res. **37,** 172–181.
Takagi, T., H. Ishii, H. Takahashi, S. Kato, F. Okuno, Y. Ebihara, H. Takamatsu, M., (1962), Kumamoto. Med. J. **15:** 43–54.
Takahashi, J., I. Johansson, S. W. French, and M. Ingelman-Sundberg, (1992), Pharmacol. Toxicol., **70**: 347–352.
Terelius, Y. and M. Ingelman-Sundberg, (1986), Eur. J. Biochem. **1661:** 303–308.
Terelius, Y. and M. Ingelman-Sundberg, (1988), in: Cytochrome P-450: Biochemistry and Biophysics, (I. Schuster, ed.), Taylor and Francis, pp. 608–611.
Terelius, Y. and M. Ingelman-Sundberg, (1991), submitted.
Terelius, Y., C. Norsten-Höög, T. Cronholm and M. Ingelman-Sundberg, (1991), Biochem. Biophys. Res. Commun. **179:** 689–694.
Thomas, P. E., S. Bandiera, S. L. Maines, D. E. Ryan, and W. Levin, (1987), Biochemistry **26:** 2280–2289.
Thompson, J. A., B. Ho, and S. L. Mastovich, (1984), Chem. Biol. Interacts **51:** 321–333.
Thummel, K. E., J. T. Slattery, and S. D. Nelson, (1988), J. Pharma. Exptl. Ther. **245:** 129–136.
Thummel, K. E. and J. B. Schenkman, (1990), Mol. Pharmacol. **37:** 119–129.
Timbrell, J. A., (1979), Drug Metab. Rev. **10:** 125–47.
Timbrell, J. A., J. R. Mitchell, W. R. Snodgrass, and S. D. Nelson, (1980), J. Pharmacol. Exp. Ther. **213:** 364–9.
Tindberg, N. and M. Ingelman-Sundberg, (1989), Biochemistry **28:** 4499–4504.
Tomasi, A., S. Billing, A. Garner, T. F. Slater, and E. Albano, (1983), Chem. Biol. Interacts. **46:** 353–368.
Tomasi, A., E. Albano, F. Biasi, T. F. Slater, V. Vannini, and M. U. Dianzani, (1985), Chem. Biol. Interact **55:** 303–316.
Trudell, J. R., B. Bosterling, and A. Trevor, (1982), Mol. Pharmacol. **21:** 710–717.
Tsukamoto, H., S. W. French, N. Benson, G. Delgado, G. A. Rao, E. C. Larkin, and C. Largman, (1985), Hepatology **5:** 224–232.
Tsukamoto, H., S. J. Towner, L. M. Ciofalo, and S. W. French, (1986), Hepatology **6:** 814–822.
Tsutsumi, R., Leo, A. M., C. Kim, M. Tsutsumi, J. Lasker, N. Lowe, and C. S. Lieber, (1990), Alcoholism. Clin. Exp. Res. **14:** 174–179.
Tynes, R. E. and E. Hodgson, (1983), Biochem. Pharmacol. **32:** 3419–3428.
Uematsu, F., H. Kikuchi, M. Motomiya, T. Abe, I. Sagami, T. Ommachi, A. Wakui, R. Kanamaru, and M. Watanabe, (1991), Jap. J. Cancer Res. **82,** 254–256.
Umeno, M., O. W. McBridge, C. S. Yang, H. V. Gelboin, and F. J. Gonzalez, (1988), J. Biol. Chem. **263:** 4956–4962.
Vadi, H. V. and R. A. Neal, (1981), Chem. Biol. Interact. **32:** 25–38.

VÄÄNÄNEN, H., M. SALASPURO, and K. O. LINDROS, (1984), Hepatology **4:** 862–866.
VAN DE STRAAT, R. J. DE VRIES, T. KULKENS, A. J. J. DEBETS, and N. P. E. VERMEULEN, (1986), Biochem. Pharmacol. **35:** 3693–3699.
VAN DE STRAAT, R., J. DE VRIES, H. J. R DE BOER, R. M. VROMANS, and N. P. E. VERMUELEN, (1987), Xenobiotica **17:** 1–9.
VAN DE STRAAT, R., R. M. VROMANS, P. BOSMAN, J. DE VRIES, and N. P. E. VERMUELEN, (1988), Chem. Biol. Interact. **64:** 267–280.
VAN DYKE, R. A., M. T. BAKER, I. JANSSON, and J. SCHENKMAN, (1988), Biochem. Pharmacol. **37:** 2357–2361.
WALKER, R. M., T. F. MCELLIGOTT, E. M. POWER, T. E. MASSEY, and W. J. RACZ, (1983), Toxicology **28:** 193–206.
WATKINS, J. B., R. A. SANDERS, and L. V. BECK, (1988), Toxicol. Appl. Pharmacol. **93:** 329–238.
WAXMAN, D. J., J. J. MORRISSEY, and G. A. LEBLANC, (1989), Endocrinology **124:** 2954–2966.
WILLIAMS, M. T. and L. C. SIMONET, (1988), Biochem. Biophys. **10:** 279–284.
WILLIAMS, A. T. and R. F. BURK, (1990), Semin. Liver Dis. **10:** 279–284.
WINTERS, D. K. and A. I. CEDERBAUM, (1990), Biochem. Pharmacol. **39:** 697–705.
WOODWARD, K. N. and J. A. TIMBRELL, (1984), Toxicology **30:** 65–74.
WRIGHTON, S. A., C. CAMPANELE, P. E. THOMAS, S. L. MAINES, P. B. WATKINS, G. PARKER, G. MENDEZ-PICON, M. HANIU, J. E. SHIVELY, W. LEVIN, and P. S. GUZELIAN, (1986), Molec. Pharmac. **29:** 405–410.
WRIGHTON, S. A., P. E. THOMAS, D. E. RYAN and W. LEVIN, (1987), Arch. Biochem. Biophys. **258:** 292–297.
YAMAUCHI, S. NAGATA, M. TASHIRO, and M. TSUCHIYA, (1983), Pharmacol. Biochem. Behav. **18:** 461–465.
YANG, C. S., Y. Y. TU, D. R. KOOP, and M. J. COON, (1985), Cancer Res. **45:** 1140–1145.
YANG, C. S., C. PATTEN, M. J. LEE, M. LI, J. S. YOO, J. PAN, and J. HONG, (1987), Iarc. Sci. Publ. **84:** 104–108.
YOO, J.-S. H., F. P. GUENGERICH, and C. S. YANG, (1988), Cancer Res. **88:** 1499–1504.
YOO, J.-S. H., H.-S. PARK, S. M. NING, M.-J. LEE, and C.. S. YANG, (1990), Biochem. Pharmacol. **39:** 519–525.
YOO, J.-S. H., S. M. NING, C. B. PANTUK, E. J. PANTUK, and C. S. YANG, (1991), J. Nutr. **121:** 959–965.
ZAMPAGLIONI, N., D. J. JOLLOW, J. R MITCHELL, B. STRIPP, M. HAMRICK, and J. R. GILLETTE, (1973), J. Pharmacol. Exp. Ther. **187:** 218–227.
ZIMMERMANN, H. J., (1978), Hepatotoxicity, Appleton-Century-Crofts, New York.
ZIMMERMAN, H. J., (1986), Alcoholism. Clin. Exptl. **10:** 3–15.

Chapter 6

Molecular Mode of Action of Organic Nitrates and their Bioactivation via Cytochrome P-450

H. SCHRÖDER

1. Introduction

Organic nitrates such as glyceryl trinitrate are among the oldest and most widely used drugs for the treatment of myocardial ischemia and its principal symptom angina pectoris. Bearing in mind that the antianginal effect of glyceryl trinitrate was described more than one hundred years ago (MURRELL, 1879) it seems astonishing that the cellular mechanism underlying this action has remained completely unknown until the seventies of this century. Since then it has become increasingly clear that the nitric oxide free radical and its second messenger cyclic GMP mediate the vasodilatory and antianginal effects of organic nitrates. Thus, organic nitrates have to be classified as prodrugs releasing the pharmacologically active metabolite nitric oxide. Although several different pathways have been suggested, recent evidence points to a cytochrome P-450 system playing an important role in the process of enzymatic nitric oxide generation from organic nitrates.

2. Nitric oxide and cyclic GMP as mediators of organic nitrate action

The antianginal action of organic nitrates is due to their capacity to relax vascular smooth muscle in veins and to a lesser degree in arteries. The resulting reduction in preload and afterload decreases the myocardial requirement for oxygen thus counteracting the inadequate oxygenation during ischemia and anginal attacks. Moreover, dilation of collaterial vessels and large coronary arteries directly improves blood flow to ischemic myocardium (for review see BASSENGE et al., 1981).

That the second messenger cyclic GMP might be involved in the smooth muscle relaxing effect of certain nitrogen-containing compounds was first suggested in the mid seventies when the smooth muscle relaxants sodium azide, hydroxylamine and sodium nitrite were shown to elevate tissue cyclic GMP levels and to activate soluble guanylate cyclase (KIMURA et al., 1975; MITTAL et al., 1975; MIKI et al., 1976). Subsequently, DIAMOND and co-workers demonstrated glyceryl trinitrate to increase cyclic GMP levels in rat myometrium and canine femoral artery (DIAMOND and HOLMES, 1975; DIAMOND and BLISARD, 1976). Further studies in several laboratories revealed that sodium nitroprusside and other vasodilators containing nitrogen-oxide constituents (nitrovasodilators, Fig. 1) produced relaxation of vascular and nonvascular smooth muscle via stimulation of cyclic GMP formation (SCHULTZ et al., 1977; ARNOLD et al., 1977; KATSUKI and MURAD, 1977; BÖHME et al., 1978; AXELSSON et al., 1979; JANIS and DIAMOND, 1979; KUKOVETZ et al., 1979; GRUETTER et al., 1981). The mechanism by which nitrovasodilators increase intracellular cyclic GMP levels became clear when the

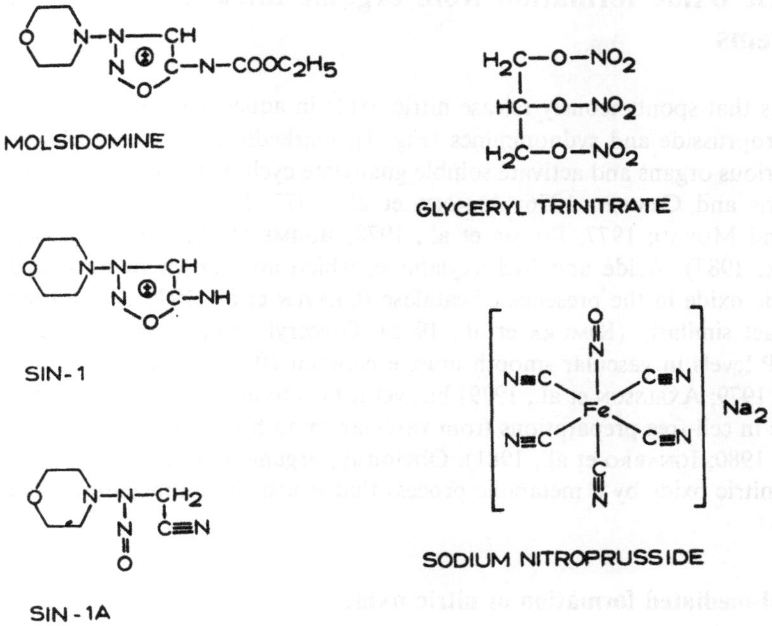

Fig. 1. Chemical structure of different nitrovasodilators. The antianginal agent molsidomine is a mesoionic compound, which is metabolized enzymatically to 3-morpholinosydnonimine (SIN-1). SIN-1 is unstable in aqueous solution and converted to N-nitroso-N-morpholinoaminoacetonitrile (SIN-1A) by pH-dependent, non-enzymatic opening of the oxdiazole ring. SIN-1A and sodium nitroprusside but not glyceryl trinitrate spontaneously release nitric oxide in aqueous solution.

unstable free radical nitric oxide was found to be a potent and reversibly relaxing agent which activated soluble guanylate cyclase (Arnold et al., 1977; Miki et al., 1977; Gruetter et al., 1979). From these findings it was concluded that prior to cyclic GMP stimulation and vasodilation, nitrovasodilator drugs have to be acted upon either enzymatically or non-enzymatically to release free nitric oxide.

Cyclic GMP-dependent vasodilation has been shown to involve activation of a cyclic GMP-dependent protein kinase followed by a decrease in cytosolic calcium levels (for reviews see Lincoln, 1989; Walter, 1990). However, it was not until recently that the physiological relevance of nitric oxide as endothelium-derived relaxing factor and endogenous activator of soluble guanylate cyclase has emerged (Palmer et al., 1987; Knowles et al., 1989; Schröder and Schrör, 1989; Ishii et al., 1989).

3. Nitric oxide formation from organic nitrates in cell-free systems

Compounds that spontaneously release nitric oxide in aqueous solution such as sodium nitroprusside and sydnonimines (Fig. 1), markedly elevate cyclic GMP levels in various organs and activate soluble guanylate cyclase in cell-free systems (DeRubertis and Craven, 1976; Arnold et al., 1977; Schultz et al., 1977; Katsuki and Murad; 1977; Böhme et al., 1978; Böhme et al., 1984; Feelisch and Noack, 1987). Azide and hydroxylamine, which are known to form and release nitric oxide in the presence of catalase (Craven et al., 1979), have been shown to act similarly (Kimura et al., 1975). Glyceryl trinitrate also elevates cyclic GMP levels in vascular smooth muscle between 10- and 130-fold (Kukovetz et al., 1979; Axelsson et al., 1979) but yet it fails to activate soluble guanylate cyclase in cell-free preparations from vascular smooth muscle (Ignarro and Gruetter, 1980; Ignarro et al., 1981). Obviously, organic nitrates require conversion to nitric oxide by a metabolic process that is lost on homogenization of the tissue.

3.1. Thiol-mediated formation of nitric oxide

As early as 1950 it was demonstrated by Heppel and Hilmoe that reduced glutathione undergoes a spontaneous reaction with glyceryl trinitrate and other organic esters of nitric acid to form inorganic nitrite and oxidized glutathione. A similar reductive hydrolysis of organic nitrates was found to take place in the presence of cysteine or cysteinylglycine (Heppel and Hilmoe, 1950). These and other findings suggesting a thiol-induced biotransformation of organic nitrates (Needleman et al., 1973) prompted Ignarro and colleagues to include cysteine in their guanylate cyclase assay. Under these conditions, glyceryl trinitrate produced a marked activation of guanylate cyclase in cell-free preparations from vascular tissues (Ignarro and Gruetter, 1980; Ignarro et al., 1981). It was proposed that free thiols such as cysteine were necessary co-factors for nitric oxide generation from organic nitrates and that this process involved the formation of S-nitrosothiols as intermediates. Cysteine was later shown to release nitric oxide from organic nitrates in a non-enzymatic chemical reaction and in correlaton to the observed degree of guanylate cyclase activation (Schröder et al., 1985; Feelisch and Noack, 1987; Schröder and Noack, 1987). The thiol dependence of guanylate cyclase activation by organic nitrates was found to be highly specific for cysteine and structurally related sulfhydryl compounds such as N-acetylcysteine and thiosalicyclic acid (Schröder et al., 1985; Schmidt et al., 1985). Whereas other thiols only release nitrite during the non-enzymatic reaction with organic nitrates, the three thiols mentioned above were shown to yield both nitrite and nitric oxide (Feelisch et al., 1988).

3.2. Cytochrome P-450-mediated formation of nitric oxide

The generation of nitric oxide upon incubation of organic nitrates with rat liver microsomal fractions was first reported by WOLF and WERRINGLOER (1987, 1988). SERVENT and co-workers (1989) demonstrated the formation of a cytochrome P-450-Fe(II)-NO complex during microsomal hepatic denitration of glyceryl trinitrate. In the same paper, it was shown that inhibitors of microsomal cytochrome P-450 such as carbon monoxide, metyrapone and miconazole lead to a marked attenuation of glyceryl trinitrate denitration. Moreover, this reaction was reported to be NADPH-dependent and sensitive to inhibition by dioxygen (SERVENT et al., 1989; MCDONALD and BENNETT, 1990; BENNETT et al., 1992) suggesting microsomal denitration of glyceryl trinitrate to be a reductive process according to the following equation (SERVENT et al., 1989):

$$\text{RO-NO}_2 + \text{P-450-Fe(II)} \overset{\text{NADPH}}{\longrightarrow} \text{ROH} + \text{P-450-Fe(II)} - \text{NO}$$

These findings are consistent with the view that organic nitrates serve as an electron acceptor for cytochrome P-450 and may compete directly with dioxygen or carbon monoxide for binding to the heme moiety of cytochrome P-450. A similar reductase activity of cytochrome P-450 has been reported for the conversion of nitro compounds to amines (GILLETTE et al., 1968), the dehalogenation of haloalkanes (AHR et al., 1982; NASTAINCZYK et al., 1982) and for other cytochrome P-450-catalyzed reactions (for review see ARCHAKOV and ZHUKOV, 1989).

3.3. Glutathione S-transferase-mediated formation of nitric oxide

The question whetheer enzymes other than cytochrome P-450 may also catalyze nitric oxide formation from organic nitrates was addressed by WOLF and WERRINGLOER (1987, 1988). Using rat liver cytosolic fractions they found that degradation of glyceryl trinitrate by the cytosolic glutathione S-transferase resulted in the formation of both nitrite and nitric oxide as metabolites. However, the ratio of nitrite to nitric oxide formed during this reaction (150 : 1) and the respective K_m value for glyceryl trinitrate (4 nM) were both rather high as compared to the cytochrome P-450-catalyzed reaction (15 : 1; $K_m < 200 \mu M$). Thus, metabolism by cytochrome P-450 but not by glutathione S-transferase exhibits a high efficiency for nitric oxide formation at low concentrations of glyceryl trinitrate. Moreover, it can be shown that the glutathione-dependent biotransformation of glyceryl trinitrate is not accompanied by activation of soluble guanylate cyclase in rat liver cytosolic fractions (Fig. 2). This finding demonstrates that although glutathione S-transferase may generate nitric oxide from organic nitrates, the amount of nitric oxide formed is not high enough to produce a significant increase in guanylate cyclase activity. Thus, biotransformation by

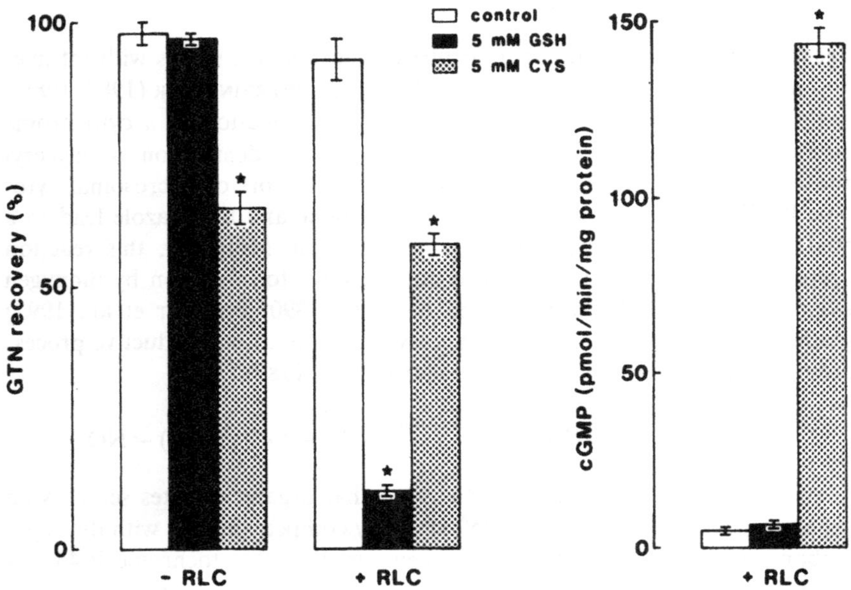

Fig. 2. Glutathione (GSH)-dependent biotransformation of glyceryl trinitrate (GTN, 0.05μM) does not promote activation of guanylate cyclase in a rat liver cytosolic fraction (RLC). Thus, nitric oxide formation during metabolism of organic nitrates by glutathione S-transferase does not yield sufficient amounts of nitric oxide for guanylate cyclase activation. In contrast, the non-enzymatic degradation of GTN by cysteine (CYS) occurs independently of added RLC and results in nitric oxide formation as demonstrated by activation of guanylate cyclase. *): $P < 0.05$, treatment vs. control, two tailed t-test. The data are mean ±SEM of n = 6 observations (data from Schröder, 1992).

glutathione S-transferase, which mainly yields the pharmacologically inactive metabolite nitrite (Oberst and Snyder, 1948; Needleman, 1975; Keen et al., 1976), can be characterized as a pathway for elimination but not bioactivation of organic nitrates.

4. Cytochrome P-450 rather than cysteine mediates nitric oxide formation from organic nitrates in intact cells

According to Ignarro and colleagues, cysteine or related sulfhydryl compounds were the only co-factors required for the non-enzymatic formation of nitric oxide from organic nitrates (Ignarro and Gruetter, 1980; Ignarro et al., 1981). This so-called nitrosothiol hypothesis was considered to hold true not only in cell-free systems but also in intact blood vessels, an assumption based on earlier observations by Needleman and colleagues. These authors had found ethacrynic acid, a known thiol alkylating agent, to antagonize glyceryl trinitrate-induced relaxation of rabbit thoracic aortic strips (NEEDLEMAN et al., 1973). However, the inhibitory effect of ethacrynic acid on vascular relaxation was not specific for organic nitrates but was also observed with isoproterenol, sodium nitroprusside, cyclic AMP, cyclic GMP and other relaxing agents (NEEDLEMAN et al., 1973; RAPOPORT and MURAD, 1988). Thus, the effect of ethacrynic acid and other thiol blocking reagents on vascular relaxation (RAPOPORT and MURAD, 1988) may rather be due to interaction with the contractile apparatus than to inhibition of organic nitrate metabolism. Moreover, the nitrosothiol hypothesis was challenged by GRUETTER and LEMKE (1985) who reported the dissociation of cysteine and glutathione levels from glyceryl trinitrate-induced relaxation in bovine coronary arteries. Using cultured porcine aortic endothelial cells, we found that the molsidomine metabolite SIN-1 but not glyceryl trinitrate markedly elevates cyclic GMP levels (SCHRÖR et al., 1989; SCHRÖDER, 1992). Even in the presence of N-acetylcysteine, the cells remained unresponsive to organic nitrates (SCHRÖDER 1992) demonstrating that in intact cells, nitric oxide formation from organic nitrates is not mediated via free thiols. These and other findings (SCHRÖDER et al., 1987, 1988; BRIEN et al., 1988; BENNETT et al., 1989; CHUNG and FUNG, 1990) argued for an enzymatic rather than a thiol-dependent mechanism for the cellular bioconversion of organic nitrates to nitric oxide.

That cytochrome P-450 is involved in cellular nitric oxide formation from organic nitrates, is suggested by recent findings demonstrating inhibition of organic nitrate-induced cyclic GMP elevation by cytochrome P-450 inhibitors in cultured cells (SCHRÖDER and SCHRÖR, 1990; SCHRÖDER, 1992, see also Fig. 3). The RFL-6 and LLC-PK_1 cells used in these investigations have been established as model systems to study organic nitrate action. Particularly with respect to the regulation of organic nitrate metabolism and cyclic GMP synthesis, these cell lines are very similar to vascular smooth muscle cells (SCHRÖDER et al., 1987, 1988; BENNETT et al., 1988, 1989). An experiment with the cytochrome P-450 inhibitor SKF 525A, proadifen, (MANNERING, 1971) is shown in Table 1. Cyclic GMP stimulation by glyceryl trinitrate and isoidide denitrate is reduced to a

Table 1. Effect of SKF 525A on cyclic GMP stimulation by glyceryl trinitrate, isoidide dinitrate, and sodium nitroprusside in RFL-6 cells

Addition	**Cyclic GMP** (pmol/10^6 cells)		
	10 μM GTN	10 μM IIDN	1 μM SNP
none (control)	33.6 ± 2.5	26.9 ± 2.0	37.2 ± 2.3
100 μM SKF 525A	10.4 ± 0.5*	7.3 ± 0.3*	32.0 ± 3.9

Cell culture conditions and incubation procedures as in SCHRÖDER et al., 1987 and SCHRÖDER, 1992. Basal cyclic GMP levels were 1.5 ± 0.1 pmol/10^6 cells. *): $P < 0.05$, treatment vs. control, two tailed t-test. The data are mean ± SEM of n = 6 observations.

similar extent by SKF 525A in cultured RFL-6 cells. In contrast, sodium nitroprusside-induced cyclic GMP stimulation remains unaltered under these conditions precluding a direct inhibitory effect of SKF 525A on activated guanylate cyclase (Table 1). In the same cell type, a 24-h preincubation with 3-methylcholanthrene augmented cyclic GMP stimulation by glyceryl trinitrate by up to 100% (SCHRÖDER, 1992) indicating the involvement of an inducible cytochrome P-450 system.

It seems unlikely that glutathione S-transferase mediates bioactivation of organic nitrates, since sulfobromophthalein, a known inhibitor of this enzyme (YALCIN et al., 1983), has no effect on intracellular cyclic GMP stimulation by glyceryl trinitrate (Fig. 3). Moreover, it has recently been shown that the potency of organic nitrates for relaxation and cyclic GMP stimulation correlates well with their respective degree of biotransformation by cytochrome P-450 but not by cysteine or glutathione S-transferase (SCHRÖDER et al., 1987; BENNETT et al., 1989; MCDONALD and BENNETT, 1990).

5. Unanswered questions

5.1. Is cytochrome P-450 relevant for the action of organic nitrates in vivo?

Although there is now considerable evidence for cytochrome P450 mediating

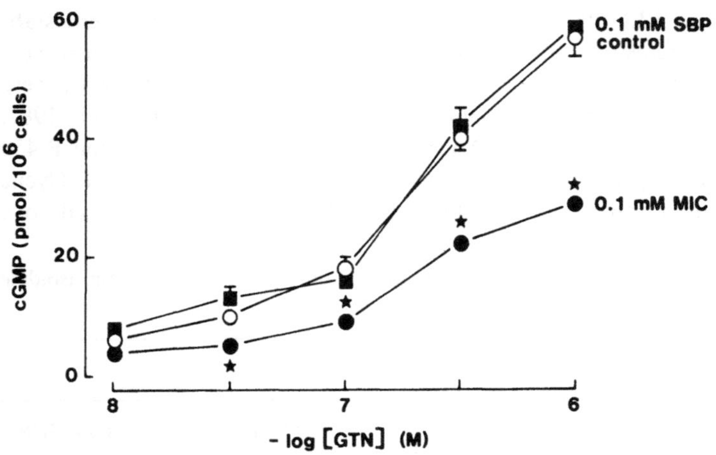

Fig. 3. Miconazole (MIC), an inhibitor of cytochrome P-450, but not the inhibitor of glutathione S-transferase sulfobromophthalein (SBP) attenuates cyclic GMP stimulation by glyceryl trinitrate (GTN) in LLC-PK_1 cells. *):$P < 0.05$, treatment vs. control, two tailed t-test. The data are mean ±SEM of $n = 6$ observations (methods as in SCHRÖDER and SCHRÖR, 1990 and SCHRÖDER, 1992).

nitric oxide formation from organic nitrates in vitro, the relevance of this pathway in vivo has still to be determined. This is particularly important considering the high sensitivity to inhibition by dioxygen. The reductive denitration of organic nitrates by cytochrome P-450 under aerobic conditions in vitro is only 20% and less of that observed under anaerobic conditions (WOLF and WERRINGLOER, 1987, 1988; SERVENT et al., 1989; MCDONALD and BENNETT, 1990). Thus, optimal expression of this pathway in vivo should be expected to occur in tissues with low oxygen tension, f.e. in the venous system (NASTAINCZYK et al., 1977; KRATZ and STAUDINGER, 1965). As a matter of fact, venous capacitance vessels are more sensitive to the relaxant effect of organic nitrates than arterial resistance vessels, a phenomenon that is well documented in vivo (IMHOF et al., 1980; KOBER et al., 1981) and in vitro (MACKENZIE and PARRATT, 1977; EDWARDS et al., 1984). Moreover, it has been suggested that the sensitivity of blood vessels to organic nitrates is related to their nitrate metabolizing capacity (FUNG et al., 1984; KAWAMOTO et al., 1987). In this context, it is interesting to note that cytochrome P-450 has been identified in vascular wall tissues (JUCHAU et al., 1976; ABRAHAM et al., 1985; SERABJIT-SINGH et al., 1985) although its isoenzyme characteristics (for review see GUENGERICH, 1989) and relevance for biotransformation of organic nitrates has yet to be examined in more detail. On the other hand, blood cells such as neutrophils, which in terms of cyclic GMP elevation are unresponsive to organic nitrates (SCHRÖDER et al., 1990), may lack the required metabolizing system, i.e. cytochrome P-450. Clearly, more research is needed, but to date most observations made in this field are well in line with the assumption that cytochrome P-450 also plays a role for the action of organic nitrates in vivo (Fig. 4).

5.2. Is cytochrome P-450 involved in nitrate tolerance?

Prolonged exposure to organic nitrates is known to induce tolerance to the cardiovascular effects of these drugs in humans and experimental animals (CRANDALL et al., 1931; SCHELLING and LASAGNA, 1967; BOGAERT and DE SCHAEPDRYVER, 1968). This tolerance to organic nitrates is associated with a reduced cyclic GMP response in vascular tissue (WIKBERG et al., 1978; KEITH et al., 1982; RAPOPORT et al., 1987). There is now substantial evidence, that the attenuation of organic nitrate induced-cyclic GMP stimulation is caused by a decrease in biotransformation of organic nitrates (BRIEN et al., 1986; BENNETT et al., 1989) and subsequent nitric oxide formation (FÖRSTER et al., 1991). According to the nitrosothiol concept, it was previously suggested that nitrate tolerance is due to intracellular depletion or oxidation of cysteine and other thiols (NEEDLEMAN and JOHNSON, 1973; for review see NOACK, 1990). However, the results of early attempts to reverse tolerance with different thiol-containing compounds were contradictory (AXELSSON et al., 1982; GRUETTER and LEMKE, 1985; TORRESI

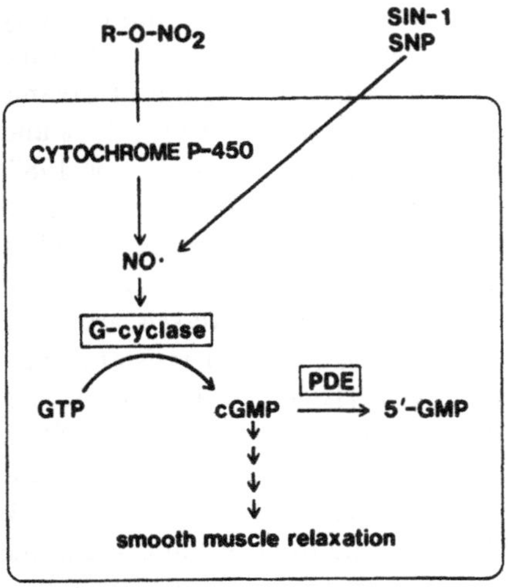

Fig. 4. Schematic diagram of possible mechanisms for nitrovasodilator-induced relaxation of vascular smooth muscle. In cell-free systems, nitric oxide (NO) formation from organic nitrates ($RO\text{-}NO_2$) has also been observed in the presence of free thiols or glutathione S-transferase. Recent evidence, however, suggests cytochrome P-450 to play a key role in cellular bioactivation of organic nitrates. Abbreviations: SIN-1, 3-morpholinosydnonimine; SNP, sodium nitroprusside; G-cyclase, guanylate cyclase; PDE, phosphodiesterase.

et al., 1985). Recently, it has become clear that nitrate tolerance cannot be reversed by cysteine or related compounds and that the apparent restoration of nitrate responsiveness observed previously can be attributed to tolerance-independent interactions between cysteine and organic nitrates (Schröder et al., 1988; Münzel et al., 1989).

Today, cytochrome P-450 has emerged as another possible site for the induction of nitrate tolerance. An exhaustion of this enzymatic pathway would be a plausible explanation for the observed decrease in nitric oxide formation during nitrate tolerance (Förster et al., 1991). So far, no studies have been published that were specifically designed to investigate the role of cytochrome P-450 in nitrate tolerance, but in a previous investigation, we found that the reversal of nitrate tolerance in intact cells requires de novo synthesis of proteins (Schröder et al., 1988). These findings may now be reinterpreted as evidence for a tolerance-induced down-regulation or irreversible inactivation of an enzyme being responsible for organic nitrate bioactivation, i.e. cytochrome P-450. In addition to down-regulation, other mechanisms leading to a decrease in enzymatic activity are also conceivable, e.g. reveersible inhibition of cytochrome P-450 by product or substrate (for reviews see Whitlock, 1986; Guengerich, 1989; Kühn-Velten et al., 1990). Future investigations will have to test the hypothesis of cytochrome P-450 being involved in nitrate tolerance and by doing so may solve this still continuing "100 year old mystery" (Packer, 1990).

6. References

Abraham, N. G., A. Pinto, K. M. Mullane, R. D. Levere, and E. Spokas, (1985), Hypertension 7, 899–904.

Ahr, H. J., L. J. King, W. Nastainczyk, and V. Ullrich, (1982), Biochem. Pharmacol. **31**, 383–390.

Archakov, A. I. and A. A. Zhukov, (1989), in: Basis and Mechanisms of Regulation of Cytochrome P-450, (K. Ruckpaul and H. Rein, eds.), Taylor and Francis, London, 151–175.

Arnold, W. P., C. K. Mittal, S. Katsuki, and F. Murad, (1977), Proc. Natl. Acad. Sci. USA **74**, 3203–3207.

Axelsson, K. L., J. E. S. Wikberg, and R. G. G. Andersson, (1979), Life Sci. **24**, 1779–1786.

Axelsson, K. L., R. G. G. Andersson, and J. E. S. Wikberg, (1982), Acta Pharmacol. Toxicol. **50**, 350–357.

Bassenge, E., J. Holtz, H. Kinadeter, and A. Kolin, (1981), in: Nitrates III, (P. Lichtlen, H. J. Engel, A. Schrey, and H. J. C. Swan, eds.), Springer, Berlin, 238–249.

Bennett, B. M., H. Schröder, L. D. Hayward, and F. Murad, (1988), Circ. Res. **63**, 693–701.

Bennett, B. M., D. C. Leitman, H. Schröder, J. H. Kawamoto, K. Nakatsu, and F. Murad, (1989), J. Pharmacol. Exp. Ther. **250**, 316–323.

Bennett, B. M., B. J. McDonald, and M. J. St. James, (1992), J. Pharmacol. Exp. Ther. **261**, 716–723.

Böhme, E., H. Graf, and G. Schultz, (1978), Adv. Cyclic Nucleotide Res. **9**, 131–143.

Böhme, E., G. Grossmann, J. Herz, A. Mülsch, C. Spies, and G. Schultz, (1984), Adv. Cyclic Nucleotide Protein Phosphor. Res. **17**, 259–266.

Bogaert, M. G. and A. F. De Schaepdryver, (1968), Arch. Int. Pharmacodyn. Ther. **171**, 221–224.

Brien, J. F., B. E. McLaughlin, T. H. Breedon, B. M. Bennett, K. Nakatsu, and G. S. Marks, (1986), J. Pharmacol. Exp. Ther. **237**, 608–614.

Brien, J. F., B. E. McLaughlin, S. M. Kobus, J. H. Kawamoto, K. Nakatsu, and G. S. Marks, (1988), J. Pharmacol. Exp. Ther. **244**, 322–327.

Chung, S.-J. and H.-L. Fung, (1990), J. Pharmacol. Exp. Ther. **253**, 614–619.

Crandall, L. A., Jr., C. D. Leake, A. S. Loevenhart, and C. W. Muehlberger, (1931), J. Pharmacol. Exp. Ther. **41**, 103–120.

Craven, P. A., F. R. DeRubertis, and D. W. Pratt, (1979), J. Biol. Chem. **254**, 8213–8222.

DeRubertis, F. R. and P. A. Craven, (1976), Science **193**, 897–899.

Diamond, J. and T. G. Holmes, (1975), Can. J. Physiol. Pharmacol. **53**, 1099–1107.

Diamond, J. and K. S. Blisard, (1976), Mol. Pharmacol. **12**, 688–692.

Edwards, J. C., L. J. Ignarro, A. L. Hyman, and P. J. Kadowitz, (1984), J. Pharmacol. Exp. Ther. **228**, 33–42.

Feelisch, M. and E. A. Noack, (1987), Eur. J. Pharmacol. **139**, 19–30.

Feelisch, M., E. Noack, and H. Schröder, (1988), Eur. Heart J. **9**, Suppl. A, 57–62.

Förster, S., I. Woditsch, H. Schröder, and K. Schrör, (1991), J. Cardiovasc. Pharmacol. **17**, 867–872.

Fung, H.-L., S. C. Sutton, A. Kamiya, (1984), J. Pharmacol. Exp. Ther. **228**, 334–341.

Gillette, J. R., H. Kamin, and H. A. Sasame, (1986), Mol. Pharmacol. **4**, 541–548.

Gruetter, C. A., B. K. Barry, D. B. McNamara, D. Y. Gruetter, P. J. Kadowitz, and L. J. Ignarro, (1979), J. Cyclic Nucleotide Res. **5**, 211–224.

Gruetter, C. A., P. J. Kadowitz, and L. J. Ignarro, (1981), Can. J. Physiol. Pharmacol. **59**, 150–156.

Gruetter, C. A. and S. M. Lemke, (1985), Eur. J. Pharmacol. **111**, 85–95.

Guengerich, F. P., (1989), Annu. Rev. Pharmacol. Toxicol. **29**, 241–264.

Heppel, L. A. and R. J. Hilmoe, (1950), J. Biol. Chem. **182**, 129–138.

Ignarro, L. J. and C. A. Gruetter, (1980), Biochim. Biophys. Acta **631**, 221–231.

IGNARRO, L. J., H. LIPPTON, J. C. EDWARDS, W. H. BARICOS, A. L. HYMAN, P. J. KADOW9ITZ, and C. A. GRUETTER, (1981), J. Pharmacol. Exp. Ther. **218**, 739–749.
IMHOF, P. R., B. OTT, P. FRANKHAUSER, L.-C. CHU, and J. HODLER, (1980), Eur. J. Clin. Pharmacol. **18**, 455–460.
ISHII, K., L. D. GORSKY, J. FÖRSTERMANN, and F. MURAD, (1989), J. Applied Cardiol. **4**, 505–512.
JANIS, R. A. and J. DIAMOND, (1979), J. Pharmacol. Exp. Ther. **211**, 480–484.
JUCHAU, M. R., J. A. BOND, and E. P. BENDITT, (1976), Proc. Natl. Acad. Sci. **73**, 3723–3725.
KATSUKI, S. and F. MURAD, (1977), Mol. Pharmacol. **13**, 330–341.
KAWAMOTO, J. H., J. F. BRIEN, G. S. MARKS, and K. NAKATSU, (1987), Can. J. Physiol. Pharmcol. **65**, 1146–1150.
KEEN, J. H., W. H. HABIG, and W. B. JAKOBY, (1976), J. Biol. Chem. **251**, 6183–6188.
KEITH, R. A., A. M. BURKMAN, T. D. SOKOLOSKI, and R. H. FERTEL, (1982), J. Pharmacol. Exp. Ther. **221**, 525–531.
KIMURA, H., C. K. MITTAL, and F. MURAD, (1975), Nature **257**, 700–702.
KNOWLES, R. G., M. PALACIOS, R. M. J. PALMER, and S. MONCADA, (1989), Proc. Natl. Acad. Sci. USA **86**, 5159–5162.
KOBER, G., R. GROBMANN, W. SCHULZ, and M. KALTENBACH, (1981), Z. Kardiol. **70**, 547–554.
KRATZ, F. and H. STAUDINGER, (1965), Hoppe Seyler's Z. Physiol. Chem. **343**, 27–34.
KÜHN-VELTEN, N., J. REDIGER, and W. STAIB, (1990), Horm. Metab. Res. **22**, 528–532.
KUKOVETZ, W. R., S. HOLZMANN, A. WURM, and G. PÖCH, (1979), Naunyn-Schmiedeberg's Arch. Pharmacol. **310**, 129–138.
LINCOLN, T. M., (1989), Pharmacol. Ther. **41**, 479–502.
MACKENZIE, J. E. and J. R. PARRATT, (1977), Br. J. Pharmacol. **60**, 155–160.
MANNERING, G. J., (1971), in: Fundamentals of Drug Metabolism and Disposition, (B. LA DU, H. MANDEL, and E. WOY, eds.), Williams and Wilkins, Baltimore, 206–252.
MCDONALD, B. J. and B. M. BENNETT, (1990), Can. J. Physiol. Pharmacol. **68**, 1552–1557.
MIKI, N., M. NAGANO, and K. KURIYAMA, (1976), Biochem. Biophys. Res. Commun. **72**, 952–959.
MIKI, N., Y. KAWABE, and Y. KURIYMA, (1977), Biochem. Biophys. Res. Commun. **75**, 851–856.
MITTAL, C. K., H. KIMURA, and F. MURAD, (1975), J. Cyclic Nucleotide Res. **1**, 261–269.
MÜNZEL, T., J. HOLTZ, A. MÜLSCH, D. J. STEWART, and E. BASSENGE, (1990), Circulation **79**, 188–197.
MURRELL, W., (1879), Lancet **1**, 80–81, 113–115, 151–152, 225–227.
NASTAINCZYK, W., V. ULLRICH, and H. SIES, (1977), Biochem. Pharmacol. **27**, 387–392.
NASTAINCZYK, W., H. J. AHR, and V. ULLRICH, (1982), Biochem. Pharmacol. **31**, 391–396.
NEEDLEMAN, P. and E. M. JOHNSON, Jr., (1973), J. Pharmacol. Exp. Ther. **184**, 709–715.
NEEDLEMAN, P., B. JAKSCHIK, and E. M. JOHNSON, Jr., (1973), J. Pharmacol. Exp. Ther. **187**, 324–331.
NEEDLEMAN, P., (1975), in: Organic Nitrates, (P. NEEDLEMAN, ed.), Springer, Berlin Heidelberg New York, 57–95.
NOACK, E., (1990), Z. Kardiol. **79**, Suppl. 3, 51–55.
OBERST, F. W. and F. H. SNYDER, (1948), J. Pharmacol. Exp. Ther. **93**, 444–450.
PACKER, M., (1990), J. Am. Coll. Cardiol. **16**, 932–935.
PALMER, R. M. J. A. G. FERRIGE, and S. MONCADA, (1987), Nature **327**, 524–526.
RAPOPORT, R. M., S. A. WALDMAN, R. GINSBERG, C. R. MOLNA, and F. MURAD, (1987), J. Cardiovasc. Pharmacol. **10**, 82–89.
RAPOPORT, R. M. and F. MURAD, (1988), Gen. Pharmacol. **19**, 61–65.
SCHELLING, J. L. and L. LASAGNA, (1967), Clin. Pharmacol. Ther. **8**, 256–260.

SCHMIDT, K., R. REICH, and W. R. KUKOVETZ, (1985), J. Cyclic Nucleotide Protein Phosphor. Res. **10**, 43–53.
SCHRÖDER, H., E. NOACK, and R. MÜLLER, (1985), J. Mol. Cell. Cardiol. **17**, 931–934.
SCHRÖDER, H. and E. NOACK, (1987), Arch. Int. Pharmacodyn. Ther. **290**, 235–246.
SCHRÖDER, H., D. C. LEITMAN, L. D. HAYWARD, B. M. BENNETT, and F. MURAD, (1987), J. Applied Cardiol. **2**, 301–311.
SCHRÖDER, H., D. C. LEITMAN, B. M. BENNETT, S. A. WALDMAN, and F. MURAD, (1988), J. Pharmacol. Exp. Ther. **245**, 413–418.
SCHRÖDER, H. and K. SCHRÖR, (1989), Naunyn-Schmiedeberg's Arch. Pharmacol. **340**, 475–477.
SCHRÖDER, H. and K. SCHRÖR, (1990), Naunyn-Schmiedeberg's Arch. Phrmacol. **342**, 616–618.
SCHRÖDER, H., P. NEY, I. WODITSCH, and K. SCHRÖR, (1990), Eur. J. Pharmacol. **182**, 211–218.
SCHRÖDER, H., (1992), J. Pharmacol. Exp. Ther. **262**, 298–302.
SCHRÖR, K., S. FÖRSTER, I. WODITSCH, and H. SCHRÖDER, (1989), J. Cardiovasc. Pharmacol. **14**, Suppl. 11, 29–34.
SCHULTZ, K.-D., K. SCHULTZ, and G. SCHULTZ, (1977), Nature **265**, 750–751.
SERABJIT-SINGH, C. J., J. R. BEND, and R. M. PHILPOT, (1985), Mol. Pharmacol. **28**, 72–79.
SERVENT, D., M. DELAFORGE, C. DUCROCQ, D. MANSUY, and M. LENFANT, (1989), Biochem. Biophys. Res. Commun. **163**, 1210–1216.
TORRESI, J., J. D. HOROWITZ, and G. J. DUSTING, (1985), J. Cardiovasc. Pharmacol. **7**, 777–783.
WALTER, U., (1990), Rev. Physiol. Biochem. Pharmacol. **113**, 41–88
WHITLOCK, J. P., Jr., (1986), Annu. Rev. Pharmacol. Toxicol. **26**, 333–369.
WIKBERG, J. E. S., K. AXELSSON, and R. G. G. ANDERSSON, (1978), in: Vascular Neuroeffector Mechanisms, (J. A. BEVAN, T. GODFRAIND, R. A. MAXWELL, P. M. VANHOUTTE, eds.), Raven, New York, 261–263.
WOLF, J. and J. WERRINGLOER, (1987), Biol. Chem. Hoppe Seyler **368**, 1142.
WOLF, J. and J. WERRINGLOER, (1988), Naunyn-Schmiedeberg's Arch. Pharmacol. **337**, Suppl., R9.
YALCIN, S., H. JENSSON, and B. MANNERVIK, (1983), Biochem. Biophys. Res. Commun. **114**, 829–834.

List of Authors

E. ALBANO
Karolinska Institute
Department of Physiological Chemistry
S-10401 Stockholm
Sweden

M. A. CORREIA
University of California
School of Pharmacy
Department of Pharmaceutical Chemistry
San Francisco CA 94143
USA

M. INGELMAN-SUNDBERG
Karolinska Institute
Department of Physiological Chemistry
S-10401 Stockholm
Sweden

B. KEMPER
University of Illinois
Department of Physiology and Biophysics
407 South Goodwin Avenue
Urbana IL 61801
USA

K. O. LINDROS
Karolinska Institute
Department of Physiological Chemistry
S-10401 Stockholm
Sweden

T. OMURA
Kyushu University
Department of Molecular Biology
Fukuoka 812
Japan

P. R. ORTIZ DE MONTELLANO
University of California
School of Pharmacy
Department of Pharmaceutical Chemistry
San Francisco CA 94143
USA

M. SAKAGUCHI
Kyushu University
Department of Molecular Biology
Fukuoka 812
Japan

H. SCHRÖDER
Stanford University School of Medicine
Division of Oncology
VA Medical Center 154-N
3801 Miranda Avenue
Palo Alto, CA 94304
USA

I. SCHUSTER
Sandoz Forschungsinstitut
Brunner Str. 59
A-1235 Wien
Austria

Y. TERELIUS
Karolinska Institute
Department of Physiological Chemistry
S-10401 Stockholm
Sweden

Subject Index

Bei Fragen zur Produktsicherheit wenden Sie sich bitte an:
If you have any questions regarding product safety,
please contact:

Walter de Gruyter GmbH
Genthiner Straße 13
10785 Berlin
productsafety@degruyterbrill.com